Fractal Binding and Dissociation Kinetics for Different Biosensor Applications

Fractal Binding and Dissociation Kinetics for Different Biosensor Applications

Ajit Sadana

Chemical Engineering Department
Composite Structures and Nanoengineering Research
University of Mississippi
University, MS, USA

2005

ELSEVIER

Amsterdam – Boston – Heidelberg – London – New York – Oxford – Paris
San Diego – San Francisco – Singapore – Sydney – Tokyo

ELSEVIER B.V.
Radarweg 29
P.O. 211, 1000 AE
Amsterdam, The Netherlands

ELSEVIER Inc.
525 B Street
Suite 1900, San Diego
CA 92101-4495, USA

ELSEVIER Ltd
The Boulevard
Langford Lane, Kidlington,
Oxford OX5 1GB, UK

ELSEVIER Ltd
84 Theobalds Road
London WC1X 8RR
UK

First edition 2005

Library of Congress Cataloging in Publication Data
A catalog record is available from the Library of Congress.

British Library Cataloguing in Publication Data
A catalogue record is available from the British Library.

ISBN: 0-444-51945-9

♾ The paper used in this publication meets the requirements of ANSI/NISO Z39.48-1992 (Permanence of Paper).
Printed in The Netherlands.

This book is dedicated to my
daughters Neeti and Richa

Foreword

The abilities of biological organisms to recognize foreign substances are exquisite and have to some extent been exploited by researchers in the development of biosensors. Using bioreceptors from biological organisms or receptors that have been patterned after biological systems, scientists have developed a new means of chemical analysis that often has the high selectivity of biological recognition systems. These biorecognition elements in combination with various transduction methods have helped to create the rapidly expanding fields of bioanalysis and related technologies known as biosensors. Nowadays, the increasing interest in biosensors is largely a result of the wise applicability of the method, and the considerable selectivity, good sensitivity, and possibility of automation for bioanalysis. Biosensors have been of greater research interest during the past 10 years and this trend is likely to continue for many years with the expanding growth of biomedical technology.

In this excellent book, fundamental principles and biochemical mechanisms on fractals and their relationships with biosensor design and applications are covered in a tutorial manner. Dr. A. Sadana also covers in a comprehensive, logical, and easily readable fashion such topics as single-fractal and dual-fractal analysis, binding and dissociation kinetics as well as important background in fractals, instrumentation, methodologies for a wide variety of application areas of fractals.

The symbols, units, terminology, and organization in this book are consistent, logical and clear throughout. This book should be of considerable use to anyone who wishes to apply fractal analysis to biosensor research, design and applications.

I am proud to have prepared a foreword for such an outstanding book and a fine friend.

Tuan Vo-Dinh
Oak Ridge National Laboratory
July 2005

Preface

Biosensors are finding increasing application in the areas of healthcare, drug design, environmental monitoring, detection of biological, chemical, and toxic agents, biotechnology, aviation, physics, oceanography, and the protection of civilian and engineering infrastructures. This book emphasizes the two areas of expanding biosensor development that include (a) detection of biological and chemical pathogens in the atmosphere, and (b) biomedical applications, especially in healthcare. In both of these cases, one needs to detect analytes at very dilute concentrations.

Modifications in the fractal mathematics is made in the appropriate chapters, and effectively applied to help model the binding and dissociation kinetics of analytes in solution to receptors immobilized on a biosensor surface. It is anticipated that the application of biosensors for the detection of analytes that have medical applications, such as thrombin, prions, glucose, and heat-shock proteins is bound to increase in the future. The early detection of analytes that lead to the onset of disease will considerably assist in its management and control, and hopefully in the alleviation of insidious and intractable diseases. For example, the early detection of markers of diseases such as cancer, systemic lupus erythomatosus (SLE), arthritis, and cardiac-related diseases is invaluable. Biosensors can assist in this effort, and this book emphasizes these viewpoints.

Chapter 1 provides an introduction to biosensors, that includes a definition of a biosensor, advantages and disadvantages of biosensors, and a brief overview of economics of biosensors. Chapter 2 outlines and examines the fractal theory used to model the binding and dissociation kinetics of analytes in solution to receptors on biosensor surfaces. Both simple and complex mechanisms are presented.

Chapter 3 analyzes the kinetics of detection of harmful pathogens on biosensor surfaces. Considering the present geopolitical environment this is an extremely important area of investigation. Large sums of money are being spent in this area of biosensor development by different governmental agencies around the world. Chapter 4 analyzes the binding and dissociation kinetics of heat shock proteins on biosensor surfaces. The levels of these proteins increase when the body is subjected to physiological stress, and they assist other proteins to fold properly. Protein misfolding is involved in quite a few diseases.

Chapter 5 analyzes the binding and dissociation kinetics of prions on biosensor surfaces. These prions are involved in diseases such as mad cow disease (bovine spongiform encephalopathy, BSE) and Alzheimers. The human form of BSE is variant Creutzfeld-Jakob disease (vCJD). Chapter 6 is the heart of the book and presents different examples of the kinetics of binding and dissociation of analytes related to human health on biosensor surfaces. Examples analyzed include the ever prominent glucose, riboflavin, Pb^{++}, prostrate specific antigen (PSA), etc.

Chapter 7 presents the kinetics of binding and dissociation of human heart fatty acid binding protein on biosensor surfaces. This is an early marker of ischemic heart disease. Chapter 8 presents a kinetic analysis of the binding and dissociation of p38α mitogen-activated protein (MAP) kinase occurring on biosensor surfaces. P38α MAP kinase is a well-characterized target for inflammatory disease therapy.

Chapter 9 presents a kinetic analysis of heparin–protein interaction studies on biosensor surfaces. Heparin binds to growth factors, and influences angiogenesis and other proliferation dependent processes. Chapter 10 analyzes the binding and dissociation kinetics of thrombin on biosensor surfaces. The prevention of clot formation due to thrombin is an important area of investigation.

Chapter 11 analyzes the binding and dissociation kinetics of Interleukin 5 (IL5) on biosensor surfaces. IL5 is a single-chain major cytokine, and exerts its hematopoietic growth factor function by binding to a membrane-bound receptor. Chapter 12 presents a kinetic analysis of the binding and dissociation of different environmental contaminants on biosensor surfaces. The kinetics of some of the pollutants analyzed include benzene, toluene, 2,4 dinitrophenol, and methanol.

Chapter 13 presents the market size and economics for biosensors. This chapter provides an overall balance for the whole book. Understandably, this type of information is very difficult to obtain in the open literature free of charge. Not unexpectedly, industries working in the area of biosensors will guard their information very carefully. This is exacerbated by the fact that this is a small and niche market, and putting it mildly, the competition is fierce.

The author wishes to express his appreciation to Dr. Kai-Fong Lee, Dean, School of Engineering at the University of Mississippi for his continued support and encouragement for research that facilitates the writing of treatises like this.

A lot of credit goes to my students who have patiently helped me develop over the years the fractal theory for biosensor applications. My present student, Mr. Atul M. Doke has helped me considerably in getting this book ready for Elsevier.

Contents

– 1 –

Introduction

1.1 DEFINITION OF BIOSENSORS

A chemical sensor is a device that transforms, for example, a concentration of a chemical into a useful analytical signal (adapted from IUPAC, International Union for Pure and Applied Chemistry definition). Walsh (2003) indicates that a biosensor may be considered as a chemical sensor with three components: (a) a receptor, a transducer, and a separator. The receptor or biological element (for example, enzymes, antigens, antibodies, tissues, whole cells, bacteria, etc.) converts the biochemical binding event to a measurable component. The transducer converts this measurable component to generally a measurable electrical or optical signal. The transducer could, for example, be an acoustical device, a calorimetric device, an optical device, or an electrochemical device. The separator (for example, a membrane) separates the transducer from the bioreceptor.

Walsh (2003) has provided some examples of biosensors that have been commercialized, and include: glucose sensors to help monitor sugar levels in diabetics, lactate biosensors, amperometric sensors for gases, and ion-selective electrode (ISE) for blood gases and electrolytes.

Figure 1.1 shows the components of a biosensor (Biowise, 2001). Simply speaking, there is a biological component and an electronic device. The biological receptor (component) reacts with the analyte of interest (binding and/or dissociation), and produces a biochemical change. This biochemical change is transduced or converted to a measurable signal. The amplifier increases the intensity of the signal enabling easier measurement. Biowise (2001) indicates that these components are housed in a single unit that may either be placed at a strategic location or made more portable.

With the advent of nanotechnology, miniaturization, and improved fabrication techniques, there is more and more emphasis on hand-held devices, especially for the detection of biological hazards and biowarfare agents. According to Check (2004), the United States Department of Homeland Security has a $41.5 million program to develop and evaluate hand-held kits to detect harmful biologicals in a possible terrorist attack situation. These are to be used by emergency workers and by first responders. However, these hand-held detection devices still have problems during use. One way around this is to use these hand-held detectors to rule out a lot of other things, and to use them along with other techniques. Some evaluators of these hand-held detectors indicated that they were too limited in their use. Hopefully, with more analysis and research these hand-held

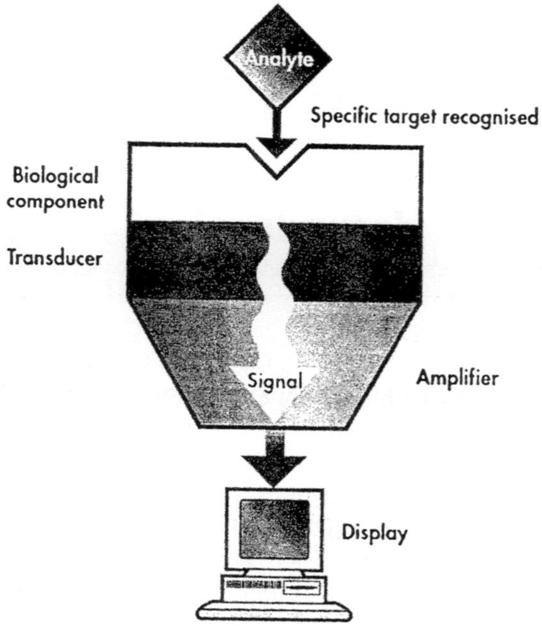

Figure 1.1 Components of a biosensor (Biowise, 2001).

detection devices will become an integral part of the arsenal used by emergency workers and by first responders.

Turner *et al.* (1987) define a biosensor as a 'compact analytical device that incorporates a biological element or a biologically-derived element that is either integrated with or intimately associated with a physicochemical transducer'. These authors further indicate that the signals from the biosensor may be either discrete or continuous. They indicate that the major demand for biosensors is in glucose monitoring for diabetics which comprises 85–90% of the total market for biosensors. This is facilitated by the biosensors providing a convenient, compact, and hygienic method of glucose measurement (Turner, 1996). The three major players involved in glucose monitoring are Abbott, Boehringer Mannheim, and Bayer.

Pharmacia in Sweden has had, and continues to have considerable commercial success with its Biacore surface plasmon resonance (SPR) biosensor. This is based on the SPR principle, and was initially described by Liedberg *et al.* (1983). It is finding increasing application since it monitors reactions in real time. This is an expensive piece of equipment in the $300,000–400,000 range. The software that comes along with it does provide values of the binding and dissociation rate coefficient(s), and affinity values. However, the model used to describe the kinetics assumes (a) the receptors are homogeneously immobilized on the sensor chip surface (no heterogeneity), and no diffusional limitations are assumed to be present if the SPR is run properly. These two assumptions could lead to errors in the estimated values of the binding and dissociation rate coefficients, and affinity values. The fractal analysis presented in this book, and which is used to analyze the binding and dissociation kinetics is an alternative method.

1.2 CURRENT AND FUTURE APPLICATIONS

Traditionally, biosensors have found increasing applications in the biomedical areas. Over the years these areas of applications have expanded to include biotechnology, physics, chemistry, medicine, aviation, food safety, oceanography, and environmental control. Recently, as noted by events occurring world wide, the emphasis has shifted to include biosensor application for the detection of biological and chemical threats, and for homeland security. Other countries, may have a different terminology for homeland security, but the emphasis remains the same: one needs to defend one's borders, and, if one may partially borrow from the motto of the police to preserve and to protect the nation's infrastructure and population.

There has been an increasing emphasis and resource allocation for biosensor research in the recent years, and in the areas of biosensor application. This is made evident, for example, in the Requests for Proposals and Program Solicitations being put out recently by United States Governmental agencies. For example, the National Science Foundation (National Science Foundation Program Solicitation, 2003) in its program solicitation NSF 03-512 entitled Sensors and Sensor Networks (where the proposal deadline was March 06, 2003) indicated the need for the development of sensors to detect biological agents, explosives, and toxic chemicals. Approximately, $34,000,000 was available for competitive research applications. Emphasis was placed on enhancing biosensor performance parameters that included robustness, fewer false alarms, sensitivity, stability, speed of response, regenerability (if possible), and reliability. This document indicated that the availability of wireless and internet communication, and miniaturization and nanotechnology/nanobiotechnology was predicted to place biosensor applications in an increasingly dominant role for sensing and for detection.

This NSF program solicitation was followed by the program solicitation NSF 04-532 entitled 'Sensors and Sensor Networks (Sensors)' (proposals due February 26, 2004) (National Science Foundation Program Solicitation, 2004) that replaced the above mentioned document, and emphasized the advancement of knowledge in materials engineering for biosensor development. Newer concepts and designs were encouraged. Sensors need to be included in engineering systems. Furthermore, it was recommended that sensor data and the analysis of such data should be more included in decision-making processes. The program solicitation emphasized that emerging technologies would impact sensor development significantly, primarily with regard to the decrease in size, weight, and cost. Similar documents, are presumably available in European (United Kingdom, France, Germany, Sweden) and other countries (such as Japan and Australia) to name a few.

The National Science Foundation has come out with a more recent program solicitation document NSF 05-522 entitled 'Sensors and Sensor Networks (Sensors)' (proposals due March 03, 2005) (National Science Foundation Program Solicitation, 2005). The solicitation seeks and if we may quote, 'to advance fundamental knowledge in new technologies for sensors and sensor networks'.

Though, in general academic institutions, can and will place a lot of emphasis on the size and weight of a biosensor, traditionally the economics or the cost of biosensor development and its market cost vis-a-vis the market size is traditionally left to the industry. As expected, one may anticipate that industrial sources would guard their knowledge with regard to biosensor development, especially the economics. Very little, if any, economic information is available in the open literature. If this information is available in the open literature, then presumably it is sparsely available, and spread out in different sources. One of the goals of this book is to provide under one cover the economic information on biosensors such as market size, cost of development, number of years required to develop and test a prototype, etc. if available in the open literature. The last chapter in the book is devoted to this area.

In order that one may obtain a better perspective of where the current applications of biosensors are (along with the research areas emphasized by the above mentioned National Science Foundation funding possibility documents), we now provide a list of recent areas of biosensor research available in the literature. This is only a partial list. Other recent biosensor examples, where the kinetics of binding (and dissociation) have also been analyzed in detail are presented in later chapters.

Some of the biosensor applications that have recently appeared in the literature include:

(a) *Acoustic Wave Chemical Sensor*: Valentine *et al.* (2004) have very recently developed an acoustic wave chemical sensor. This is based on the microelectromechanical systems (MEMS) approach. Binding of target molecules to a functionalized surface are determined by these types of sensors. These authors indicate that a sensor should be sensitive, easy to use, fast and be reusable. They emphasize that their approach does satisfy all of the above requirements. Besides, since their sensor has a higher surface area to mass ratio than other sensor designs, such as the cantilevers, their approach exhibits potential for increased sensitivity compared to the other sensor designs.

(b) *Sensing Biomolecules and Cells*: Haddock *et al.* (2003) have recently using tapered fibers to develop a rapid, convenient, and accurate sensor for biomolecules and cells. Their sensor uses volumes of cells around 150 μl. They emphasize that the sensing of biomolecules and cells is important in clinical, pharmaceutical, and in cellular applications (Chuang *et al.*, 2001; Cullum *et al.*, 2000; Ferreira *et al.*, 2001). Using their developed biosensor and an analytical grade spectrofluorometer Haddock *et al.* (2003) were available to detect and measure nicotinamide adenine dinucleotide (NADH), nicotinamide adenine dinucleotide phosphate (NADPH), and Chinese Hamster Ovary (CHO) cells at different concentrations. They indicate that their results show that the sensitivity obtained with their tapered fibers is at least an order of magnitude more than that obtained with a cuvette arrangement.

(c) *Drug Screening*: Borch and Roepstorff (2004) have very recently developed a novel strategy to help identify enzyme inhibitors. They indicate that the activities of some medical drugs are based on their inhibitory action on specific enzyme(s). For example, the anticancer drug, Imatinib (Glivec) that inhibits tyrosine kinases (Capdeville *et al.*, 2002), and HIV protease inhibitors that act against the HIV virus (Molla *et al.*, 1998).

The protocol designed by Borch and Roepstorff (2004) is simple. An enzyme is immobilized on a sensor chip. The activity of the enzyme is noted by incubating the enzymes with model substrates and testing by mass spectrometry for the products. Potential enzyme inhibitors are passed over the sensor chip containing the enzyme. The binding kinetics (if any) is noted by SPR. Then, model substrates are passed over the sensor chip again, and mass spectrometric analysis determines if the enzyme activity has been inhibited by the compounds been tested for possible therapeutic usage. Enzyme inhibitors apparently exhibit an increasing potential for use as therapeutic agents, thus screening procedures, such as those proposed by Borch and Roepstorff (2004) are bound to gain increasing importance in the future.

Skretas and Wood (2004) have recently indicated the need for a variety of drug-screening assays to help test different compounds and protein targets for potential drugs. This needs to be done in a high throughput fashion. These authors have engineered hormone sensitive bacteria for efficient drug screening. Their method is based on ligand binding of *in vivo* sensors. Their *in vivo* sensor was a hormone, and they used it to analyze ligand binding in *Escherichia Coli*. By changing the parameters of their assays and by observing the changes in cell growth these authors were able to report the presence of active compounds. This procedure permitted these authors to help identify drug compounds from a wide range of test molecules.

(d) *Diagnostic Biomarkers*: May *et al.* (2004a,b) very recently indicate that over a million people are diagnosed with cancer each year. It would be extremely beneficial to be able to detect cancer at an early stage. Growth of cancer may be broadly classified into three stages: first (latent phase), second (intermediate) phase, and a third ('blast') phase. In order to improve survivability, it is essential to be able to detect cancer at the earlier stages. May *et al.* (2004a,b) indicate that vascular endothelial growth factor (VEGF) is a potential cancer biomarker. It is present in the normal human blood in very small quantities. These authors indicate that correlations have been obtained between large quantities of VEGF in the serum and in the plasma of cancer patients. They have developed a whole-cell based biosensor for the detection of VEGF *in vivo*. Their biosensor comprises of a monolayer of human umbilical vein endothelial cells (HUVECs) attached to a cellulose triacetate (CTA) membrane on an ISE. These authors were able to optimize the detection limit as a function of exposure time. This increased the sensitivity of their whole-cell based biosensor.

(e) *Pathogen Detection*: Fitch *et al.* (2003) have very recently provided an overall perspective of the detection and identification of chemical and biological agents that may be considered as 'terrorism' threats. They indicate the need for increased sensitivity, greater automation, and fewer false alarms. Furthermore, on a more practical note they indicate the attempts being made to make these systems more cost effective as well as reducing the complexity of these systems in order that they may be more effectively employed in the field. They emphasize the need for *early intervention*.

For example, Inglesby (2000) indicates that plague (caused by *Yersinia pestis*) symptoms occur within 1–6 days after exposure. Fitch *et al.* (2004) indicate that antibiotics are most effective when administered within 24 h of exposure. In order that early intervention may be facilitated in the case of an inadvertent or deliberate (terrorism) exposure to a chemical or a biological agent, Fitch *et al.* (2004) indicate that environmental monitoring systems are in place at major United States cities (Cole, 2003).

Hostadler *et al.* (News, 2004) have developed the triangular identification for genetic evaluation of risk (TIGER) to identify both known and uncharacterized pathogens. This method has the capacity to identify viruses, bacteria, fungi, and parasitic protozoa. The authors claim that TIGER is able to detect mixtures of organisms in the same sample. They anticipate the use of their technique in infectious disease epidemics, biowarfare, food contamination, and human forensics. Using their technique Hostadler and his colleagues were able to identify the SARS virus as a new member of the coronavirus family.

Bae *et al.* (2004) have recently used imaging ellipsometry (IE) to detect *Yersinia enterocolitica*. These authors indicate that this is an optical technique that involves measuring the change of a polarization state of an elliptically polarized beam reflected from thin films (Azzam and Bashara, 1997). Bae *et al.* (2004) point out that the advantage of using the IE for biosensor applications is that it permits label-free detection, it is simple to operate, and it is highly sensitive. Durisin *et al.* (1997) have indicated that *Y. enterocolitica* is a human pathogenic species and causes yersiniosis. This disease is characterized by fever, diarrhea, and abdominal pain. Using their developed immunosensor Bae *et al.* (2004) were able to detect *Y. enterocolitica* concentrations in the range of $10^3 - 10^7$ cfu/ml.

Joshi *et al.* (2004) have recently used a carbon nanotube based biosensor to detect a VX analog and its degradation products. These authors indicate that sarin, soman, and VX are highly toxic nerve agents. They indicate that their degradation products are more stable than the original compounds. Thus, their detection in the atmosphere can be used to (a) prove the existence of the use of these toxic nerve agents, and (b) assist in monitoring the destruction of these harmful compounds. Using electrochemical detection and the use of carbon nanotubes (CNT) these authors were able to (a) detect VX degradation products and (b) with a modification, the detection of the VX analog, Demton-S.

May *et al.* (2004a,b) have very recently developed a whole-cell based biosensor to detect histamine as a model toxin. These authors indicate that histamine resides in seafood, and in patients with severe allergic reactions (Niwa *et al.*, 2000). May *et al.* (2004a,b) emphasize that their biosensor could find applications in homeland security, food and medical areas, and in environmental monitoring. A monolayer of HUVECs was attached to a CTA membrane of an ISE. These authors indicate that histamine alters the permeability of HUVECs. In the absence of toxic agents, and in the presence of potassium (K^+) ions, the monolayer blocks the interface, yielding no response from the ISE. In the presence of toxins, the permeability of the cells is affected, K^+ reaches the ISE, which gives rise to a change in the potential of the ISE.

(f) *Homeland Security*: Viswanathan and Staples (2004) indicate that virtual chemical sensors and odor profiling can be combined to yield effective virtual chemical sensors. They indicate that chemical signatures and electronic odor profiles permits one to quickly recognize and identify the presence of hazardous materials. They emphasize that cargo and port security are very important with regard to preserving homeland security. According to them approximately 20,000 containers enter the United States daily, and screening methods are urgently required that are rapid, and cost-effective. They describe an electronic nose wherein a single solid-state sensor is able to create an unlimited number of chemical sensors. These authors indicate that their method permits them to speciate chemical vapors in less than 10 s with picogram sensitivity using a solid-state surface acoustic wave (SAW) sensor with electronic variable sensitivity. They provided examples to detect chemical and biological compounds from the odors released. These included explosives, contraband drugs, hazardous chemicals, and biologicals.

(g) *Water Safety*: Acha *et al.* (2004) recently indicate that environmental water pollutants such as atrazine (a pesticide) are persistent and can remain in the aquatic environment for years. They indicate that atrazine levels as low as 0.1 ppb (μg/l) are known to cause hermaphroditism in frogs, affect the health of humans, and are responsible for ecological damage. These authors developed a sensitive fiber optic biosensor that contained a two-layer detection element: (a) a cellular layer that contained the detection enzyme, and (b) a pH-sensitive fluorophore. These layers were attached to the distal end of an optical fiber. These authors were able to detect atrazine at sub-ppb concentrations, and their atrazine biosensors had a life time of the order of days.

(h) *Biomimetic Imprinted Polymers*: Bolisay *et al.* (2004) very recently indicate that molecularly imprinting, an emerging technology, has permitted the synthesis of materials with highly specific receptor sites for different analytes (target compounds). These authors have used hydrogels with imprinted cavities to bind to select plant and insect viruses. In spite of the swelling of hydrogels in water these authors indicate that the affinity for the viruses remains high. They indicate that their hydrogel imprinted virus cavities could find application in national security, biologicals production, crop deterioration prevention, and in human and animal health.

Wilson *et al.* (2004) state that molecularly imprinted devices have been used for drug delivery and in chemical detection. They emphasize that molecularly imprinted polymers may be used as a robust substitute for antibodies. For example, they state that conditions in the gastrointestinal (GI) tract would denature antibodies, making them unsuitable as drug delivery devices. They have also formulated lab-on-a-chip microfluidic platforms for the binding and detection of cells.

Lauten and Peppas (2004) very recently indicate that naturally occurring biologicals are not only expensive but are also unstable. Thus, the need to generate synthetic biomaterials that mimic natural recognition properties. These authors have developed the configurational biomimesis process whereby they are able to generate surfaces and polymeric recognition networks that have stereo-specific three-dimensional binding cavities based on a given molecule. They indicate that their technique has the potential to generate synthetic biomaterials with molecular recognition properties that may be applied in the therapeutic and diagnostic areas.

Hilt *et al.* (2004) very recently indicate that biomimetic networks are more robust and cost effective than biological compounds for use as recognition elements as biosensors. These authors have developed methods to integrate biomimetic networks onto silicon substrates. For example, they have micropatterned polymer networks onto silica substrates to recognize D-glucose amongst similar molecules. They have also analyzed the binding and dissociation kinetics, as well as affinities.

(i) *Sol–Gels*: Rayss and Sudolski (2002) indicate that the sol–gel method may be readily employed for transducer immobilization. A glass-like porous structure is created at room temperature, and unstable transducers (typically organic compounds) may be entrapped in a rigid network of silica. Changing the sol–gel composition, gelation conditions, as well as the gel treatment process permits one to tailor-make the properties of the sol–gel matrix (Klein, 1988). Rayss and Sudolski (2002) showed that due to the relationship between the refractive index of a silica film and pH, a sol–gel film deposited on an optical fiber core could be used as a pH-transducing element in a pH biosensing system.

(j) *Food Pathogens*: McLeish (2000) indicates that more than 75 million people become ill every year in the United States due to food poisoning. This is 5-year old data. The food poisoning is due to pathogens such as Salmonella and *E. coli*. Out of these 75 million people about 1–1.5% are hospitalized (~325,000), and about 0.05–0.1% (~5000) of these cases are fatal. This author indicates that Rand, Letcher, and Brown at the University of Rhode Island have developed a fiber optic probe along with immobilized Salmonella antibodies that bind to the pathogen (Salmonella) cell. The Salmonella antibodies are labeled with a fluorescent dye. Rand indicates that the binding of the pathogen cells to the antibodies takes about an hour ($\sim$60 min), and the processing of the concentration signal occurs in about $1\frac{1}{2}$ min. The aim was to extend the application of their biosensor to the detection of pathogens in seafood and a (hand-held) scanning system for supermarket checkout.

Taylor *et al.* (2004) have recently used the SPR biosensor to detect food pathogens and toxins in complex media. They used a multichannel SPR biosensor for the quantitative and simultaneous detection of food pathogens and toxins. For example, using their biosensor these authors were able to detect *E. coli* at levels as low as 104 cfu/ml. They emphasize that this level is two orders of magnitude lower than that obtained and reported by standard SPR or ELISA methods.

Pal *et al.* (2004) have very recently developed a membrane-based immunofiltration assay that is able to detect T2 toxin in wheat and poultry feed. The limits of detection are 12.5 and 25 μg/kg, respectively. This is a competitive analysis method wherein the labeled analyte is T-2 toxin-horseradish peroxidase (T-2 toxin-HRP) and the substrate is 4-chloro-1-naphthol (4CN). These authors indicate that the ELISA method is an order of magnitude more sensitive than their membrane-based methods. They state that soil-fungi contaminate food grains in temperate climates. These fungi produce mycotoxins, such as the T2 toxin which causes alimentary toxicity. ELISA is an appropriate way to do this, but it is time consuming and sophisticated equipment is required. Furthermore, laborious and elaborate procedures are required to remove interfering substances from the matrix of the sample prior to detection of the T2 toxin (Sukhadin, 2003; Langseth and Rundberget, 1998; Pascale *et al.*, 2003).

(k) *Microcantilevers*: Bottomley *et al.* (2004) indicate that an emerging class of chemo-mechanical sensors are microcantilevers (Barnes *et al.*, 1994; Thundat *et al.*, 1995; Chen *et al.*, 1995). A differential stress results when there is adsorption of molecules on one side of the microcantilever. This leads to bending of the microcantilever. Bottomley *et al.* (2004) indicate that this cantilever bending can be measured with angstrom resolution using optical resolution, capacitance and piezoelectric measurements. These authors analyzed the influence of nano- and mesoscale particles on the performance of microcantilever sensors. They noted that the direct injection of, for example, biological fluids without the removal of particles ($>$0.7 μm) may cause problems when cantilevers are used. These authors suggest that the presence of particles in the fluid produces scattering of the laser beam used to measure the cantilever deflection. This may significantly influence the results, thus particle sizes greater than 0.7 μm should be removed from fluids prior to injection to a cantilever.

(l) *Self-assembling arrays*: LaBauer *et al.* (2004) indicate that although protein micro-arrays may be used for high throughput interactions, they are still not widely used. These authors suggest a method, which overcomes the limitations of the currently used methods by making proteins directly on the microarray slide. Their method is the nucleic acid programmable protein array (NAPPA) method. They do indicate that their method, however, still needs refining; for example, a third protein may be necessary to bridge the interaction.

(m) *Nitrogen Monoxide Sensors*: Nitrogen monoxide (NO) is a pollutant that needs to be monitored. Liu *et al.* (2004) indicate that NO also plays a critical role in biochemical processes (Lewis *et al.*, 1995; Palmer *et al.*, 1987). Liu *et al.* (2004) have very recently designed an electrochemical sensor for the selective detection of NO. These authors immobilized a polyoxometalate (POM) cluster on an electrode through a polyelectrolyte matrix. They suggest that apparently the POM electrocatalyzes the reduction of NO. Liu *et al.* (2004) further suggest that the reduction current is proportional to the NO concentration in the range analyzed from 1 nM to 10 μM.

(n) *Catecholamines*: Stoica *et al.* (2004) have very recently developed a biosensor to detect catecholamines using cellobiose dehydrogenase. Henriksson *et al.* (2000) and Cameron and Aust (2004) indicate that cellobiose dehydrogenase is an extracellular hemoflavooxidoreductase that catalyzes the oxidation of cellobiose, cellodextrins, and a few low molecular saccharides. A cellobiose dehydrogenase-modified electrode was used for the amperometric detection of catecholamines in the flow-injection mode. Stoica *et al.* (2004) indicate the need to detect catecholamines (a biogenic amine) as they are involved in a wide range of neural pathways. These biogenic amines may act as neurotransmitters and as hormones. These authors further indicate that the concentration of these biogenic amines is in the sub-nanomolar range, and very sensitive methods with low detection levels are required. Their sensor was able to detect catecholamines at levels lower than 1 nM.

(o) *Antibody Nanoarrays*: Klenerman *et al.* (2004) have developed a technique that attaches antibodies to a nanoscale surface. This permits the authors to create antibody nanoarrays. A nanosurface was created by using a gallium focus ion beam microscope. Regularly spaced holes were on a thin gold film of thickness 50 nm. A self-assembled monolayer of 3-mercaptopropionic acid permitted the immobilization of IgG antibodies on the array surface via electrostatic interactions. The authors were also able to minimize nonspecific adsorption.

(p) *SPR and Improved SPR*: Inherent diffusional limitations are present in the SPR biosensor. Furthermore, Knoll *et al.* (2004) indicate that an additional limitation is the thin metal layer on the SPR chip surface may quench the fluorescence signal. This is especially true if the fluorophore is near the surface. One way of overcoming this fluorescence signal limitation, these authors suggest, is to keep the protein interaction 'far away' from the surface. They used long molecules as a scaffold on which the target proteins reside. This permitted these authors to obtain complete fluorescence detection.

(q) *Sol–gel Particle Polyurethane Glucose Biosensors*: Shin *et al.* (2004) recently indicate that mild synthesis conditions involved in sol–gel synthesis and the chemical flexibility involved has stimulated research on sol–gels for biosensor applications. They emphasize that sol–gels are porous in nature, and diffusional limitations may be minimized if these sol–gels are used as thin coatings. Sol–gel biosensors have been developed (Chen *et al.*, 2002; Pandey *et al.*, 1999) due to the ambient and aqueous processing conditions required that are favorable for enzymes (which may be used as receptors or as biocatalysts, as required, for example, in glucose biosensors). Shin *et al.* (2004) have developed a nitric-oxide releasing sol–gel particle/polyurethane glucose biosensor. These authors indicate that *in vivo* glucose biosensors still remain a challenge due to the poor incompatibility which leads to scar formation and infection. These authors indicate that nitric oxide is a potent inhibitor of platelet adhesion (Radomski *et al.*, 1987), and an antibacterial agent (Nablo *et al.*, 2001). Nitric oxide has also been identified as an angiogenic factor (Ziche *et al*., 1994). Thus, the hybrid sol–gel/polyurethane glucose biosensor that releases nitric oxide is perhaps a step in the right direction.

(r) *Wireless Glucose Biosensor*: Gai *et al*. (2004) have recently developed a wireless and remote query biosensor. They used a pH-sensitive polymer. A ribbon-like mass-sensitive magnetoelastic sensor is used as a transducer. The magnetoelastic ribbon was coated with a pH-sensitive polymer followed by a layer containing glucose oxidase. These authors indicate that the enzymatic oxidation of glucose decreases the pH. The decrease in pH is detected by the pH-sensitive polymer, which shrinks. This results in a reduction in the mass load on the magnetoelastic transducer. This decrease in the mass load leads to an increase in the resonance frequency of the magnetoelastic sensor.

The oxidation of β-D-glucose results in the production of β-D-gluconic acid. The dissociation of gluconic acid produces H^+. This leads to the shrinking of the polymer after the H^+ has diffused to the bulk solution.

(s) *Multianalyte Sensors*: Misiakos *et al*. (2004) have recently developed an optical real-time affinity sensor. This sensor uses a monolithic silicon optoelectronic transducer and a microfluidic module. These authors indicate that some of the features that permit the application of biosensors to a variety of fields include miniaturization, portability, multianalyte potential, and interfacing with electronic functions. Turner (2000) has emphasized that optical detection in biosensors is superior to other sensing approaches since optical transducers are versatile, and a large variety of labels (such as fluorescent tags) could be used. This real-time affinity biosensor developed by Misiakos *et al*. (2004) was able to detect, for example, gold nanoparticle labeled streptavidin at 3.8 pM. Furthermore, these authors demonstrated the multianalyte capabilities of their biosensor by simultaneously monitoring in real time the binding of (a) streptavidin to biotinylated bovine serum albumin, and (b) antimouse IgG to mouse IgG. Streptavidin and antimouse IgG were in solution, whereas the biotinylated bovine serum albumin and antimouse IgG were immobilized on adjacent fibers of the same chip.

1.3 BIOSENSOR ECONOMICS

In this section we briefly provide some economic numbers on the biosensor market, and the estimated growth worldwide. These are estimates, and should be treated as such. More details about the biosensor market, and other factors involved therein are provided in the last chapter of the book.

Biosensor and bioelectronic devices may be characterized into the following areas: agriculture, food analysis, medical analysis, high throughput screening, and nano-biotechnology (Talukder, 2002). According to this author these categories may be subdivided further into specific applications. Kelzai (2004) emphasizes the four major driving forces for the development of biosensors:

(a) Increasing rate of obesity and the rising rate of diabetes. This necessitates the monitoring of diabetic patients' glucose levels.
(b) The pharmaceutical industry is continuously looking for methods to be able to screen for new drugs. Biosensors are a method to provide these rapid assays required.

(c) The newly emerging war on terror and biowarfare is bound to gain increasing importance worldwide with an increasing investment in standard, unconventional, and innovative biosensors. The need for efficient and accurate handheld devices for the field diagnoses of harmful biologicals is now more and more evident.
(d) Other—environmental, food safety, and in general, improving the quality of life.

Kelzai (2004) indicated that the worldwide market for biosensors in the year 2003 was projected to be \$7.3 billion. At a conservative projected growth rate of 10.4% for biosensors, the estimated market is provided in Table 1.1a.

Projections have been made using the 10.4% growth factor for 5 years hence presuming the present day geopolitical worldwide conditions. The growth rate may become more significant (increase more than the assumed 10.4% if the geopolitical conditions demand it), if, presumably, there is another incident of the magnitude of September 11, 2002 that occurred in New York, NY.

Technical advances would also very significantly impact the market size for biosensors. The estimated cost of developing a biosensor is around \$20 million (Walsh, 2003). This is an older report, but a 10% increase in cost per year puts the present estimated cost of development around \$40 million (once again, using the factor of 72). Needless to say, this type of investment requires important marketing choices (Walsh, 1998), since one really has to look for a 'niche'. It is estimated that around 90% of the market is in the medical area, and testing for diabetes is the major market there. However, and as expected, the more established companies already have diagnostic tests for diabetes (glucose testing), and one might expect fierce competition for market share.

Perhaps, one needs to pick up on another ailment, and develop a biosensor to detect it before it gets to be a full-blown disease. For example, cancer markers, or markers for the onset of ischemic heart disease. As expected, the earlier one is able to detect these types of diseases, the better is the quality of life once treatment starts, and perhaps, there is an improvement in the prognosis of the disease. Testing for the onset of autoimmune diseases, such as arthritis, systemic lupus erythomaosus (SLE), etc. is bound to gain

Table 1.1a

Estimated worldwide market in \$ billions for biosensors. Base year 2003. Estimated growth rate per year 10.4% (adapted from Kelzai, 2004)

Year	Estimated market \$, billions
2003	7.3
2004	8.06
2005	8.90
2006	9.83
2007	10.85
2008	11.98
2009	13.23
2010	14.60

importance in the future. Incidentally, diabetes is also an autoimmune disease. For SLE, one needs to estimate levels of different analytes (such as creatinine, autoantibodies, etc.) to be able to predict the onset of a 'flare', and thus help better control the disease. This was quite evident at the 11th International Congress of Immunology held in July/August 2001 in Stockholm, Sweden, where quite a few companies displayed the state-of-the-art biosensors for the early detection of quite a few autoimmune diseases. However, this type of a market is speciality type of a market, and the demand for these types of biosensors will be nowhere near that of diabetes testing. Finally, hopefully, as technologies develop the cost of manufacturing these integrated devices will decrease, and their reliability will improve to the extent to make biosensors a competitive product with regard to other existing methods of detection.

As expected, the worldwide estimates for the biosensor market differs when projected by different authors and sources. It is perhaps useful to provide another estimate, from a very reliable source (*www.cranfield.ac.uk/biomark.htm*) even though the estimate may be rather dated (for the year 1996). The estimate was for one billion British pounds. Using an exchange rate of one British pound equal to 1.8269 US $ (Exchange checked on October 21, 2004). This works out to $1.82 billion. Using a 10.4% growth rate in the worldwide biosensor market, Table 1.1b shows the estimated worldwide market going back a few years (this is unusual), and then projecting ahead till the year 2010.

It is of interest to compare the estimated worldwide markets for biosensors from the two different sources (Kelzai, 2004) and *www.cranfield.ac.uk/biomark.* The initial estimate from Cranfield University is 1.0 billion British pounds for the year 1996, and this translates to 1.827 billion $. This translates to 3.656 billion $ for the year 2003, which may be compared to the estimate for 7.3 billion $ for the year 2003 by Kelzai (2004). The two

Table 1.1b

Estimated worldwide market in $ billions for biosensors (Base year 1996) (www.cranfield.ac.uk/biomark.htm). Estimated growth rate per year 10.4% (adapted from Kelzai, 2004)

Year	Estimated market, $ billions
1996	1.827
1997	2.017
1998	2.227
1999	2.459
2000	2.715
2001	3.00
2002	3.312
2003	3.656
2004	4.037
2005	4.446
2006	4.91
2007	5.42
2008	5.985
2009	6.61
2010	7.30

Table 1.2

Estimated worldwide market in $ billions for patient monitoring devices. Base year 2002. Estimated growth rate 10.0% (adapted from McWilliams, 2003)

Year	Estimated market $ billions
2002	6.25
2003	6.88
2004	7.57
2005	8.33
2006	9.16
2007	10.1
2008	11.1
2009	12.2
2010	13.4

estimates differ by only a factor of two, which is remarkable considering the way estimates are made, and these are from two different sources.

McWilliams (2003) has recently analyzed the market for patient monitoring devices. This author defines these as 'products that measure, display, and document physiological information obtained at regular intervals of time from sensors or other devices attached to a patient.' The regular intervals phrase distinguishes patient monitoring devices from diagnostic kits and devices. Some common measurements made by these devices include electrocardiogram (ECG), noninvasive blood pressure, body temperature, and respiration rate. This author mentions that estimates for the monitoring market though expanding do differ widely from different sources. In the year 2002, the estimated worldwide market according to McWilliams (2003) was estimated to be $2.5–10.0 billion. Assuming a 10% increase in the worldwide market for monitoring devices Table 1.2 indicates the estimated worldwide market till the year 2010 for patient monitoring devices. As expected, factors such as technological breakthroughs, economic and regulatory considerations would presumably greatly impact the size and structure of the market. For the base year (2002) we start with the average of $2.5–10.0 billion = $6.25 billion.

If one were to look at the numbers presented in Table 1.1a,b and Table 1.2, one would perhaps come to the conclusion that the estimated worldwide market for the year 2010 is roughly the same order of magnitude for the biosensor market and for the patient monitoring devices. This is of course considering the nature of the estimates made, the assumptions used, and the inherent variability of economic estimates.

1.4 OVERVIEW

The material to be presented in the chapters to follow is now briefly presented. In Chapter 2 we present modeling and theory involved in the binding and in the dissociation phases. The theory behind single-, double-, and triple-fractal analysis is presented. In Chapter 3 we present examples of the binding of different types of pathogens (food, involved in biological warfare, etc.) to biosensor surfaces, and how their binding and dissociation may be modeled using fractal analysis. The detection of pathogens (especially those that have a 'terrorist' potential) is becoming more and more critical.

In Chapter 4 we analyze the binding and dissociation of heat shock proteins. These become important in the human body, when the body is 'stressed'. They are also important with regard to the proper protein folding process. Continuing along the line of protein folding, Chapter 5 analyzes the binding and dissociation of prions on biosensor surfaces. Prions are 'misfolded' proteins, that lead to intractable diseases, such as Alzheimers, etc. As expected, these prions have generated, and will presumably continue to generate a considerable amount of interest in the research community. In Chapter 6 we analyze the binding and dissociation kinetics of analytes related to human health on biosensor surfaces. Initially, the medical area has been the major driving force in the development of biosensors. Testing for sugar levels (for the onset or presence of diabetes, and in the control of this aliment) has been a predominant market for biosensors. However, presently, more and more sensors are becoming available in the market for determining the levels of analytes, other than sugar levels. Also, in some intractable, persistent, and perhaps difficult to diagnose diseases, one needs to analyze for more than one particular analyte, and quite frequently. For example, systemic lupus erythomatosus (SLE), where one needs to determine levels of creatinine, autoantibodies, etc. Once these analyte levels are out of their 'normal range', then one can *quickly* take corrective action, for example, the 'flare' in SLE diagnosed individuals.

Continuing along in the medical vein, in Chapter 7 we analyze for and present the binding and dissociation kinetics for human heart fatty acid binding protein. This is an early marker for ischemic heart disease. One of the major themes of this book is the medial slant, and the preponderance of examples of fractal analysis of binding and dissociation kinetics of analytes that have medical implications. The basic idea is to help provide fresh physical insights into these analytes of medical relevance. In Chapter 8 we analyze the binding kinetics of p38α mitogen-activated protein kinase occurring on biosensor surfaces.

In Chapter 9 we present a fractal analysis for heparin–protein interactions on biosensor surfaces. Heparin is another compound that has medical significance. Heparin is an anticoagulant, and is used to decrease the clotting ability of blood. It helps prevent the formation of harmful clots in blood vessels. In Chapter 10 we present an analysis of the binding and dissociation of thrombin on biosensor surfaces. Thrombin is not a normal constituent of circulating blood. It is generated by the cleavage of its precursor in plasma, pro-thrombin. After injury to a particular area, blood flow is restricted to that area (vascular constriction). Then, platelets become activated by thrombin, and aggregate at the site of the injury. In Chapter 11 we present the binding and dissociation kinetics of interleukin to biosensor surfaces. Interleukin-2 (IL-2) is a protein that is manufactured in the body. It stimulates the immune system, and has been approved by the Federal Drug Administration (FDA, USA) for the treatment of certain types of cancer.

The analysis of environmental pollution is an important area of ongoing investigation. Chapter 12 presents the fractal analysis of the binding and dissociation kinetics of different environmental contaminants on biosensor surfaces.

Biosensors are gradually becoming more and more important in our daily applications. On the practical side it would be useful to know what is the present market size for biosensors worldwide, and perhaps also in different geographical locations, for example in different continents, or in sections of the world, such as Europe, the Americas, Far East including Australia, etc. Also, what is the investment required to possibly set-up a biosensor industry, etc. This sort of economic information is difficult to obtain in the open literature, and the industrial sources who presumably have access to this type of data, will needless to say guard it very carefully. Nevertheless, this type of information is very valuable to possess, and the last chapter is an attempt to address this critical need. Besides, the last chapter, provides, and if I may quote a reviewer of this book's proposal, 'a balance to the book'. In Chapter 13 we try to address the market size and economics for biosensors. Information is gleaned from different sources, and placed together (pieced together if I may so) in one chapter. Hopefully, the information is accurate. Predictions of market size, etc. should, needless to say, be viewed with caution. An appropriate caveat, and this is fair to indicate up-front, that besides some consultancy experience, the author has not been involved in a biosensor start-up company. On the other hand, tongue and cheek one might add, that those who have been involved in biosensor start-up companies may not want to reveal their methods or 'secrets'.

REFERENCES

Acha, V, CD Jensen, J Jaeger, DS Dandy, KL Lear and KF Reardon, *Field Measurements of Atrazine in Groundwater Using Fiber Optic Biosensors*, paper 41d, Annual American Institute of Chemical Engineers Meeting, Austin, Texas, November 7–12, 2004.

RMA Azzam and NM Bashara, *Ellipsometry and Polarized Light*. 1st Edn., Chapter 3, North-Holland Publishing Company, Amsterdam, 1997.

Bae, YM, BK Oh, WH Lee and JW Choi, Immunosensor for detection of *Yersinia enterocolitica* based on imaging ellipsometry. *Analytical Biochemistry*, 2004, **76**, 1799–1803.

Barnes, JR, CH Gerber, E Mayer and RR Schlitter, *Chemical and Physics Letters*, 1994, **217**, 79.

Biowise, *Biotechnology at Work: Biosensors for Industrial Applications. A Review of Biosensor Technology*, June, Department of Trade & Industry, Didiot, Oxfordshire, UK, 2001.

Bolisay, LDV, WE Bentley, JN Culvera and P Kofinas, *Recognition of Viruses Using Molecularly Imprinted Polymers*, paper 65a, Annual American Chemical Engineers Meeting, Austin, Texas, November 7–12, 2004.

Borch, J and P Roepstorff, Screening of enzyme inhibitors by surface plasmon resonance combined with mass spectrometry. *Analytical Chemistry*, 2004, **76**, 5243–5248.

Bottomley, LA, MA Poggi and S Shen, Impact of nano- and mesoscale particles on the performance of microcantilever-based sensors. *Analytical Chemistry*, 2004, **76**, 5685–5689.

Cameron, MD and SD Aust, *Enzyme & Microbial Technology*, 2004, **28**, 129–138.

Capdeville, R, E Buchdunger, J Zimmermann and A Matter, *Nat. Reviews Drug Discovery*, 2002, **1**, 493–502.

Check, E, Bioterror tester kits trouble federal agencies. *Nature*, 2004, **1**, 454.

Chen, GY, T Thundat, EA Wachter and RJ Warmack, *Journal of Applied Physics*, 1995, **77**, 3618.

Chen, X, Y Hu and GS Wilson, *Biosensors & Bioelectronics*, 2002, **17**, 1005–1013.

Chuang, H, P Macuch and MB Tobacco, Optical sensors for detection of bacteria. Part 1. General concepts and initial development. *Analytical Chemistry*, 2001, **73**, 462–466.

Cole, S, *Homeland Security Solutions Magazine*, May, 2003, See www.omeda.com/cgiwin/hss.cgi?ADD (online journal).

Cullum, BM, GD Griffin, GH Miller and T Vo-Dinh, Intracellular measurements in mammary carcinoma cells using fiber-optic nanosensors. *Analytical Biochemistry*, 2000, **277**, 25–32.

Durisin, MD, A Ibrahim and MW Griffiths, *International Journal of Food Technology*, 1997, **37**, 103–112.

Ferreira, AP, MM Werneck and RM Ribero, Development of an evanescent-field fibre optic sensor for *E. coli* O157:H7. *Biosensors & Bioelectronics*, 2001, **16**, 399–408.

Fitch, JP, E Raber and DR Imbro, Technology challenges in responding to biological or chemical attacks in the civilian sector. *Science*, 2003, **21**, 1350–1354.

Gai, Q, K Zeng, C Ruan, TA Desai and CA Grimes, A wireless, remote query glucose biosensor based on a pH-sensitive polymer. *Analytical Chemistry*, 2004, **76**, 4038–4043.

Haddock, HS, PM Shankar and R Mutharasan, Evanescent sensing of biomolecules and cells. *Sensors & Actuators B*, 2003, **88**, 67–74.

Henriksson, G, G Johannson and G Petersson, *Journal of Biotechnology*, 2000, **78**, 93–113.

Hilt, JZ, NA Peppas and MA Bryne, *Biomimietic Polymers in Drug Delivery and Sensing Applications: Effect of Network Molecular Structure on Recognition Properties*, paper 66d, Annual American Chemical Engineers Meeting, Austin, Texas, November 7–12, 2004.

Inglesby, TV, *Journal of the American Medical Association*, 2000, **283**, 2281.

Joshi, KA, A Mulchandani and W Chen, *A Biosensor for Sensitive and Direct Determination of VX Analog and Degradation Products Based on Carbon Nanotubes*, paper 36f, Annual American Chemical Engineers Meeting, Austin, Texas, November 7–12, 2004.

Kelzai, F, April, 2004, US & worldwide biosensor market, R7D and commercial implication, http://www.mindbranch.com/catalog/product.jsp?code + R98-105&psrc + i060704, Internet site accessed August 20, 2004.

Klein, LC (ed.), *Sol–gel Technology*. Noyes Publication, New Jersey, 1988.

Klenerman, D, A Brockbauer, D Zhou, D Kang, YE Korchev and CA Bell, *Journal of the American Chemical Society*, 2004, **126**, 6508–6509.

Knoll, W, F Yu, B Persson and S Loefs, *Journal of the American Chemical Society*, 2004, **126**, 8902–8903.

LaBaer, J, N Ramachandran, E Hainsworth, B Bhullar, S Eisenstein, B Rosen, AY Lau and JC Walter, *Science*, 2004, **305**, 86–90.

Langseth, W and T Rundberget, *Journal of Chromatography A*, 1998, **815**, 103–121.

Lauten, H and NA Peppas, *Incorporation of Degradable Components into a Configurational Biomimietic Imprinted Polymer*, paper 66e, Annual American Chemical Engineers Meeting, Austin, Texas, November 7–12, 2004.

Lewis, RS, S Tamir, SR Tannenbaum and WM Deen, *Journal of Biological Chemistry*, 1995, **270**, 29350–29355.

Liedberg, B, C NyLander and I Lundstrom, *Sensors and Actuators*, 1983, **4**, 299–304.

Liu, S, D Volkmer and DG Kurth, Smart polyoxometalate-based nitrogen monoxide sensors. *Analytical Chemistry*, 2004, **76**, 4579–4582.

May, KML, KW Anderson, LG Bachas, *Vascular Endothelial Growth Factor as a Diagnostic Biomarker for the Early Detection of Cancer with a Whole Cell-Based Biosensor*, paper 509c, Annual American Chemical Engineers Meeting, Austin, Texas, November 7–12, 2004a.

May, KML, Y Wang, LG Bachas and KW Anderson, Development of a whole-cell based biosensor for detecting histamine as a model toxin. *Anaytical Chemistry*, 2004b, **76**, 4156–4161.

McLeish, T, New biosensor makes detection of food pathogens quick, easy, 2000, http://www.advance.uri.edu/pacer/october2000/story15.htm, Internet site accessed August 20, 2004.

McWilliams, A, *B-186 Patient Monitoring Devices*, Business Communications Company, Inc., Connecticut, 2003, http://www.bccresearch.com/biotech/B186.html.

Misiakos, K, SE Kakabos, PS Petrou and HS Ruf, A monolithic silicon optoelectronic transducer as a real-time affinity biosensor. *Analytical Chemistry*, 2004, **76**, 1366–1373.

Molla, A, GR Granneman, E Sun and D Kempf, *Antiviral Research*, 1998, **39**, 1–23.

Nablo, BJ, TY Chen and MH Schoenfisch, *Journal of the American Chemical Society*, 2001, **123**, 9712–9713.

National Science Foundation Program Solicitation, 2003, NSF 03-512, Sensors and Sensor Networks, National Science Foundation Funding Opportunity Document, National Science Foundation, Arlington, Virginia, USA, Proposals due March 06, 2003.

National Science Foundation Program Solicitation, 2004, NSF 04-522, Sensors and Sensor Networks (Sensors), National Science Foundation Funding Opportunity Document, National Science Foundation, Arlington, Virginia, USA, Proposals due February 26, 2004.

National Science Foundation Program Solicitation, 2005, NSF 05-526, Sensors and Sensor Networks (Sensors), National Science Foundation Funding Opportunity Document, National Science Foundation, Arlington, Virginia, USA, Proposals due March 03, 2005.

News, Detecting newly emerging pathogens by MS, *Analytical Chemistry*, 2004, 261A.

Niwa, O, R Kunta, K Hayashi, T Horiuchi, K Torimitsu, K Maeyama and K Tanizawa, *Sensors & Actuators B*, 2000, **67**, 43–51.

Pal, A, D Acharya, D Saha and TK Dhar, Development of a membrane-based immunofiltration assay for the detection of T-2 toxin. *Analytical Chemistry*, 2004, **76**, 4237–4240.

Palmer, RMJ, AG Ferrige and S Moncada, *Nature*, 1987, **327**, 524–526.

Pandey, PC, S Upadhay and HC Pathak, *Sensors & Actuators B*, 1999, **60**, 83–89.

Pascale, M, M Haidukowski and A Visconti, *Journal of Chromatography*, 2003, **989**, 257–264.

Radomski, MW, RMJ Palmer and S Moncada, *British Journal of Pharmacology*, 1987, **92**, 639–646.

Rayss, J and G Sudolski, Ion adsorption in the porous sol–gel silica layer in the fiber optic pH sensor. *Sensors & Actuators B*, 2002, **87**, 397–405.

Shin, JH, SM Marxer and MH Schoenfisch, Nitric oxide-releasing sol–gel particle/polyurethane glucose biosensors. *Analytical Chemistry*, 2004, **76**, 4543–4549.

Skretas, G and DW Wood, *Engineering Hormone-Sensitive Bacteria for Efficient Drug Screening*, paper 509a, Annual American Chemical Engineers Meeting, Austin, Texas, November 7–12, 2004.

Stoica, L, A Lindgren-Sjolander, T Ruzgas and L Gorton, Biosensor based on cellulose dehydrogenase for detection of catecholamines. *Analytical Chemistry*, 2004, **76**, 4690–4696.

Sukhadin, DL, *Technology Letters*, 2003, 97–107.

Talukder, G, *Biosensors & Bioelectronics*, Business Communications Company, Inc., Connecticut, 2002, http://www.buscom/biotech/B159.htm, Internet site accessed August 20, 2004.

Taylor, A, J Ladd, Q Yu, S Chen, J Homola and S Jiang, *Surface Plasmon Resonance (SPR) Biosensor Detection for Food Safety*, paper 36c, Annual American Chemical Engineers Meeting, Austin, Texas, November 7–12, 2004.

Thundat, T, GY Chen, RJ Warmack, DP Allison and EA Wachter, *Analytical Chemistry*, 1995, **77**, 3618.

Turner, APF, *Biosensors: past, present, and future*, 1996, http://www.cranfield.ac.uk/biotech/chinap.htm.

Turner, APF, *Science*, 2000, **87**, 397–405.

Turner, APF, I Karube and GS Wilson, *Biosensors: Fundamentals and Applications.* Oxford University Press, Oxford, 1987.

Valentine, JE, TM Przybycien, S Hauan, TM Przybycien and S Hauan, *Response Surface Determination of a Multi-Target MEMS Sensor*, paper 509f, Annual American Chemical Engineers Meeting, Austin, Texas, November 7–12, 2004.

Viswanathan, S and EJ Staples, Homeland Security, *Olfactory Images, and Virtual Chemical Sensors*, paper 41a, Annual American Chemical Engineers Meeting, Austin, Texas, November 7–12, 2004.

Walsh, M, Nano- and MEMS technologies for chemical biosensors, *Advanced Technology Program*, updated April 29, 2003, http://www.atp.nist.gov/atp/focus/98wp-nan.htm.

Wilson, TM, H He, J Guan and LJ Lee, *Molecularly Imprinted Polymers Microdevices for Cellular Recognition*, paper 36e, Annual American Chemical Engineers Meeting, Austin, Texas, November 7–12, 2004.

Ziche, M, E Morbidelli, E Masini, S Amerini, HJ Granger, CA Maggi, P Geppetti and F Ledda, *Journal of Clinical Investigation*, 1994, **94**, 2036–2044.

– 2 –

Modeling and Theory

2.1 INTRODUCTION

In a biosensor based assay the molecule to be detected (analyte) is present in solution and the appropriate receptor is immobilized on a solid surface. The interaction between the analyte and the receptor on the solid biosensor surface is detected either by a change in the refractive index (in SPR instruments) or by changes in the fluorometric intensity, ultraviolet light intensity, etc. The SPR biosensor protocol analyzes the binding (and dissociation where applicable) kinetic curves using classical saturation models involving analyte–receptor binding using 1:1, 1:2, etc. ratios, generally under diffusion-free conditions and assuming that the receptors are homogeneously distributed over the sensor surface. Computer programs and software that come with the equipment provide values of the binding (and the dissociation) rate coefficients. Though a careful analysis and experimental protocol may eliminate or minimize the influence of diffusional limitations; realistically speaking, it is more appropriate to include a heterogeneous distribution on the sensing surface. Heterogeneity on the sensing surface and in the biosensor systems itself may be due to other reasons, such as for example, nonspecific binding, inherent irregularities on the sensing surface, mixture of receptors on the surface, and mixture of analytes in solution which includes the analyte of interest.

Two factors need to be addressed whilst analyzing the analyte–receptor binding and dissociation kinetics. The system by its design is heterogeneous. For example, and as indicated above, the receptors immobilized on the biosensor surface may exhibit some heterogeneity, that is, surface roughness. No matter how careful one is in immobilizing the receptors on the biosensor surface, there will be some degree of heterogeneity on the surface. For the analyte–receptor binding (and dissociation) to take place the analyte by the diffusion process must come within the ‘proximity’ of the active site on the receptor. Mass transport limitations may be minimized or eliminated if the system is either properly designed or properly operated or both. In most cases, however, both diffusional effects and heterogeneity aspects will be present in biosensor systems, and their influence on binding and dissociation kinetics need to be determined. Ideally, one would like to determine the influence of each of these separately on the binding and dissociation kinetics. In the theoretical analysis to be presented below (the Havlin analysis, 1989) the effects of diffusion and of heterogeneity are presented coupled together. One possible way of accounting for the presence of diffusional limitations and the heterogeneity that exists on the surface is by using fractals. Ideally, and as indicated above, one would prefer to decouple the influence diffusion and heterogeneity. Presumably, an approach other than fractal analysis is required to decouple these two effects.

A characteristic feature of fractals is self-similarity at different levels of scale. Fractals exhibit dilatational symmetry. Fractals are disordered systems, and the disorder is described by nonintegral dimensions (Pfeifer and Obert, 1989). Fractals have nonintegral dimensions, and are smaller than the dimension they are embedded in. In other words, the highest value that a fractal can have is three. In our case, an increase in the degree of heterogeneity on the biosensor surface would lead to an increase in the value of the fractal dimension. Another way of looking at the fractal dimension is its 'space filling' capacity. The more the space a surface fills, the higher is its fractal dimension. The fractal dimension cannot have a negative value, and very low values of the fractal dimension on the surface indicate that the surface exists as a Cantor like dust.

Kopelman (1988) indicates that surface diffusion-controlled reactions that occur on clusters or islands are expected to exhibit anomalous and fractal-like kinetics. These kinetics exhibit anomalous reaction orders and time-dependent (for example, binding) rate coefficients. As long as surface irregularities show scale invariance they can be characterized by a single number, the fractal dimension. Later on in the book we will characterize the surfaces of the biosensors used in the different examples by a fractal dimension. More specifically, we will characterize the heterogeneity present on these biosensor surfaces by a fractal dimension.

The fractal dimension is a global property, and it is insensitive to structural or morphological details (Pajkossy and Nyikos, 1989). Markel *et al.* (1991) indicate that fractals are scale self-similar mathematical objects that possess nontrivial geometrical properties. Furthermore, these authors indicate that rough surfaces, disordered layers on surfaces, and porous objects all possess fractal structure. A consequence of the fractal nature is a power-law dependence of a correlation function (in our case the analyte–receptor on the biosensor surface) on a coordinate (for example, time).

The binding of an analyte in solution to a receptor attached to a solid (albeit flow cell or biosensor surface) is a good example of a low dimension reaction system in which the distribution tends to be 'less random' (Kopelman, 1988), and a fractal analysis would provide novel physical insights into the diffusion-controlled reactions occurring at the surface. Also, when too many parameters are involved in a reaction, which is the case for these analyte–receptor reactions on a solid (for example, biosensor surface), a fractal analysis provides a useful lumped parameter. It is appropriate to pay particular care to the design of such systems and to explore new avenues by which further insight or knowledge may be obtained in these biosensor systems. The fractal approach is not new and has been used previously in analyzing different phenomena on lipid membranes.

Fatin-Rouge *et al.* (2004) have recently presented a summary of cases where the analysis of diffusion properties in random media have led to significant theoretical and experimental interest. These cases include soils (Sahimi, 1993), gels (Starchev *et al.*, 1997; Pluen *et al.*, 1999), bacteria cytoplasm (Berland *et al.*, 1995; Schwille *et al.*, 1999), membranes (Saffman and Delbruck, 1975; Peters and Cherry, 1982; Ghosh and Webb, 1988), and channels (Wei *et al.*, 2000). In this chapter, and in this book as a whole, we are extending the analysis to analyte–receptor binding (and dissociation) on biosensor surfaces.

Fatin-Rouge *et al.* (2004) indicate that in most real systems disorder may exist over a finite range of distances. Harder *et al.* (1987) and Havlin (1989) indicate that in this range the diffusion process cannot be characterized by the classical Fick's law. In this range, anomalous diffusion applies. Fatin-Rouge *et al.* (2004) emphasize that at larger distances than in the above window range, the effects of disorder on diffusion may be very small due to statistical effects, and may cancel each other.

Prior to presenting the Havlin (1989) analysis modified for the analyte–receptor binding occurring on biosensor surfaces, it is appropriate to present briefly the analysis presented by Fatin-Rouge *et al.* (2004) on size effects on diffusion processes within agarose gels, and apply it to analyte–receptor binding and dissociation for biosensor kinetics. This analysis provides some insights into general fractal-related processes. Fatin-Rouge *et al.* (2004) have considered diffusion within a fractal network of pores. They indicate that fractal networks such as percolating clusters may be characterized by a power law distribution (Havlin, 1989):

$$M \propto (L)^{D_f} \tag{2.1}$$

Here M is the average number of empty holes in the (gel) space characterized by a linear size, L. The exponent, D_f is the mass fraction dimension. Fatin-Rouge *et al.* (2004) emphasize that in the general case of fractals, D_f is smaller than the dimension of space of interest. Furthermore, the independence of D_f on scale is also referred to as self-similarity, and is an important property of rigorous fractals.

Havlin and Ben-Avraham (1987) indicate that the diffusion behavior of a particle within a medium can be characterized by its mean-square displacement, $r^2(t)$ versus time, t, which is written as:

$$r^2(t) = \Gamma t^{(2/D_w)} \tag{2.2a}$$

Here Γ is the transport coefficient, and D_w is the fractal dimension for diffusion. Normal or regular diffusion occurs when D_w is equal to 2. In this case, $r^2(t)$ is equal to Γt. In other words,

$$r^2(t) = 2dDt \tag{2.2b}$$

Here d is the dimensionality of space, and D is the diffusion coefficient.

Harder *et al.* (1987) and Havlin (1989) describe *anomalous diffusion* wherein the particles sense obstructions to their movement. This is within the fractal matrix, or in our case due to heterogeneities on the biosensor surface, perhaps due to irregularities on the biosensor surface. Fatin-Rouge *et al.* (2004) are careful to point out that anomalous diffusion may also occur due to nonelastic interactions between the network and the diffusing particles in a gel matrix (Saxton, 2001). Furthermore, Fatin-Rouge *et al.* (2004) indicate that *anomalous diffusion* is different from *trapped diffusion* wherein the particles are permanently trapped in holes, and are unable to come out of these holes. When the particles (analyte in our case) are in these trapped holes, then as time, $t \rightarrow \infty$, the mean-square displacement, $r^2(t)$ tends to a constant value. Fatin-Rouge *et al.* (2004) emphasize that in real heterogeneous porous media anomalous diffusion of particles occurs over a limited length- or time-scales since the structure is only fractal over a limited size scale. In other words, there is a lower bound and an upper bound over which the fractal structure applies. Similarly, in our case, the anomalous diffusion of the analyte on the biosensor surface occurs over a limited range of length- or time-scales.

For anomalous diffusion, one may combine the right-hand sides of eqs. 2.2a and 2.2b. Then, the diffusion coefficient, D is given by (Fatin-Rouge *et al.*, 2004):

$$D(t) = (1/4)\Gamma t^{[(2/D_w)-1]} \tag{2.3}$$

Due to the temporal nature of $D(t)$, it is better to characterize the diffusion of the analyte in our case by D_w. If we were still talking about the medium and gels, then D_w would refer to the diffusing medium.

We will now develop the theory for the analyte–receptor binding and dissociation on biosensor surfaces. We will use the (Havlin, 1989) approach.

2.2 THEORY

We present now a method of estimating fractal dimension values for analyte–receptor binding and dissociation kinetics observed in biosensor applications. The following chapters will present the different examples of data that have been modeled using the fractal analysis. The selection of the binding and dissociation data to be analyzed in the later chapters is constrained by whatever is available in the literature.

2.2.1 Variable rate coefficient

Kopelman (1988) has indicated that classical reaction kinetics are sometimes unsatisfactory when the reactants are spatially constrained on the microscopic level by either walls, phase boundaries, or force fields. Such heterogeneous reactions, for example, bioenzymatic reactions, that occur at interfaces of different phases, exhibit fractal orders for elementary reactions and rate coefficients with temporal memories. In such reactions, the rate coefficient exhibits a form given by:

$$k_1 = k' t^{-b} \qquad 0 \leq b \leq 1 \ (t \geq 1) \tag{2.4}$$

In general, k_1 depends on time whereas $k' = k_1$ $(t = 1)$ does not. Kopelman (1988) indicates that in three dimensions (homogeneous space) $b = 0$. This is in agreement with the results obtained in classical kinetics. Also, with vigorous stirring, the system is made homogeneous and b again equals zero. However, for diffusion-limited reactions occurring in fractal spaces, $b \geq 0$; this yields a time-dependent rate coefficient. The diffusion-limited binding kinetics of antigen (or antibody or analyte or substrate) in solution to antibody (or antigen, or receptor, or enzyme) immobilized on a biosensor surface has been analyzed within a fractal framework (Sadana and Beelaram, 1994; Sadana *et al.*, 1995). One of the findings, for example, is that an increase in the surface roughness or fractal dimension leads to an increase in the binding rate coefficient. Furthermore, experimental data presented for the binding of HIV virus (antigen) to the antibody immobilized on a surface displays characteristic ordered ‘disorder’ (Anderson, 1993). This indicates the possibility of a fractal-like surface.

A biosensor system (wherein either the antigen, antibody, analyte, or substrate is attached to the surface), along with its different complexities, which include heterogeneities on the surface and in solution, diffusion-coupled reaction, time-varying adsorption or binding rate coefficients, etc. can be characterized as a fractal system.

The diffusion of reactants towards fractal surfaces has been analyzed (De Gennes, 1982; Pfeifer *et al.*, 1984a,b; Nyikos and Pajkossy, 1986). Havlin (1989) has briefly reviewed and discussed these results. The diffusion is in the Euclidean space surrounding the fractal surface (Giona, 1992). Havlin (1989) presents an equation that may be utilized to describe the build-up of the analyte–receptor on a biosensor surface during the binding reaction. The receptor is immobilized on the biosensor surface. This equation is given below. In all fairness, at the outset, it is appropriate to indicate that the biosensor surface is assumed to be fractal, or possibly so.

In general, to demonstrate fractal-like behavior log–log plots of distribution of molecules $M(r)$ as a function of the radial distance (r) from a given molecule are required. This plot should be close to a straight line. The slope of $\log M(r)$ *versus* $\log(r)$ plot determines the fractal dimension. In our case, one could try to obtain a log–log plot of two variables, k and time, t and perform a least squares fit in this parameter space to find the slope of the curve. A regression coefficient at this stage could be beneficial in understanding the efficacy of this metric. However, an easier method, without the use of the required log–log plots, is presented below. This is the equation developed by Havlin (1989) for diffusion of analytes towards fractal surfaces.

2.2.2 Single-fractal analysis

In the literature some authors refer to binding as comprising of two phases, an association phase and a dissociation phase. In this chapter and in the book, we will be refer to binding as just binding. The dissociation phase is separate.

Binding rate coefficient

Havlin (1989) indicates that the diffusion of a particle (analyte) from a homogeneous solution to a solid surface (e.g. receptor-coated surface) on which it reacts to form a product (analyte–receptor complex) is given by:

$$(\text{Analyte}\cdot\text{Receptor}) \sim \begin{cases} t^{(3-D_{\mathrm{f,bind}})/2} = t^{p} & (t < t_{\mathrm{c}}) \\ t^{1/2} & (t > t_{\mathrm{c}}) \end{cases} \tag{2.5a}$$

where the analyte–receptor represents the association (or binding) complex formed on the surface. Here $p = -b$, and D_{f} is the fractal dimension of the surface. Havlin (1989) states that the crossover value may be determined by $r_{\mathrm{c}}^2 \sim t_{\mathrm{c}}$. Above the characteristic length, r_{c}, the self-similarity of the surface is lost and the surface may be considered homogeneous. Eq. 2.5a indicates that the concentration of the product [analyte–receptor] on a solid fractal surface scales at short and intermediate times as analyte–receptor $\sim t^{p}$ with the coefficient $p = (3 - D_{\mathrm{f}})/2$ at short time scales and $p = 1/2$ at intermediate time scales. Note that D_{f}, $D_{\mathrm{f,assoc}}$, and $D_{\mathrm{f,bind}}$ are used interchangeably. This equation is associated with the short-term diffusional properties of a random walk on a fractal surface. Note that, in perfectly stirred kinetics on a regular (nonfractal) structure (or surface), the binding rate coefficient, k_1 is a constant, that is, it is independent of time. In other words, the limit of regular structures (or surfaces) and the absence of diffusion-limited kinetics leads to k_1 being independent of time. In all other situations, one would expect a scaling behavior given by $k_1 \sim k' t^{-b}$ with $-b = p < 0$. Also, the appearance of the coefficient, p different from $p = 0$ is the consequence of two different phenomena, that is, the heterogeneity (fractality) of the surface and the imperfect mixing (diffusion-limited) condition. Finally, for a homogeneous surface where D_f is equal to 2, and when only diffusional limitations are present, $p = 1/2$ as it should be. Another way of looking at the $p = 1/2$ case (where $D_{f,bind}$ is equal to two) is that the analyte in solution views the fractal object, in our case the receptor-coated biosensor surface, from a 'large distance'. In essence, in the association process, the diffusion of the analyte from the solution to the receptor surface creates a depletion layer of width $(Đt)^{1/2}$ where $Đ$ is the diffusion constant. This gives rise to the fractal power law, (Analyte·Receptor) $\sim t^{(3-D_{f,bind})/2}$.

The values of the parameters k (binding rate coefficient), p, and D_f in eq. 2.5a may be obtained for analyte–receptor association kinetics data. This may be done by a regression analysis using, for example, Corel Quattro Pro (1997) along with eq. 2.5a where (analyte·receptor) $= kt^{\mathrm{p}}$ (Sadana and Beelaram, 1994; Sadana *et al.*, 1995). The fractal dimension may be obtained from the parameter p. Since $p = (3 - D_{f,bind})/2$. $D_{f,bind}$ is equal to $(3 - 2p)$. In general, low values of p would lead to higher values of the fractal dimension, $D_{f,bind}$. Higher values of the fractal dimension would indicate higher degrees of 'disorder' or heterogeneity or inhomogeneity on the surface.

Another way of looking at the diffusive process is that it inherently involves fluctuations at the molecular level that may be described by a random walk (Weiss, 1994). This author indicates that the kinetics of transport on disordered (or heterogeneous) media itself needs to be described by a random-walk model. When both of these are present, that is the diffusion phenomena as well as a fractal surface, then one needs to analyze the interplay of both of these fluctuations. In essence, the disorder on the surface (or a higher fractal dimension, D_f) tends to slow down the motion of a particle (analyte in our case) moving in such a medium. Basically, according to Weiss (1994) the particle (random walker analyte) is trapped in regions in space, as it oscillates for a long time before resuming its motion.

Havlin (1989) indicates that the crossover value may be determined by $r_c^2 \sim t_c$. Above the characteristic length, r_c, the self-similarity of the surface of the surface is lost. Above t_c, the surface may be considered homogeneous, and 'regular' diffusion is now present. One may consider the analysis to be presented as an intermediate 'heuristic' approach in that in the future one may also be able to develop an autonomous (and not time-dependent) model of diffusion-limited kinetics in disordered media.

It is worthwhile commenting on the units of the association and the dissociation rate coefficient(s) obtained for the fractal analysis. In general, for SPR biosensor analysis, the unit for the analyte·receptor complex on the biosensor surface is RU (resonance unit). One thousand resonance units is generally 1 ng/(mm)^2 (of surface), or one resonance unit is 1 pg/(mm)^2. Here, ng and pg are nanogram and picogram, respectively. Then, to help determine the units for the binding coefficient, k, from eq. 2.5a:

$$\text{(Analyte·Receptor)}, \mathrm{pg}/(\mathrm{mm})^2 = kt^p = kt^{(3-D_{f,bind})/2}$$

This yields a unit for the binding rate coefficient, k as (pg)(mm)$^{-2}$(s)$^{(D_{f,bind}-3)/2}$. Note that the unit of dependence in time exhibited by the association (or binding) rate coefficient, k changes slightly depending on the corresponding fractal dimension obtained in the binding phase, $D_{f,bind}$. The fractal dimension value is less than or equal to three. Three is the highest value of the fractal dimension, since the system is embedded in a three-dimensional system. k and k_{bind}, and D_f, and $D_{f,bind}$ are used interchangeably in this chapter and in the book.

It should be indicated that different laboratories use different technologies or different experimental designs to analyze the binding affinity of ligands to target proteins (or analytes) of interest (or to determine the rate coefficients for association and dissociation kinetics for binding). The comparison of data between different technologies and experimental designs and conclusions thereof should be made with great caution. The fractal analysis is of value in that it provides the pros and cons of different *in vitro technologies* (or more precisely, in this case, analysis procedures). It makes the user of the technology aware of the quality of data generated and what can be done to improve the analysis.

One might very reasonably question the utility of the approach considering the different dimensions, and subsequently the units one may obtain even for the same interactions. It would be difficult to compare this technique with other approaches for different interactions. Nevertheless, the inclusion of the surface effects is essential, albeit difficult. This is especially true, if the rate coefficients for association and dissociation for binding are very significantly dependent on the nature of the surface. Unless, a simpler and alternate approach is suggested that includes the surface effects, it is reasonable for the present, to follow the present approach. Hopefully, modifications, to this approach may be suggested that permit the comparison for different interactions as well as with other approaches.

It would be useful to specify what the carrier of fractal properties is. It could either be the analyte surface, the receptor surface, or the immobilizing (in our case, the biosensor) surface. There is a considerable body of work on fractal surface properties of proteins (Li *et al.*, 1990; Federov *et al.*, 1999; Dewey and Bann, 1992; Le Brecque, 1992). Le Brecque (1992) indicates that the active sites (in our case the receptors on the biosensor surface) may themselves form a fractal surface. Furthermore, the inclusion of nonspecific association sites on the surface would increase the degree of heterogeneity on the surface, thereby leading to an increase in the fractal dimension of the surface. At present, we are unable to specify what the carrier of the fractal properties is. This is exacerbated by our re-analysis of kinetic data available in the literature. Presumably, it is due to a composite of some or all of the factors mentioned above. No evidence of fractality is presented.

Dissociation rate coefficient

The diffusion of the dissociated particle (receptor or analyte) from the solid surface (e.g. analyte–receptor complex coated surface) into solution may be given, as a first approximation by:

$$\begin{aligned}(\text{Analyte}\cdot\text{Receptor}) &\sim -t^{(3-D_{f,diss})/2}, \qquad (t > t_{diss})\\ &= -k_{diss}t^{(3-D_{f,diss})/2}\end{aligned} \tag{2.5b}$$

Here $D_{f,diss}$ is the fractal dimension of the surface for the dissociation step. t_{diss} represents the start of the dissociation step. This corresponds to the highest concentration of the analyte–receptor complex on the surface. Henceforth, its concentration only decreases. $D_{f,bind}$ may or may not be equal to $D_{f,diss}$. k_d and k_{diss}, and $D_{f,d}$ and $D_{f,diss}$ are used interchangeably in this chapter and in this book.

One may obtain a unit for the dissociation rate coefficient, k_d in a similar manner as done for the binding rate coefficient. In this case, the units for the binding and the dissociation rate coefficient are the same. The unit for the dissociation rate coefficient, k_d is $(pg)(mm)^2(s)^{(D_{f,diss}-3)/2}$. Once again, note that the unit dependence on time exhibited by k_d changes slightly due to the dependence on $D_{f,diss}$.

2.2.3 Dual-fractal analysis

Binding rate coefficient

The single-fractal analysis we have just presented is extended to include two fractal dimensions. At present, the time ($t = t_1$) at which the first fractal dimension "changes" to the second fractal dimension is arbitrary and empirical. For the most part, it is dictated by the data analyzed and the experience gained by handling a single-fractal analysis. The r^2 (regression coefficient) value obtained is also used to determine if a single-fractal analysis is sufficient, or one needs to use a dual-fractal analysis to provide an adequate fit. Only if the r^2 value is less than 0.97 for a single-fractal analysis, do we use a dual-fractal model. In this case, the analyte–receptor complex is given by:

$$(\text{Analyte}\cdot\text{Receptor}) \sim \begin{cases} t^{(3-D_{f1,bind})/2} = t^{p_1} & (t < t_1) \\ t^{(3-D_{f2,bind})/2} = t^{p_2} & (t_1 < t < t_2 = t_c) \\ t^{1/2} & (t > t_c) \end{cases} \tag{2.5c}$$

In analyte–receptor binding the analyte–receptor binds with the active site on the surface and the product is released. In this sense the catalytic surface exhibits an unchanging fractal surface to the reactant in the absence of fouling and other complications. In the case of analyte–receptor association the biosensor surface exhibits a changing fractal surface to the analyte in solution. This occurs since as each association reaction takes place, smaller and smaller amounts of 'association' sites or receptors are available on the biosensor surface to which the analyte may bind. Furthermore, as the reaction proceeds, there is an increasing degree of heterogeneity on the biosensor surface for some reaction systems. This is manifested by two degrees of heterogeneity, or two fractal dimensions on the biosensor surface. In the theoretical limit one might envisage a temporal fractal dimension wherein there is a continuous change in the degree of heterogeneity on the surface; though of course, such situations would be very rare, if at all.

Surfaces exhibit roughness, or a degree of heterogeneity at some scale. This degree of heterogeneity on the surface may be due to fracture or erosion. In our case of biosensors, this may arise due to (a) the inherent roughness of the biosensor surface, or (b) due to the immobilization or deposition of the receptors on the biosensor surface. The method of deposition of the receptors on the surface would also lead to different degrees of heterogeneity on the surface. The binding reaction takes place between the analyte in solution and the receptors on the surface through chemical bond formation and subsequent molecular association. The geometric nature (or parameter) of the surface will significantly influence these reactions. The influence of surface morphology and structure has been analyzed (Lee and Lee, 1994; Chaudhari *et al.*, 2002, 2003). It would be of interest to determine the scale of these roughness heterogeneities. Are these at the Angstrom level or lower? With the current emphasis on nano-technology and nano-biotechnology these types of questions are becoming more and more relevant and of significance. The nature of surfaces in general, and of biosensors in particular (our case) should exhibit a fractal nature at the molecular level. Furthermore, one of the reasons for the emphasis on nano-technology is that as one goes down in scale, the properties of some substances change, sometimes for the better. It is these beneficial changes that one wishes to exploit in nano-technology and nano-biotechnology. Hopefully, similar parallels can be drawn on analyzing the fractal nature of biosensor surfaces. Do they exhibit self-similarity; and if they do what are their limits? In other words, what are their lower and upper bounds.

Furthermore, each binding event, need not result in the formation of an analyte–receptor on the biosensor surface. All of the receptors on the biosensor surface are presumably not, and do exhibit the same activity. In other words, their active sites should comprise of presumably a probability distribution in 'activity.' *In lieu* of any prior information, it is reasonable to assume a bell-shaped Gaussian (or normal) distribution of active sites on the surface. A probabilistic approach is more realistic here. Such sort of an analyses have presumably not been performed (at least this author is unaware of this) for analyte–receptor reactions occurring on biosensor surfaces. Thus, the fractal analysis is a convenient method of providing a lumped parameter analysis of analyte–receptor reactions occurring on biosensor surfaces.

Note that, at present, the dual-fractal analysis does not have a basis at the molecular level. This represents two different levels of heterogeneity on the biosensor surface. But, in some of the examples presented, a single-fractal analysis is clearly inadequate to model the data. Only in these cases does one resort to a dual-fractal analysis. The binding rate coefficients, k_1 and k_2 in the dual-fractal analysis have the same units $(\mathrm{pg})(\mathrm{mm})^{-2} \times (\mathrm{s})^{(D_{\mathrm{f1,bind}}-3)/2}$ and $(\mathrm{pg})(\mathrm{mm})^{-2}(\mathrm{s})^{(D_{\mathrm{f2,bind}}-3)/2}$, respectively, as the association rate coefficient, k, in the single-fractal analysis.

Dissociation rate coefficient

In this case the dissociation rate coefficient is given by:

$$(\text{Analyte}\cdot\text{Receptor}) \sim \begin{cases} -t^{(3-D_{\mathrm{f1,diss}})/2} & (t_{\mathrm{diss}} < t < t_{\mathrm{d1}}) \\ -t^{(3-D_{\mathrm{f2,diss}})/2} & (t_{\mathrm{d1}} < t < t_{\mathrm{d2}}) \end{cases} \tag{2.5d}$$

Here $D_{\mathrm{f,diss}}$ is the fractal dimension of the surface for the dissociation step. t_{diss} represents the start of the dissociation step. This corresponds to the highest concentration of the analyte·receptor on the surface. Henceforth, its concentration only decreases. $D_{\mathrm{f,bind}}$ or $D_{\mathrm{f,assoc}}$ may or may not be equal to $D_{\mathrm{f,diss}}$. The dissociation rate coefficients, k_{d1} and k_{d2} in the dual-fractal analysis have the same units $(\mathrm{pg})(\mathrm{mm})^{-2}(\mathrm{s})^{(D_{\mathrm{fd1}}-3)/2}$ and $(\mathrm{pg})(\mathrm{mm})^{-2} \times (\mathrm{s})^{(D_{\mathrm{fd2}}-3)/2}$, respectively, as the dissociation rate coefficient, k_{d}, in the single-fractal analysis.

2.2.4 Triple-fractal analysis

One resorts to a triple-fractal analysis if the dual-fractal analysis does not provide a reasonable fit. As will be shown later on in the book, one resorts to a triple-fractal analysis when the dual-fractal analysis does not provide an adequate fit. The equation for the fractal analysis equation is generic in nature, and one may easily extend the single-and the dual-fractal analysis equations (eqs. 2.5a and 2.5c) to describe the binding (and/or the dissociation) kinetics for a triple fractal analysis. In fact, in the extreme case, *n* fractal dimensions may be present. In this case, the degree of heterogeneity, D_f or the fractal dimension is continuously changing on the biosensor surface, and the surface needs to be represented by D_{fi} where *i* goes from 1 to *n*. Similarly, we have *n* binding rate coefficients on the biosensor surface. A similar representation may also be or made for the dissociation phase.

It is perhaps appropriate to at least mention one more approach that has been used to model the binding kinetics on surfaces.

2.2.5 Pfeifer's fractal binding rate theory

Pfeifer (1989) has suggested an alternate form of the binding rate theory. In the equation given in this reference *N* is the number of complexes, N_0 is the number of receptors on the solid surface, *D* is the diffusion coefficient of the analyte, *L* is the receptor diameter, and λ is the mean distance between two neighboring receptors. This equation may also be used to analyze the analyte–receptor binding kinetics. The problem, however, is that it may not be possible in all instances to estimate *a priori* all of the parameters described in the equation (not given here). In that case, one may have to approximate or assume certain values, and this will affect the accuracy and reliability of the analysis. The suggested equation does have an advantage compared to the fractal analysis described above in that it does include a prefactor necessary to convert the time interval over which fractal scaling is observed into a length interval. It also provides an expression for $t_c(= L^2/D)$, which separates the short-term regime from the long term regime. The short-term regime is the one in which the anomalous diffusion applies. At the end of the short term interval ($t = t_c$), the self-similarity of the system is lost, the surface is homogeneous, and regular diffusion applies.

Pfeifer (1989) states that the application of the above equation is contingent on:

(a) The analyte is uniformly distributed in the solution at time, *t* equal to zero.
(b) Binding is irreversible and first-order (*N* equals the number of analyte particles that have reached the receptors).
(c) Binding occurs whenever an incoming analyte particle hits a receptor surface for the first time. In other words, the 'sticking' probability is one.

It is very difficult to imagine perhaps any one or all of the these conditions being satisfied for analyte–receptor binding interactions occurring in continuous-flow reactors.

Given the extremely small volume of the flow channels there is a high probability that the mixing of the analyte is not proper. This in turn may lead to analyte depletion in the flow channel. Also, the binding cannot be assumed to be irreversible in all instances. There may be cases of extremely fast binding and dissociation, especially for analytes with low affinity which can dissociate in the continuously flowing buffer without any regeneration reagent.

Condition (c) may be satisfied, however, it does not include the 'sticking' probability, in that each collision leads to a binding event. Also, the presence of nonspecific binding, avidity effects, and binding with reactions or binding of dissociated analytes may interfere with condition (c) being satisfied. Also, the equation makes assumptions about the number of active sites, and the immobilized receptors. For example it states that the analyte binds to one specific active site. The receptor cannot bind to more than one analyte molecule at one time (1:1 binding).

The equilibrium dissociation rate coefficient, $K_D = k_{diss}/k_{assoc}$ can be calculated using the above models. The K_D value is frequently used in analyte–receptor reactions occurring on biosensor surfaces. The ratio besides providing physical insights into the analyte·receptor system is of practical importance since it may be used to help determine (and possibly enhance) the regenerablility, reusability, stability, and other biosensor performance parameters. K_D has the unit $(s)^{[D_{f,diss}-D_{f,assoc}]/2}$. This applies to both the single- as well as the dual-fractal analysis. For example, for a single-fractal analysis, K_D has the units $(s)^{[D_{fd}-D_f]/2}$. Similarly, for a dual-fractal analysis, the affinity, K_{D1} has the units $(s)^{[D_{fd1}-D_{fassoc1}]/2}$ and K_{D2} has the units $(s)^{[D_{fd2}-D_{fassoc2}]/2}$. Note the difference in the units of the equilibrium dissociation rate coefficient obtained for the classical as well as the fractal-type kinetics. Though the definition of the equilibrium dissociation rate coefficient is the same in both types of kinetics (ratio of the dissociation rate coefficient to the association rate coefficient), the difference(s) in the units of the different rate coefficients eventually leads to a different unit for the equilibrium dissociation rate coefficient in the two types of kinetics. This is not entirely unexpected since the classical kinetic analysis does not include the characteristics of the surface in the definition of the equilibrium dissociation rate coefficient whereas the present fractal analysis does. Thus, one may not be able to actually compare the equilibrium dissociation rate coefficient affinities in these two types of systems. This is a significant difference in the kinetic analysis of binding and dissociation reactions on biosensor surfaces from what is available in the literature.

REFERENCES

Anderson, J, NIH Panel Review Meeting, Case Western Reserve University, Cleveland, Ohio, July 1993.

Berland, KM, PTC So and E Gratton, Two-photon fluorescence correlation spectroscopy: method and application to the intracellular environment. *Biophys. J.*, 1995, **68**, 694–701.

Chaudhari, A, CC Yan and SL Lee, *Chem. Phys. Lett.*, 2002, **207**, 220.

Chaudhari, A, CC Yan and SL Lee, *J. Phys. A*, 2003, **36**, 3757.

Corel Quattro Pro, Corel Corporation Limited, Ottawa, Canada, 1997.

De Gennes, PG, Diffusion-controlled reactions. *Polymer Melts. Radiat. Phys. Chem.*, 1982, **22**, 193.

Dewey, TG and JG Bann, Diffusion-controlled reaction in polymer melts. *Biophys. J.*, 1992, **63**, 594.

Fatin-Rouge, N, K Starchev and J Buffle, Size effects on diffusion process with agarose gels. *Biophys. J.*, 2004, **86**, 2710–2719.

Federov, BA, BB Federov and PW Schmidt, An analysis of the fractal properties of globular proteins. *J. Chem. Phys.*, 1999, **99**, 4076–4083.

Giona, M, First-order reaction-diffusion in complex fractal media. *Chem. Eng. Sci.*, 1992, **47**, 1503–1515.
Ghosh, RN and WW Webb, Results of automated tracking of low density lipoprotein receptors on cell surfaces. *Biophys. J.*, 1988, **53**, A352.
Harder, FH, S Havlin and A Bunde, Diffusion in fractals with singular waiting-time distribution. *Phys. Rev. B*, 1987, **36**, 3874–3879.
Havlin, S, Molecular diffusion and reaction, in *The Fractal Approach To Heterogeneous Chemistry: Surfaces, Colloids, Polymers*. (ed. D Avnir), Wiley, New York, 1989, pp. 251–269.
Havlin, S and D Ben-Avraham, Diffusion in disordered media. *Adv. Phys.*, 1987, **36**, 695–798.
Kopelman, R, Fractal reaction kinetics. *Science*, 1988, **241**, 1620–1624.
Le Brecque, M, *Mosaic*, 1992, **23**, 12–15.
Lee, CK and SL Lee, *Chem. Phys. Lett.*, 1994, **228**, 539.
Li, HL, S Chen and H Zhao, Fractal mechanisms for the allosteric effects of proteins and enzymes. *Biophys. J.*, 1990, **58**, 1313–1320.
Markel, VA, LS Muratov, MI Stockman and TF George, *Phys. Rev. B*, 1991, **43**(10), 8183.
Nyikos, L and T Pajkossy, Diffusion to fractal surfaces. *Electrochim. Acta*, 1986, **31**, 1347.
Pajkossy, T and L Nyoikos, Diffusion to fractal surfaces II. Verification of theory. *Electrochim. Acta*, 1989, **34**, 171.
Peters, R and RJ Cherry, Lateral and rotational diffusion of bacteriorhodopsin in lipid bilayers: experimental test of the Saffman–Delbruck equations. *Proc. Natl Acad. Sci., USA*, 1982, **79**, 4317–4321.
Pfeifer, P and M Obert, in *The Fractal Approach to Heterogeneous Chemistry: Surfaces, Colloids, Polymers*. (ed. D Avnir), Wiley, New York, 1989, pp. 251–269.
Pfeifer, P, YJ Wu, MW Cole and J Krim, Multilayer adsorption on a fractally rough surface. *Phys. Rev. Lett.*, 1989, **62**(17), 1997–2000.
Pfeifer, P, D Avnir and DJ Farin, Molecular fractal surfaces. *Nature (London)*, 1984a, **308**, 261.
Pfeifer, P, D Avnir and DJ Farin, Surface geometric irregularity of particulate materials. The fractal approach. *J. Colloid and Interface Sci.*, 1984b, **103**(1), 112.
Pluen, A, PA Netti, KJ Rakesh and DA Berk, Diffusion of macromolecules in agarose gels: comparison of linear and globular configurations. *Biophys. J.*, 1999, **77**, 542–552.
Sadana, A and A Beelaram, Fractal analysis of antigen–antibody binding kinetics. Biosensor applications. *Biotechnol. Progr.*, 1994, **9**, 45.
Sadana, A, JP Alarie and T Vo-Dinh, A β-cyclodextrin based fiber-optic chemical sensor: a fractal analysis. *Talanta*, 1995, **42**, 1567.
Saffman, PG and M Delbruck, Brownian motion in biological membranes. *Proc. Natl Acad. Sci., USA*, 1975, **72**, 3111–3113.
Sahimi, M, Flow phenomena in rocks: From continuum models to fractals, percolation, cellular automata and simulated annealing. *Rev. Mod. Phys.*, 1993, **65**, 1393–1534.
Saxton, MJ, Anomalous diffusion due to binding: a Monte Carlo study. *Biophys. J.*, 2001, **77**, 2251–2265.
Schwille, P, U Haupts, S Maiti and WW Webb, Molecular dynamics in living cells observed by fluorescence correlation spectroscopy with one- and two-photon excitation. *Biophys. J.*, 1999, **77**, 2251–2265.
Starchev, K, J Sturm, G Weill and CH Brogen, Brownian motion and electrophoretic transport in agarose gels studied by epifluorescence microscopy and simple particle tracking analysis. *J. Phys. Chem.*, 1997, **101**, 5659–5663.
Wei, QH, C Bechinger and P Leiderer, Single-file diffusion of colloids in one-dimensional channels. *Science*, 2000, **287**, 625–627.
Weiss, GH, *Fractals in Science*, Springer-Verlag, p. 119, Berlin, 1994.

– 3 –

Fractal Analysis of Pathogen Detection on Biosensors

3.1 INTRODUCTION

The development of biosensors was until recently propelled mainly by medical applications. Recent events on a global level have thrusted the applications of biosensors to the detection of chemical and biological agents of mass destruction. Note that in both medical and in national security applications one needs to detect viruses and pathogens at rather low levels and in dilute concentrations. The principles of application and development in both areas exhibit similarities, with the major difference being that in most medical applications, pathogens are not immediately life-threatening, in contrast to security applications when there may be merely minutes to hours (Alwing, 2002) to (a) take corrective action to remove or to neutralize the pathogen, and (b) to remove the endangered civilian and military personnel from the affected areas. The identification of exotic and lethal diseases most frequently associated with biological weapons programs is a slow, uncertain, and fragmented process.

There was and continues to be increasing pressure to enhance biosensor performance characteristics such as stability, sensitivity, response time, robustness, selectivity, and regeneration. The biosensor system for destructive biological agents requires not only the detection of a harmful biological agent, but also its identification. A three-step process may be necessary that includes a detection step, an identification step, and finally a communication step for alarm and/or control systems Furthermore, this agent is often present in a very dilute solution, and there is an urgent need to minimize false positives. Biological and chemical agents can be present in different types of situations, such as military applications, forensics, law enforcement, and medical diagnostics (Fitch *et al.*, 2003). Each of these situations will have common themes as well as characteristics that are unique to their particular application. In a broad sense, one can present some underlying principles which may have general applicability with the caveat that these may have to be modified to one's particular application. It is imperative to develop fast, accurate, and reliable detection systems for pathogens. The ultimate aim is to provide a continuous monitoring system.

Fitch *et al.* (2003) have recently outlined the technology challenges one may expect in the civilian sector when responding to chemical and biological attacks. As expected, there is a plethora of activity in the governmental, industrial, and academic areas to promote the development of biosensors not only in the chemical and biological detection area for military applications, but also, for example, making edible food safe. Some of the techniques used for the detection of biological agents include optical biosensors (Ligler, 2001), structure–function based biosensors (Swanson, 2001), and time-of-flight Mass Spectrometry and Laser-Induced Fluorescence (Stowers, 2001). As expected, nanotechnology has played (Salamone, 2003), and will continue to play a major role in the detection of pathogenic agents. Most of these detection systems should be portable. The Defense Advanced Research Projects Agency (DARPA) (2004) in its Sensor Integration and Modeling for Biological Agent Detection (SIMBAD) program is developing tools and methodologies that will result in a significant increase in its understanding of biological warfare detection technologies. The program encompasses a complete system that includes monitoring capabilities and a reporting aspect that indicates the presence or absence of a threat.

As an example, the detection of pathogens may be underscored by the very recent mad-cow disease event that occurred in the North-Western part of the United States, and the implications it has (or may have) on the economy of certain areas, and the fall-out of such an event. The present detection technique involves immunodiagnostic techniques, and its expense per test is partly to blame. Sveshnikov (2001) has indicated the need for rapid and sensitive immunoassays for pathogen detection. Cheaper and more reliable techniques, such as involving biosensors with more frequent testing would help minimize such events that tend to create instability locally, as well as perhaps globally to some extent.

The kinetics of binding and dissociation of analytes present in solution or in the environment to (complementary) receptors immobilized on a biosensor surface is an important step that provides information not only on the biomolecular interaction taking place on the biosensor surface, but also assists significantly in improving biosensor performance parameters such as stability, selectivity, response time, sensitivity, etc.

A number of biowarfare agents such as botulinum toxin, ricin, and Staphylococcus enterotoxins are proteins (Schneider, 2002). The binding kinetics of cholera toxin (CT) in solution to fluorophore-labeled ganglioside GM1 immobilized on a biomimetic membrane surface (Song *et al.*, 2000) has been analyzed using fractal methods (Sadana and Ramakrishnan, 2002). Rider *et al.* (2003) have recently presented data for the rapid detection of pathogens such as *Francisella tularensis*, *Yersinia pestis* (the bacterium that causes plague), *Bacillus anthracis*, Venezuelan equine encephalitis (VEE) virus, and *Escherichia coli*. Note that the first four have defense or military applications whereas the last one helps to keep edible food safe. They used a B lymphocytes cell-based sensor named CANARY (cellular analysis and notification of antigens risks and yields). No kinetic analysis for either the binding or the dissociation phase was available. In this chapter, we re-analyze the data by Rider *et al.* (2003) using a fractal analysis to provide kinetic rate coefficients for the binding and the dissociation phase. Affinity values are also provided. The fractal dimension values presented provide an indication of the degree of heterogeneity present on the biosensor surface. In no way are we indicating that the present analysis is in any way better than the original analysis by Rider *et al.* (2003). The present analysis should just be considered as an alternate analysis. A unique feature of the analysis of fractals to the present system is the very dilute nature of the analyte(s) present in the atmosphere. The kinetics provided in the manuscript should complement and enhance the original Rider *et al.* (2003) analysis.

3.2 THEORY

Havlin (1989) has reviewed and analyzed the diffusion of reactants towards fractal surfaces. The details of the theory and the equations involved for the binding and the dissociation phases for analyte–receptor binding are available (Sadana, 2001). The details are not repeated here; except that just the equations are given to permit an easier reading. These equations have been applied to other biosensor systems (Ramakrishnan and Sadana, 2001; Sadana, 2001). Here we will attempt to apply these equations to the rapid detection of pathogens using a B cell-based biosensor (Rider *et al.*, 2003). We recognize that these systems are very dilute. To accommodate this very dilute nature of these systems, a triple-fractal analysis may be used if necessary. For most applications, a single- or a dual-fractal analysis is often adequate to describe the binding and the dissociation kinetics. Peculiarities in the values of the binding and the dissociation rate coefficients, as well as in the values of the fractal dimensions with regard to the dilute analyte systems being analyzed will be carefully noted, if applicable.

3.2.1 Single-fractal analysis

Binding rate coefficient

Havlin (1989) indicates that the diffusion of a particle (analyte [Ag]) from a homogeneous solution to a solid surface (e.g. receptor [Ab]-coated surface) on which it reacts to form a product (analyte–receptor complex; (Ab · Ag)) is given by:

$$(\text{Ab}\cdot\text{Ag}) \approx \begin{cases} t^{(3-D_{\text{f,bind}})/2} = t^{p} & t < t_{\text{c}} \\ t^{1/2} & t > t_{\text{c}} \end{cases} \tag{3.1a}$$

Here $D_{\text{f,bind}}$ or D_{f} (used later on in the manuscript) is the fractal dimension of the surface during the binding step. t_{c} is the cross-over value. Havlin (1989) indicates that the cross-over value may be determined by $r_{\text{c}}^2 \sim t_{\text{c}}$. Above the characteristic length, r_{c}, the self-similarity of the surface is lost and the surface may be considered homogeneous. Above time t_{c}, the surface may be considered homogeneous, since the self-similarity property disappears, and 'regular' diffusion is now present. For a homogeneous surface where D_{f} is equal to two, and when only diffusional limitations are present, $p = 1/2$ as it should be. Another way of looking at the $p = 1/2$ case (where $D_{\text{f,bind}}$ is equal to two) is that the analyte in solution views the fractal object, in our case, the receptor-coated biosensor surface, from a 'large distance'. In essence, in the association process, the diffusion of the analyte from the solution to the receptor surface creates a depletion layer of width $(Ɖt)^{1/2}$ where $Ɖ$ is the diffusion constant. This gives rise to the fractal power law, (Analyte·Receptor) $\sim t^{(3-D_{\text{f,bind}})/2}$. For the present analysis, t_{c} is arbitrarily chosen and we assume that the value of the t_{c} is not reached. One may consider the approach as an intermediate 'heuristic' approach that may be used in the future to develop an autonomous (and not time-dependent) model for diffusion-controlled kinetics.

Dissociation rate coefficient

The diffusion of the dissociated particle (receptor [Ab] or analyte [Ag]) from the solid surface (e.g. analyte [Ag]–receptor [Ab]) complex coated surface) into solution may be given, as a first approximation by:

$$(\text{Ab}\cdot\text{Ag}) \approx -k' t^{(3-D_{\text{f,diss}})/2} \qquad t > t_{\text{diss}} \tag{3.1b}$$

Here $D_{\text{f,diss}}$ is the fractal dimension of the surface for the dissociation step. This corresponds to the highest concentration of the analyte–receptor complex on the surface. Henceforth, its concentration only decreases. The dissociation kinetics may be analyzed in a manner 'similar' to the binding kinetics.

3.2.3 Dual-fractal analysis

Binding rate coefficient

Sometimes, the binding curve exhibits complexities and two parameters (k, D_f) are not sufficient to adequately describe the binding kinetics. This is further corroborated by low values of r^2 factor (goodness-of-fit). In that case, one resorts to a dual-fractal analysis (four parameters; k_1, k_2, D_{f1}, and D_{f2}) to adequately describe the binding kinetics. The single-fractal analysis presented above is thus extended to include two fractal dimensions. At present, the time $(t = t_1)$ at which the 'first' fractal dimension 'changes' to the 'second' fractal dimension is arbitrary and empirical. For the most part, it is dictated by the data analyzed and experience gained by handling a single-fractal analysis. A smoother curve is obtained in the 'transition' region, if care is taken to select the correct number of points for the two regions. In this case, the product (antibody–antigen; or analyte–receptor complex, Ab·Ag or analyte–receptor) is given by:

$$(\text{Ab}\cdot\text{Ag}) \approx \begin{cases} t^{(3-D_{\text{f1,bind}})/2} = t^{p1} & t < t_1 \\ t^{(3-D_{\text{f2,bind}})/2} = t^{p2} & t_1 < t < t_2 \\ t^{1/2} & t > t_c \end{cases} \tag{3.1c}$$

In some cases, as mentioned above, a triple-fractal analysis with six parameters (k_1, k_2, k_3, D_{f1}, D_{f2}, and D_{f3}) may be required to adequately model the binding kinetics. This is when the binding curve exhibits convolutions and complexities in its shape due perhaps to the very dilute nature of the analyte (in some of the cases to be presented) or for some other reasons. Also, in some cases, a dual-fractal analysis may be required to describe the dissociation kinetics.

3.3 RESULTS

The fractal analysis will be applied to the rapid identification of pathogens using a B-cell based biosensor (Rider *et al.*, 2003).

Alternate expressions for fitting the data are available that include saturation, first-order reaction, and no diffusion limitations, but these expressions are apparently deficient in describing the heterogeneity that inherently exists on the surface. One might justifiably argue that appropriate modeling may be achieved by using a Langmuirian or other approach. The Langmuirian approach may be used to model the data presented if one assumes the presence of discrete classes of sites (for example, double exponential analysis as compared with a single-fractal analysis). Lee and Lee (1995) indicate that the fractal approach has been applied to surface science, for example, adsorption and reaction processes. These authors emphasize that the fractal approach provides a convenient means to represent the different structures and morphology at the reaction surface. These authors also emphasize using the fractal approach to develop optimal structures and as a predictive approach. Another advantage of the fractal technique is that the analyte–receptor association (as well as the dissociation reaction) is a complex reaction, and the fractal analysis via the fractal dimension and the rate coefficient provide a useful lumped parameter(s) analysis of the diffusion-limited reaction occurring on a heterogeneous surface.

In a classical situation, to demonstrate fractality, one should make a log–log plot, and one should definitely have a large amount of data. It may be useful to compare the fit to some other forms, such as exponential, or one involving saturation, etc. At present, we do not present any independent proof or physical evidence of fractals in the examples presented. It is a convenient means (since it is a lumped parameter) to make the degree of heterogeneity that exists on the surface more quantitative. Thus, there is some arbitrariness in the fractal model to be presented. The fractal approach provides additional information about interactions that may not be obtained by conventional analysis of biosensor data.

There is no nonselective adsorption of the analyte. The present system (pathogens) being analyzed is typically very dilute. Nonselective adsorption would skew the results obtained very significantly. In these types of systems, it is imperative to minimize this nonselective adsorption. We also do recognize that, in some cases, this nonselective adsorption may not be a significant component of the adsorbed material and that this rate of association, which is of a temporal nature, would depend on surface availability. If we were to accommodate the nonselective adsorption into the model, there would be an increase in the heterogeneity on the surface, since, by its very nature, nonspecific adsorption is more homogeneous than specific adsorption. This would lead to higher fractal dimension values since the fractal dimension is a direct measure of the degree of heterogeneity that exists on the surface.

Rider *et al.* (2003) have used a B-cell based biosensor to rapidly analyze pathogens. They emphasize that there is a need to rapidly identify bioterrorism agents compared to present immunoassays (Rowe *et al.*, 1999). They have used the CANARY biosensor to rapidly identify pathogens such as *Y. pestis*, *F. tularensis*, *B. Anthracis*, and the VEE virus. Figure 3.1a shows the binding of 500 cfu *Y. Pestis* to pathogen-specific B cells immobilized on a biosensor surface. A dual-fractal analysis is required to adequately describe the binding kinetics. The dissociation kinetics is adequately described by a single-fractal analysis. The values of (a) the binding rate coefficient, k, for a single-fractal analysis, (b) the binding rate coefficients, k_1 and k_2, for a dual-fractal analysis, and (c) the dissociation rate coefficient, k_d, for a single-fractal analysis are presented in Table 3.1a. The values of (a) the fractal dimension, D_f, for binding for a single-fractal analysis, (b) the fractal dimensions, D_{f1} and D_{f2}, for binding for a dual-fractal analysis, and (c) the fractal dimension, D_{fd}, for the dissociation phase for a single-fractal analysis are presented in Table 3.1b. Due to the 'convex' nature (S-shaped curve) of the binding curve at time, t, close to zero, the fractal dimension value is estimated to be equal to zero. This indicates that the surface acts as a 'Cantor' like dust.

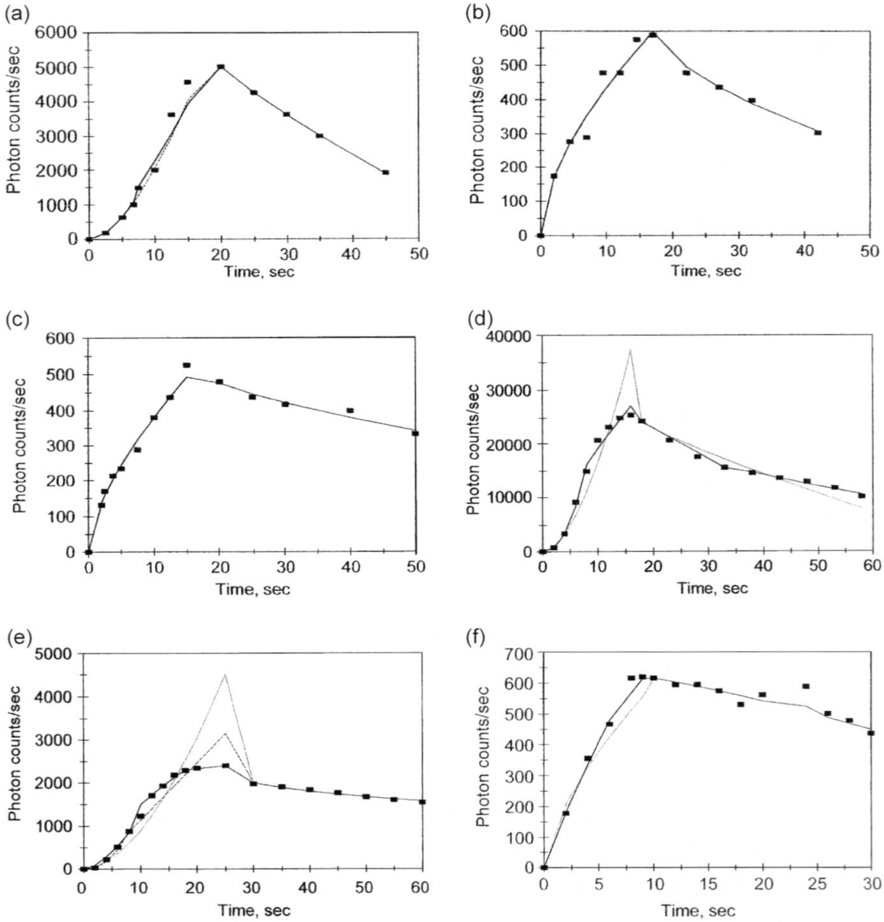

Figure 3.1 Binding and dissociation curves for the detection of different pathogens at different concentrations by B lymphocyte-based sensor (Rider *et al.*, 2003): (a) 500 cfu *Y. pestis*, (b) 50 cfu *Y. pestis*, (c) 50 cfu *Y. pestis* + 5×10^5 cfu *Francisela tularensis*, (d) 5000 pfu Vaccinia virus, (e) 500 pfu Vaccinia virus, (f) 1000 cfu *E. coli* 0157:H7, (g) 500 cfu *E. coli* 0157:H7, (h) 5×10^7 pfu Venezuelan equine encephalitis (VEE TC 83), (i) 5×10^6 pfu Venezuelan equine encephalitis (VEE TC 83), (j) 5×10^5 pfu Venezuelan equine encephalitis (VEE TC 83), (k) 10,000 cfu *B. anthracis* spores, (l) 1000 cfu *B. anthracis* spores (When only a solid line (—) is used, then a single-fractal analysis applies. When a dashed (--) and a solid line (—) is used, then the dashed line represents a single-fractal analysis, and the solid line represents a dual-fractal analysis. When a dashed (--), a dotted (···) and a solid line (—) is used, then the dashed line represents a single-fractal analysis, the dotted line represents a dual-fractal analysis, and the solid line represents a triple-fractal analysis.

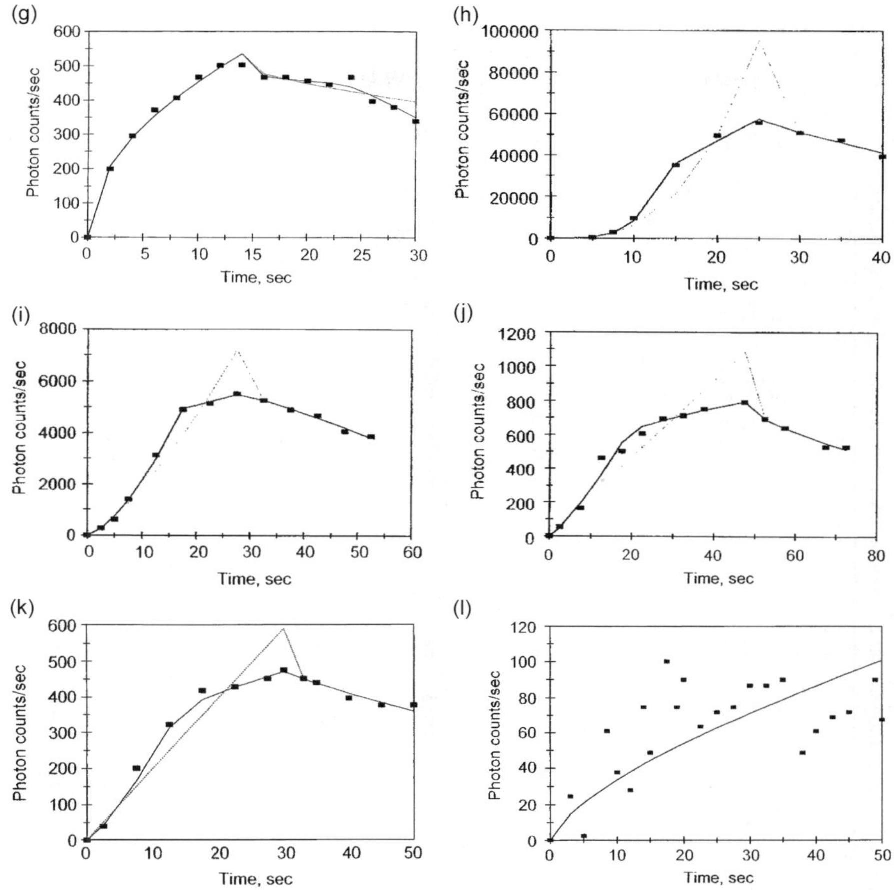

Figure 3.1 Continued.

Define affinity, K_1 as k_1/k_d and K_2 as k_2/k_d. Then, K_1 is equal to 4.057, and K_2 is equal to 0.545. The affinity value decreases by a factor of 7.44 on going from the first phase (K_1) of the binding curve to the second phase (K_2) of the binding curve.

Figure 3.1b shows the binding of 50 cfu *Y. Pestis* to pathogen-specific B cells immobilized on a biosensor surface. Once again, a dual-fractal analysis is required to adequately describe the binding kinetics. The dissociation kinetics is adequately described by a single-fractal analysis. The values of (a) the binding rate coefficient, k, for a single-fractal analysis, (b) the binding rate coefficients, k_1 and k_2, for a dual-fractal analysis, and (c) the dissociation rate coefficient, k_d, for a single-fractal analysis are presented in Table 3.1a. The values of (a) the fractal dimension, D_f, for binding for a single-fractal analysis, (b) the fractal dimensions, D_{f1} and D_{f2}, for binding for a dual-fractal analysis, and (c) the fractal dimension, D_{fd}, for the dissociation phase are presented in Table 3.1b. In this case, since the binding curve does not exhibit a S-shaped curve at time, t close to zero, the fractal dimension value, D_{f1} is equal to 1.8414 when the dual-fractal analysis

Table 3.1a

Binding and dissociation rate coefficients for different pathogens on a B lymphocyte-based sensor (Rider *et al.*, 2003)

Pathogen concentration	k	k_1	k_2	k_3	k_d	k_{d1}	k_{d2}
(a) 500 cfu *Yersinia pestis*	43.99 ± 7.72	36.21 ± 3.95	97.28 ± 18.4	na	178.49 ± 2.33	na	na
50 cfu *Yersinia pestis*	110.64 ± 12.93	113.02 ± 16.68	109.73 ± 16.38	na	36.15 ± 3.49	na	na
50 cfu *Yersinia pestis* + 5×10^5 cfu *Francisella tularensis*	88.27 ± 5.76	na	na	na	16.68 ± 2.11	na	na
(b) Vaccinia virus, 5000 pfu	304.31 ± 100.40	167.98 ± 14.67	3344.32 ± 287.5	na	838.55 ± 161.9	651.11 ± 99.26	2340.55 ± 66.87
Vaccinia virus, 500 pfu	16.14 ± 7.99	31.17 ± 1.99	268.6 ± 5.57	1529.6 ± 2.19	200.68 ± 11.62	na	na
(c) *E. coli* 0157:H7 1000 cfu	133.65 ± 22.71	98.38 ± 4.68	145.22 ± 9.17	na	4.188 ± 1.21	na	na
E. coli 0157:H7 500 cfu	147.44 ± 6.32	na	na	na	15.40 ± 8.08	25.94 ± 3.78	0.7785 ± 0.376
(d) Venezuelan equine encephalitis (VEE TC83) virus, 5×10^7 pfu	7.1750 ± 4.36	1.229 ± 0.197	2946.9 ± 216.3	na	890.56 ± 167.6	na	na
Venezuelan equine encephalitis (VEE TC83) virus, 5×10^6 pfu	85.08 ± 20.07	62.94 ± 10.61	2341.91 ± 34.93	na	38.35 ± 3.85	na	na
Venezuelan equine encephalitis (VEE TC83) virus, 5×10^5 pfu	26.46 ± 6.47	15.90 ± 3.90	278.68 ± 3.04	na	33.82 ± 2.61	na	na
B. anthracis spores 10,000 cfu	20.88 ± 6.19	13.66 ± 2.61	149.25 ± 2.41	na	9.878 ± 1.60	na	na
B. anthracis spores 1000 cfu	7.008 ± 5.97	7.008 ± 5.97	34.769 ± 4.589	0.2492 ± 0.036	na	na	na

Table 3.1b

Fractal dimensions for binding and dissociation for different pathogens on a B lymphocyte-based sensor (Rider *et al.*, 2003)

Pathogen concentration	D_f	D_{f1}	D_{f2}	D_{f3}	D_{fd}	D_{fd1}	D_{fd2}
(a) 500 cfu *Yersinia pestis*	0 + 0.1845	0 + 0.240	0.2656 ± 0.4624	na	1.2194 ± 0.022	na	na
50 cfu *Yersinia pestis*	1.8084 ± 0.1944	1.8414 ± 0.1944	1.8306 ± 0.2736	na	1.7032 ± 0.1272	na	na
50 cfu *Yersinia pestis* + 5×10^5 cfu *Francisella tularensis*	1.7311 ± 0.0639	na	na	na	1.6513 ± 0.1554	na	na
(b) Vaccinia Virus, 5000 pfu	0 + 0.3064	0 + 0.160	1.4872 ± 0.301	na	1.3790 ± 0.1282	1.0796± 0.1492	2.0138± 0.1102
Vaccinia Virus, 500 pfu	0 + 0.3350	0 + 0.1854	1.5084 ± 0.1359	2.7204 ± 0.0669	2.2654 ± 0.0610	na	na
(c) *E. coli* 0157:H7 1000 cfu	1.7036 ± 0.1846	1.2292 ± 0.0892	1.6870 ± 0.1658	na	0.5650 ± 0.1782	na	na
E. coli 0157:H7 500 cfu	2.0220 ± 0.050	na	na	na	1.6084 ± 0.4534	2.3412 ± 0.2614	0 + 1.4388
(d) Venezuelan equine encephalitis (VEE TC83) virus, 5×10^7 pfu	0 + 0.6948	0 + 0.3702	1.1522 ± 0.3910	na	0.9930 ± 0.439	na	na
Venezuelan equine encephalitis (VEE TC83) virus, 5×10^6 pfu	0.3214 ± 0.1990	0 + 0.2648	2.4864 ± 0.0924	na	0.6240 ± 0.1506	na	na
Venezuelan equine encephalitis (VEE TC83) virus, 5×10^5 pfu	1.0770 ± 0.1900	0.5908 ± 0.2980	2.4578 ± 0.0800	na	1.6910 ± 0.1176	na	na
(e) *B. Anthracis* spores 10,000 cfu	1.0344 ± 0.2402	0.5202 ± 0.2374	2.3250 ± 0.1534	na	1.3600 ± 0.1922	na	na
B. Anthracis spores 1000 cfu	1.6382 ± 0.3578	1.0882 ± 0.4962	2.494 ± 0.4182	0.0424 + 1.0896	na	na	na

applies. Also, in this case, K_1 is equal to 3.126 and K_2 is equal to 3.035. At this low value of the analyte concentration (50 cfu *Y. Pestis*), there is a very small decrease (2.91%) in the affinity value as one goes from the first- to the second-phase of binding.

Figure 3.1c shows the binding of 50 cfu *Y. pestis* and 5×10^5 cfu *F. tularensis* to pathogen-specific B cells immobilized on a biosensor surface. In this case, both the binding as well as the dissociation kinetics are adequately described by a single-fractal analysis. The values of (a) the binding rate coefficient, k, and the dissociation rate coefficient, k_d, for a single-fractal analysis are presented in Table 3.1a. The values of the fractal dimension, D_f, for binding for a single-fractal analysis and the fractal dimension, D_{fd}, for the dissociation phase for a single-fractal analysis are presented in Table 3.1b. In this case, since the binding curve does not exhibit a S-shaped curve at time, t close to zero, the fractal dimension value, D_f, is equal to 1.7311 for the single-fractal analysis. Since a single-fractal analysis applies to both the binding and the dissociation phases, the affinity, K $(= k/k_d)$, value is 5.29.

Figure 3.1d shows the binding of 5000 pfu Vaccinia virus to pathogen-specific B cells immobilized on a biosensor surface. In this case, a dual-fractal analysis is required to describe the binding as well as the dissociation phases. The values of (a) the binding rate coefficient, k, for a single-fractal analysis, (b) the binding rate coefficients, k_1 and k_2, for a dual-fractal analysis, (c) the dissociation rate coefficient, k_d, for a single-fractal analysis, and (d) the dissociation rate coefficients, k_{d1} and k_{d2}, for a dual-fractal analysis are presented in Table 3.1a. The values of (a) the fractal dimension, D_f, for binding for a single-fractal analysis, (b) the fractal dimensions, D_{f1} and D_{f2}, for binding for a dual-fractal analysis, (c) the fractal dimension, D_{fd}, for the dissociation phase for a single-fractal analysis, and (d) the fractal dimensions, D_{fd1} and D_{fd2}, for dissociation for a dual-fractal analysis, are presented in Table 3.1b. In this case, since a dual-fractal analysis is required to adequately describe both the binding as well as the dissociation phases, let us define the affinity, K_1 as k_1/k_{d1} and K_2 as k_2/k_{d2}. Then, using this definition, K_1 is equal to 0.258 and K_2 is equal to 1.423.

Figure 3.1e shows the binding of 500 pfu Vaccinia virus to pathogen-specific B cells immobilized on a biosensor surface. In this case, due to the complexities exhibited by the binding curve at this lower or dilute Vaccinia virus concentration, a triple-fractal analysis is required to adequately describe the binding kinetics. A single-fractal analysis is, however, sufficient to adequately describe the dissociation phase. The values of (a) the binding rate coefficient, k, for a single-fractal analysis, (b) the binding rate coefficients, k_1, k_2, and k_3, for a triple-fractal analysis, and (c) the dissociation rate coefficient, k_d, for a single-fractal analysis are presented in Table 3.1a. The values of (a) the fractal dimension, D_f, for binding for a single-fractal analysis, (b) the fractal dimensions, $D_{f1,}$ D_{f2} and D_{f3}, for binding for a triple-fractal analysis, and (c) the fractal dimension, D_{fd}, for the dissociation phase for a single-fractal analysis are presented in Table 3.1b. Once again, define affinity, K_1 as k_1/k_d, K_2 as k_2/k_d, and K_3 as k_3/k_d. Then, K_1 is equal to 0.155, K_2 is equal to 1.428, and K_3 is equal to 7.62. As one goes from the first phase to the second phase and then on to the third phase of binding, the affinity value keeps on increasing. It increases by a factor of 9.21 on going from the first phase to the second phase, and by a factor of 5.34 on going from the second phase to the third phase.

Figure 3.1f shows the binding of 1000 cfu *E. coli* 0157:H7 to pathogen-specific B cells immobilized on a biosensor surface. Once again, a dual-fractal analysis is required to adequately describe the binding kinetics. The dissociation kinetics is adequately described by a single-fractal analysis. The values of (a) the binding rate coefficient, k, for a single-fractal analysis, (b) the binding rate coefficients, k_1 and k_2, for a dual-fractal analysis, and (c) the dissociation rate coefficient, k_d, for a single-fractal analysis are presented in Table 3.1a. The values of (a) the fractal dimension, D_f, for binding for a single-fractal analysis, (b) the fractal dimensions, D_{f1} and D_{f2}, for binding for a dual-fractal analysis, and (c) the fractal dimension, D_{fd}, for the dissociation phase for a single-fractal analysis are presented in Table 3.1b. In this case, K_1 is equal to 23.49 and K_2 is equal to 34.68. There is an increase by a factor of 1.48 in the affinity value as one goes from the first- to the second-phase of binding.

Figure 3.1g shows the binding of 500 cfu *E. coli* 0157:H7 to pathogen-specific B cells immobilized on a biosensor surface. A single-fractal analysis is required to adequately describe the binding kinetics. The dissociation kinetics is adequately described by a dual-fractal analysis. The values of (a) the binding rate coefficient, k, for a single-fractal analysis, (b) the dissociation rate coefficient, k_d, for a single-fractal analysis, and (c) the dissociation rate coefficient, k_{d1} and k_{d2}, for a dual-fractal analysis are presented in Table 3.1a. The values of (a) the fractal dimension, D_f, for binding for a single-fractal analysis, (b) the fractal dimension, D_{fd}, for the dissociation phase for a single-fractal analysis, and (c) the fractal dimensions, D_{fd1} and D_{fd2}, for dissociation for a dual-fractal analysis are presented in Table 3.1b. In this case, K_1 is equal to 5.68 and K_2 is equal to 189.38. There is an increase by a factor of 33.34 in the affinity value as one goes from the first- to the second-phase of binding.

Figure 3.1h shows the binding of 5×10^7 pfu of the Venezuelan equine encephalitis (VEE TC83) virus to pathogen-specific B cells immobilized on a biosensor surface. A dual-fractal analysis is required to adequately describe the binding kinetics. The dissociation kinetics is adequately described by a single-fractal analysis. The values of (a) the binding rate coefficient, k, for a single-fractal analysis, (b) the binding rate coefficients, k_1 and k_2, for a dual-fractal analysis, and (c) the dissociation rate coefficient, k_d, for a single-fractal analysis are presented in Table 3.1a. The values of (a) the fractal dimension, D_f, for binding for a single-fractal analysis, (b) the fractal dimensions, D_{f1} and D_{f2}, for binding for a dual-fractal analysis, and (c) the fractal dimension, D_{fd}, for the dissociation phase for a single-fractal analysis are presented in Table 3.1b. In this case, K_1 is equal to 0.00138 and K_2 is equal to 3.31. There is an increase by a factor of 2398.5 in the affinity value as one goes from the first- to the second-phase of binding.

Figure 3.1i shows the binding of 5×10^6 pfu of the Venezuelan equine encephalitis (VEE TC83) virus to pathogen-specific B cells immobilized on a biosensor surface. A dual-fractal analysis is required to adequately describe the binding kinetics. The dissociation kinetics is adequately described by a single-fractal analysis. The values of (a) the binding rate coefficient, k, for a single-fractal analysis, (b) the binding rate coefficients, k_1 and k_2, for a dual-fractal analysis, and (c) the dissociation rate coefficient, k_d, for a single-fractal analysis are presented in Table 3.1a. The values of (a) the fractal dimension, D_f, for binding for a single-fractal analysis, (b) the fractal dimensions, D_{f1} and D_{f2}, for binding for a dual-fractal analysis, and (c) the fractal dimension, D_{fd}, for the dissociation phase for a single-fractal analysis are presented in Table 3.1b. In this case, K_1 is equal to 1.641 and K_2 is equal to 61.85. There is an increase by a factor of 37.69 in the affinity value as one goes from the first- to the second-phase of binding.

Figure 3.1j shows the binding of 5×10^5 pfu of the Venezuelan equine encephalitis (VEE TC83) virus to pathogen-specific B cells immobilized on a biosensor surface. A dual-fractal analysis is required to adequately describe the binding kinetics. The dissociation kinetics is adequately described by a single-fractal analysis. The values of (a) the binding rate coefficient, k, for a single-fractal analysis, (b) the binding rate coefficients, k_1 and k_2, for a dual-fractal analysis, and (c) the dissociation rate coefficient, k_d, for a single-fractal analysis are presented in Table 3.1a. The values of (a) the fractal dimension, D_f, for binding for a single-fractal analysis, (b) the fractal dimensions, D_{f1} and D_{f2}, for binding for a dual-fractal analysis, and (c) the fractal dimension, D_{fd}, for the dissociation phase for a single-fractal analysis are presented in Table 3.1b. In this case, K_1 is equal to 0.47 and K_2 is equal to 8.24. There is an increase by a factor of 17.53 in the affinity value as one goes from the first- to the second-phase of binding.

Figure 3.1k shows the binding of 10,000 cfu of *B. anthracis* spores to pathogen-specific B cells immobilized on a biosensor surface. A dual-fractal analysis is required to adequately describe the binding kinetics. The dissociation kinetics is adequately described by a single-fractal analysis. The values of (a) the binding rate coefficient, k for a single-fractal analysis, (b) the binding rate coefficients, k_1 and k_2, for a dual-fractal analysis, and (c) the dissociation rate coefficient, k_d, for a single-fractal analysis are presented in Table 3.1a. The values of (a) the fractal dimension, D_f, for binding for a single-fractal analysis, (b) the fractal dimensions, D_{f1} and D_{f2}, for binding for a dual-fractal analysis, and (c) the fractal dimension, D_{fd}, for the dissociation phase for a single-fractal analysis are presented in Table 3.1b. In this case, K_1 is equal to 1.383 and K_2 is equal to 15.11. There is an increase by a factor of 10.93 in the affinity value as one goes from the first- to the second-phase of binding.

Figure 3.1l shows the binding of 1000 cfu of *B. anthracis* spores to pathogen-specific B cells immobilized on a biosensor surface. At this low concentration of *B. Anthracis* spores only a binding phase is exhibited. However, a triple-fractal analysis is required to adequately describe the binding kinetics. The values of (a) the binding rate coefficient, k, for a single-fractal analysis, and (b) the binding rate coefficients, k_1, k_2, and k_3, for a triple-fractal analysis are presented in Table 3.1a. The values of (a) the fractal dimension, D_f, for binding for a single-fractal analysis, and (b) the fractal dimensions, D_{f1}, D_{f2}, and D_{f3}, for binding for a triple-fractal analysis are presented in Table 3.1b. Since no dissociation phase was exhibited here, no affinity values are presented.

Figure 3.2a shows that the binding rate coefficient, k_1, increases with an increase in the fractal dimension, D_{f1}. The binding rate coefficient, k_1, is given by:

$$k_1 = (46.59 \pm 17.51) D_{f1}^{1.842 \pm 0.306} \quad (3.2a)$$

The binding rate coefficient, k_1, is quite sensitive to the degree of heterogeneity on the biosensor surface as noted by the 1.842 order of dependence on the fractal dimension, D_{f1}. Only four data points were available. More data points are required to make the curve more reliable.

Figure 3.2b shows that the affinity, K_1 ($= k_1/k_d$), increases with an increase in the fractal dimension ratio, D_{f1}/D_{fd}. The affinity binding rate coefficient, K_1, is given by:

$$K_1 = (4.4221 \pm 3.7486)(D_{f1}/D_{fd})^{1.7975 \pm 0.404} \quad (3.2b)$$

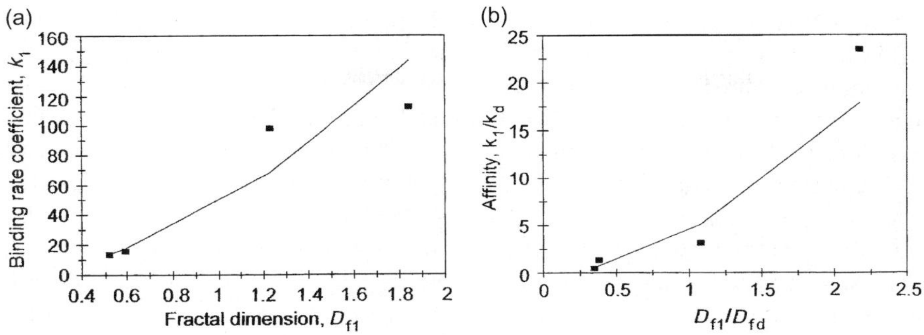

Figure 3.2 (a) Increase in the binding rate coefficient, k_1 with an increase in the fractal dimension, D_{f1}. (b) Increase in the affinity, K_1 ($= k_1/k_d$) with an increase in the fractal dimension ratio, D_{f1}/D_{fd}.

The affinity, K_1, is quite sensitive to the degree of heterogeneity on the biosensor surface as noted by the 1.7975 order of dependence on the fractal dimension ratio, D_{f1}/D_{fd}. Only four data points were available. More data points are required to make the curve more reliable.

Figure 3.3a shows the binding of 25 μl of B cells + 5 μl pRMC strain at 1.4×10^8 pfu/ml (wt A12FMDV) to pathogen-specific B cells immobilized on a biosensor surface. A triple-fractal analysis is required to adequately describe the binding kinetics. No dissociation curve is exhibited. The values of (a) the binding rate coefficient, k, for a single-fractal analysis, (b) the binding rate coefficients, k_1, k_2, and k_3 for a triple-fractal analysis are presented in Table 3.2a. The values of (a) the fractal dimension, D_f, for binding for a single-fractal analysis, and (b) the fractal dimensions, D_{f1}, D_{f2}, and D_{f3}, for binding for a triple-fractal analysis, are presented in Table 3.2b.

Figure 3.3b shows the binding of 20 μl of B cells + VEE virus strain subtype 1A TC83 to pathogen-specific B cells immobilized on a biosensor surface. A dual-fractal analysis is required to adequately describe the binding as well as the dissociation kinetics. The values of (a) the binding rate coefficient, k, for a single-fractal analysis, (b) the binding rate coefficients, k_1 and k_2, for a dual-fractal analysis, (c) the dissociation rate coefficient, k_d, for a single-fractal analysis, and (d) the dissociation rate coefficients, k_{d1} and k_{d2} for a dual-fractal analysis are presented in Table 3.2a. The values of (a) the fractal dimension, D_f, for binding for a single-fractal analysis, (b) the fractal dimensions, D_{f1} and D_{f2}, for binding for a dual-fractal analysis, (c) the fractal dimension, D_{fd}, for the dissociation phase for a single-fractal analysis, and (d) the fractal dimensions, D_{fd1} and D_{fd2}, for dissociation for a dual-fractal analysis are presented in Table 3.2b. In this case, K_1 ($= k_1/k_{d1}$) is equal to 3.81 and K_2 ($= k_2/k_{d2}$) is equal to 5.87. The affinity value increases by a factor of 1.54 as one goes from the first- to the second-phase of binding.

Figure 3.3c shows the binding of 20 μl of B cells + VEE virus strain subtype 1BPTF-39 to pathogen-specific B cells immobilized on a biosensor surface. A dual-fractal analysis is required to adequately describe the binding as well as the dissociation kinetics. The values of (a) the binding rate coefficient, k, for a single-fractal analysis, (b) the binding rate coefficients, k_1 and k_2, for a dual-fractal analysis, (c) the dissociation rate coefficient, k_d, for a single-fractal analysis, and (d) the dissociation rate coefficients, k_{d1} and k_{d2}, for a dual-fractal analysis are presented in Table 3.2a. The values of (a) the fractal

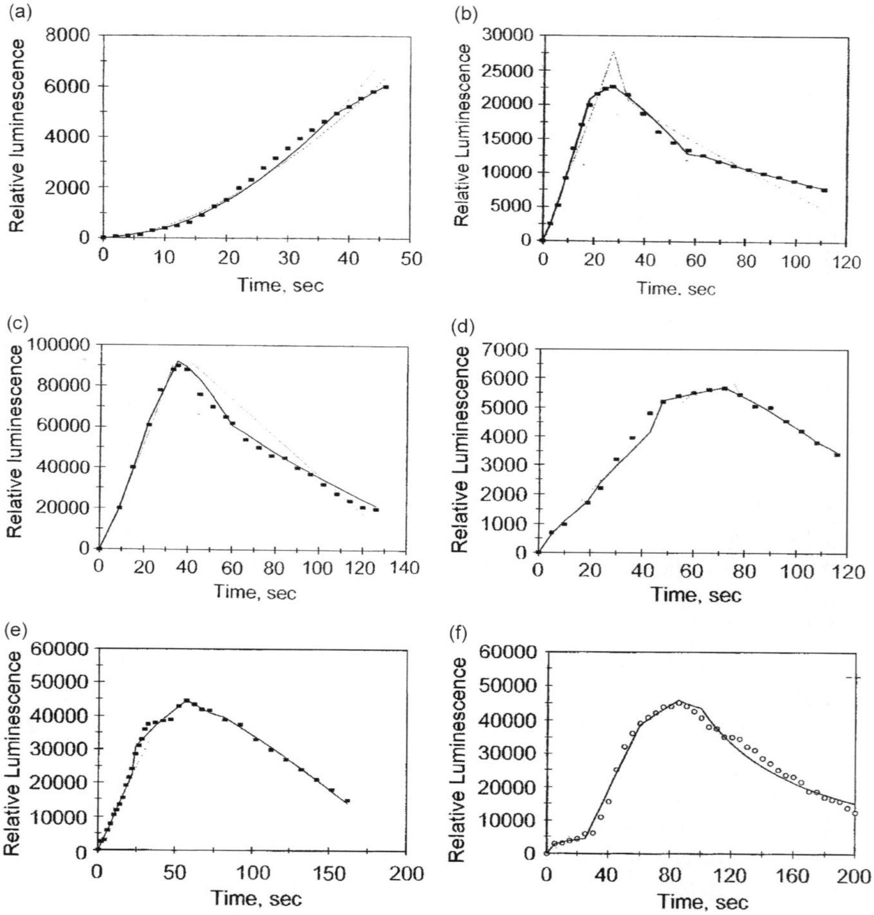

Figure 3.3 Binding and dissociation curves for the detection of different pathogens at different concentrations by B lymphocyte-based sensor (Rider *et al.*, supplement, 2003): (a) 25 μl of B cells + 5 μl pRMC35 strain at 1.4×10^8 pfu/ml (wt A12 FMDV), (b) 20 μl of B cells + Venezuelan equine encephalitis (VEE) virus strain subtype 1A TC-83, (c) 20 μl of B cells + VEE virus strain subtype 1BPTF-39, (d) 20 μl of B cells + VEE virus strain subtype 1c p676, (e) 20 μl of B cells + VEE virus strain 3880 subtype 1d, (f) 20 μl of B cells + VEE virus strain subtype 1e (Mena II).

dimension, D_{f1}, for binding for a single-fractal analysis, (b) the fractal dimensions, D_{f1} and D_{f2}, for binding for a dual-fractal analysis, (c) the fractal dimension, D_{fd}, for the dissociation phase, and (d) the fractal dimensions, D_{fd1} and D_{fd2} for dissociation for a dual-fractal analysis, are presented in Table 3.2b. In this case, K_1 is equal to 3.23 and K_2 is equal to 1.265. The affinity value decreases by a factor of 2.55 as one goes from the first- to the second-phase of binding.

Figure 3.3d shows the binding of 20 μl of B cells + VEE virus strain subtype 1cp676 to pathogen-specific B cells immobilized on a biosensor surface. A dual-fractal analysis is required to adequately describe the binding kinetics. The dissociation kinetics is adequately described by a single-fractal analysis. The values of (a) the binding rate coefficient, k, for a single-fractal analysis, (b) the binding rate coefficients, k_1 and k_2, for

Table 3.2a

Binding and dissociation rate coefficients for different pathogens on a B lymphocyte-based sensor (Rider *et al.*, science supplement, 2003)

Pathogen concentration	k	k_1	k_2	k_3	k_d	k_{d1}	k_{d2}
25 μl of B cells + 5 μl pRMC35 strain at 1.4×10^8 pfu/ml (wt A12 FMDV)	9.738 ± 2.499	19.689 ± 6.136	5.011 ± 0.549	114.60 ± 0.808	na	na	na
20 μl of B cells + Venezuelan equinine encephalitis (VEE) virus strain subtype 1A TC-83	843.72 ± 115.87	653.99 ± 36.21	11810.9 ± 96.53	na	441.89 ± 107.55	171.67 ± 37.06	2013.48 ± 16.64
20 μl of B cells+ Venezuelan equinine encephalitis (VEE) virus strain subtype 1B PTF-39	1834.40 ± 141.98	1293.41 ± 60.12	4825.63 ± 211.30	na	978.17 ± 277.66	400.26 ± 132.03	3814.72 ± 100.17
20 μl of B cells+ Venezuelan equinine encephalitis (VEE) virus strain subtype 1c p676	148.69 ± 20.08	206.28 ± 19.23	125.44 ± 16.63	na	24.38 ± 3.44	na	na
20 μl of B cells + Venezuelan equinine encephalitis (VEE) virus strain 3880 subtype 1d	1095.06 ± 176.56	999.35 ± 138.35	7744.63 ± 391.14	na	159 ± 18.3	na	na
20 μl of B cells + Venezuelan equinine encephalitis (VEE) virus strain subtype 1e (Mena II)	178.53 ± 84.54	1799.10 ± 150.1	4.145 ± 0.921	5146.31 ± 97.32	$4.82 \times 10^7 \pm 0.45 \times 10^7$	na	na

Table 3.2b

Fractal dimensions for the binding and dissociation phase for different pathogens on a B lymphocyte-based sensor (Rider *et al.*, science supplement, 2003)

Pathogen concentration	D_f	D_{f1}	D_{f2}	D_{f3}	D_{fd}	D_{fd1}	D_{fd2}
25 μl of B cells + 5 μl pRMC35 strain at 1.4×10^8 pfu/ml (wt A12 FMDV)	0	0.4598 ± 0.0324	0 + 0.107	0.9246 ± 0.094	na	na	na
20 μl of B cells + Venezuelan equinine encephalitis (VEE) virus strain subtype 1A TC-83	0.8790 ± 0.126	0.6044 ± 0.073	2.6014 ± 0.092	na	1.3302 ± 0.156	0.6170 ± 0.2368	2.0978 ± 0.0202
20 μl of B cells + Venezuelan equinine encephalitis (VEE) virus strain subtype 1B PTF-39	0.7662 ± 0.128	0.493 ± 0.143	1.3408 ± 0.235	na	1.0196 ± 0.149	0.2904 ± 0.338	1.7058 ± 0.0466
20 μl of B cells + Venezuelan equinine encephalitis (VEE) virus strain subtype 1c p676	1.2316 ± 0.094	1.5498 ± 0.146	1.1402 ± 0.1864	na	0.5946 ± 0.157	na	na
20 μl of B cells + Venezuelan equinine encephalitis (VEE) virus strain 3880 subtype 1d	1.0528 ± 0.076	0.9892 ± 0.110	2.1334 ± 0.1086	na	0.720 ± 0.068	na	na
20 μl of B cells + Venezuelan equinine encephalitis (VEE) virus strain subtype 1e (Mena II)	0.5052 ± 0.243	2.4224 ± 0.154	0 + 0.4328	2.0162 ± 0.0978	0 + 0.1568	na	na

a dual-fractal analysis, and (c) the dissociation rate coefficient, k_d, for a single-fractal analysis are presented in Table 3.2a. The values of (a) the fractal dimension, D_f, for binding for a single-fractal analysis, (b) the fractal dimensions, D_{f1} and D_{f2}, for binding for a dual-fractal analysis, and (c) the fractal dimension, D_{fd}, for the dissociation phase for a single-fractal analysis are presented in Table 3.1b. In this case, K_1 is equal to 8.46 and K_2 is equal to 5.15. The affinity value decreases by a factor of 0.61 as one goes from the first- to the second-phase of binding.

Figure 3.3e shows the binding of 20 μl of B cells + VEE virus strain 3880 subtype 1d to pathogen-specific B cells immobilized on a biosensor surface. Once again, a dual-fractal analysis is required to adequately describe the binding kinetics. The dissociation kinetics is adequately described by a single-fractal analysis. The values of (a) the binding rate coefficient, k, for a single-fractal analysis, (b) the binding rate coefficients, k_1 and k_2, for a dual-fractal analysis, and (c) the dissociation rate coefficient, k_d, for a single-fractal analysis are presented in Table 3.2a. The values of (a) the fractal dimension, D_f, for binding for a single-fractal analysis, (b) the fractal dimensions, D_{f1} and D_{f2}, for binding for a dual-fractal analysis, and (c) the fractal dimension, D_{fd}, for the dissociation phase are presented in Table 3.2b. In this case, K_1 is equal to 6.27 and K_2 is equal to 48.59. The affinity value increases by a factor of 7.75 as one goes from the first- to the second-phase of binding.

Figure 3.3f shows the binding of 20 μl of B cells + VEE virus strain subtype 1e (Mena II) to pathogen-specific B cells immobilized on a biosensor surface. Once again, a dual-fractal analysis is required to adequately describe the binding kinetics. The dissociation kinetics is adequately described by a single-fractal analysis. The values of (a) the binding rate coefficient, k, for a single-fractal analysis, (b) the binding rate coefficients, k_1 and k_2, for a dual-fractal analysis, and (c) the dissociation rate coefficient, k_d, for a single-fractal analysis are presented in Table 3.2a. The values of (a) the fractal dimension, D_f, for binding for a single-fractal analysis, (b) the fractal dimensions, D_{f1} and D_{f2}, for binding for a dual-fractal analysis, and (c) the fractal dimension, D_{fd}, for the dissociation phase for a single-fractal analysis are presented in Table 3.2b. In this case, K_1 is equal to 3.73×10^{-5} and K_2 is equal to 8.6×10^{-8}. The affinity value decreases by a factor of 433.7 as one goes from the first- to the second-phase of binding.

Figure 3.4a shows the binding of 5000 cfu *E. coli* strain 50 0157:H7 to pathogen-specific B cells immobilized on a biosensor surface. A dual-fractal analysis is required to adequately describe the binding kinetics. Only the dual-fractal analysis is presented in the figure. The dissociation kinetics is adequately described by a single-fractal analysis. The values of (a) the binding rate coefficient, k, for a single-fractal analysis, (b) the binding rate coefficients, k_1 and k_2, for a dual-fractal analysis, and (c) the dissociation rate coefficient, k_d, for a single-fractal analysis are presented in Table 3.3a. The values of (a) the fractal dimension, D_f, for binding for a single-fractal analysis, (b) the fractal dimensions, D_{f1} and D_{f2} for binding for a dual-fractal analysis, and (c) the fractal dimension, D_{fd}, for the dissociation phase are presented in Table 3.3b. In this case, K_1 is equal to 6.21×10^{-4} and K_2 is equal to 3.59×10^{-3}. There is an increase by a factor of 5.78 in the affinity value as one goes from the first- to the second-phase of binding.

Due to the 'convex' nature (S-shaped curve) of the binding curve at time, t, close to zero, the fractal dimension value is estimated to be equal to zero (either D_f or D_{f1}). However, D_{f2} (for a dual-fractal analysis) does have a nonzero value. D_{f2} equals 0.7108. Note that it does, however, have a large error estimate, equal to 1.06. Only a positive error

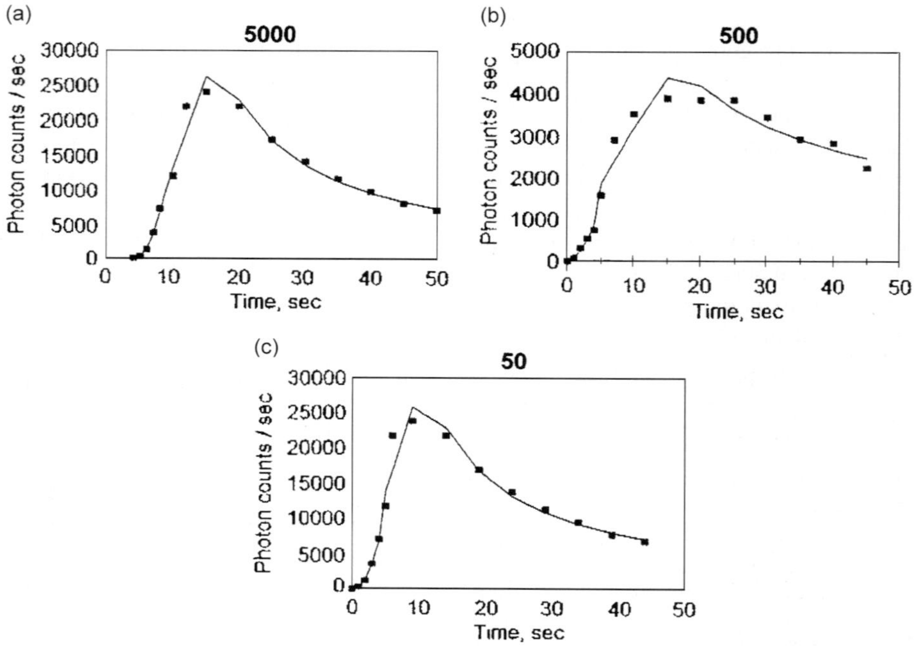

Figure 3.4 Binding and dissociation curves for the detection of *E. coli* strain 50 0157:H7 at different concentrations by the B Lymphocyte-Based Biosensor (Rider *et al.*, supplement, 2003): (a) 5000 cfu, (b) 500 cfu, (c) 50 cfu.

(or range) is given, that is $D_{f2} = 0.7108 + 1.06$, since a negative fractal dimension value has no physical significance. Similarly, the start of the dissociation curve also exhibits an 'S-shaped' curvature. Thus, D_{fd} is also estimated to be equal to zero.

Figure 3.4b shows the binding of 500 cfu *E. coli* strain 50 0157:H7 to pathogen-specific B cells immobilized on a biosensor surface. A dual-fractal analysis is required to adequately describe the binding kinetics. Only the dual-fractal analysis is presented. The dissociation kinetics is adequately described by a single-fractal analysis. The values of (a) the binding rate coefficient, k, for a single-fractal analysis, (b) the binding rate coefficients, k_1 and k_2, for a dual-fractal analysis, and (c) the dissociation rate coefficient,

Table 3.3a

Binding and dissociation rate coefficients for *E. coli* on a B lymphocyte-based sensor (Rider *et al.*, science supplement, 2003)

Pathogen concentration	k	k_1	k_2	k_d
E. coli 5000 cfu strain 50 0157:H7	na	292.2 ± 30.69	1686 ± 429.77	470246 ± 18189.7
E. coli 500 cfu strain 50 0157:H7	na	72.39 ± 15.30	515.50 ± 118.40	29839 ± 2670.9
E. coli 50 cfu strain 50 0157:H7	328.09 ± 101.42	327417 ± 14530.1	292.2 ± 30.69	2674.22 ± 986.91

Table 3.3b

Fractal dimensions for the binding and the dissociation phase for *E. coli* on a B lymphocyte-based sensor (Rider *et al.*, science supplement, 2003)

Pathogen concentration	D_f	D_{f1}	D_{f2}	D_{fd}
E. coli 5000 cfu strain 50 0157:H7	na	0	0.7108 + 1.06	0
E. coli 500 cfu strain 50 0157:H7	na	0	1.4194 ± 0.5058	0
E. coli 50 cfu strain 50 0157:H7	0	0	0	0.9310 + 1.4764

k_d, for a single-fractal analysis are presented in Table 3.3a. The values of (a) the fractal dimension, D_f, for binding for a single-fractal analysis, (b) the fractal dimensions, D_{f1} and D_{f2}, for binding for a dual-fractal analysis, and (c) the fractal dimension, D_{fd}, for the dissociation phase are presented in Table 3.3b. In this case, K_1 is equal to 0.00243 and K_2 is equal to 0.0173. There is an increase by a factor of 7.12 in the affinity value as one goes from the first- to the second-phase of binding.

Once again, due to the 'convex' nature (S-shaped curve) of the binding curve at time, t, close to zero, the fractal dimension curve is estimated to be equal to zero (either D_f or D_{f1}). However, D_{f2} (for a dual-fractal analysis) does have a nonzero value. D_{f2} equals 1.4194 ± 0.5058. Similarly, the start of the dissociation curve also exhibits an 'S-shaped' curvature. Thus, D_{fd} is also estimated to be equal to zero.

Figure 3.4c shows the binding of 50 cfu *E. coli* strain 50 0157:H7 to pathogen-specific B cells immobilized on a biosensor surface. A dual-fractal analysis is required to adequately describe the binding kinetics. The dissociation kinetics is adequately described by a single-fractal analysis. The values of (a) the binding rate coefficient, k, for a single-fractal analysis, (b) the binding rate coefficients, k_1 and k_2, for a dual-fractal analysis, and (c) the dissociation rate coefficient, k_d, for a single-fractal analysis are presented in Table 3.3a. The values of (a) the fractal dimension, D_f, for binding for a single-fractal analysis, (b) the fractal dimensions, D_{f1} and D_{f2}, for binding for a dual-fractal analysis, and (c) the fractal dimension, D_{fd}, for the dissociation phase for a single-fractal analysis are presented in Table 3.3b. In this case, K_1 is equal to 122.43 and K_2 is equal to 0.109. There is a decrease by a factor of 1123.21 in the affinity value as one goes from the first- to the second-phase of binding.

Here too, due to the 'convex' nature (S-shaped curve) of the binding curve at time, t, close to zero, the fractal dimension curve is estimated to be equal to zero (either D_f, D_{f1}, or D_{f2}). Here, D_{fd} has a nonzero value. In this case, since the start of the dissociation curve does not exhibit an 'S-shaped' curve, the D_{fd} value is equal to 0.9301 + 1.4764. There is a large error estimate, thus only the positive error (or range) is presented.

He *et al.* (2002) indicate that worldwide tuberculosis is the largest cause of death from a single infectious agent. Approximately, one-third of the people in the world are infected by *Mycobacterium tuberculosis (M. TB)*. These authors indicate that thickness shear mode (TSM) immunosensors could be used to detect *M. TB*, especially since it is a simple and cheap technique that is sensitive to small mass changes on the quartz crystal surface. These small changes are detected by the Sauerbrey equation (Sauerbrey, 1996).

He *et al.* (2002) emphasize that other diagnostic methods such as DNA and RNA probes are sensitive methods for the detection of *M. TB*, especially the application of nucleic acid

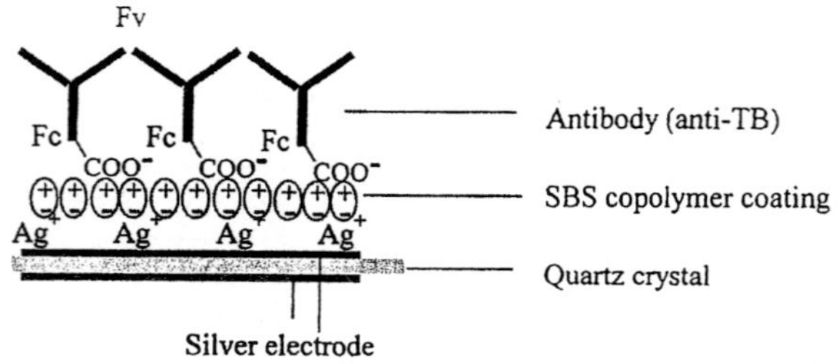

Figure 3.5 Immobilization scheme for anti-TB antibody on a FSM thickness shear mode (TSM) immunosensor (He *et al.*, 2002).

by the polymerase chain reaction (PCR). However, these methods are (a) too expensive, and (b) yield a significant amount of false positive results. Especially in developing countries (and also in developed countries) an inexpensive diagnostic method for the detection of *M. TB* is urgently required.

He *et al.* (2002) immobilized rabbit IgG against *M. TB* (anti-TB) on a styrene–butadiene–styrene (SBS) block copolymer. The author emphasizes that these copolymers are inexpensive and provide good performance. This SBS copolymer was coated on a quartz crystal. Figure 3.5 shows the procedure for the immobilization of the antibody. The Ag^+ delocalized the unsaturated covalent bonds of the SBS copolymer. The positive side of the SBS copolymer binds to the Fc section of the antibody.

Figure 3.6a shows the binding of the 0.25 mg/ml antibody (anti-TB) in solution to the SBS copolymer coated on the quartz crystal surface. This is not the actual biosensor detection application, but leads up to it. This is just the binding of the anti-TB (receptor) and the polymer-coated quartz surface. It is instructive to analyze the binding of the 'receptor' on the TSM immunosensor surface.

A dual-fractal analysis is required to adequately describe the binding kinetics. The values of (a) the binding rate coefficient, k, and the fractal dimension, D_f for a single-fractal analysis, and (b) the binding rate coefficients, k_1 and k_2, and the fractal dimensions, D_{f1} and D_{f2} for a dual-fractal analysis are given in Table 3.4a.

Figure 3.6b shows the binding of 0.50 mg/ml antibody (anti-TB) to the SBS copolymer coated on the quartz crystal surface. Once again, a dual-fractal analysis is required to adequately describe the binding kinetics. The values of (a) the binding rate coefficient, k, and the fractal dimension, D_f for a single-fractal analysis, and (b) the binding rate coefficients, k_1 and k_2, and the fractal dimensions, D_{f1} and D_{f2} for a dual-fractal analysis are given in Table 3.4a.

It is of interest to note that as the antibody concentration in solution increases by a factor of two from 0.25 to 0.5 mg/ml for a dual-fractal analysis, the binding rate coefficient, k_1, decreases by a factor of 0.413 from a value of k_1 equal to 0.0290–0.0120. Note that the estimated value of the fractal dimension, D_{f1} is equal to zero for both cases. This indicates that the TSM immunosensor surface exists as a Cantor-like dust. In the second phase of binding, an increase in the antibody concentration in solution by a factor of two from 0.25 to 0.50 mg/ml leads to (a) a decrease in the fractal dimension, D_{f2} by a factor of 0.533 from a value of 2.6525 to 1.4140, and (b) to a decrease in the binding rate coefficient, k_2 by a factor of 0.051 from a value of 71.924 to 3.6682. Changes in the binding rate coefficient, k_2 and in the fractal dimension, D_{f2} are in the same direction.

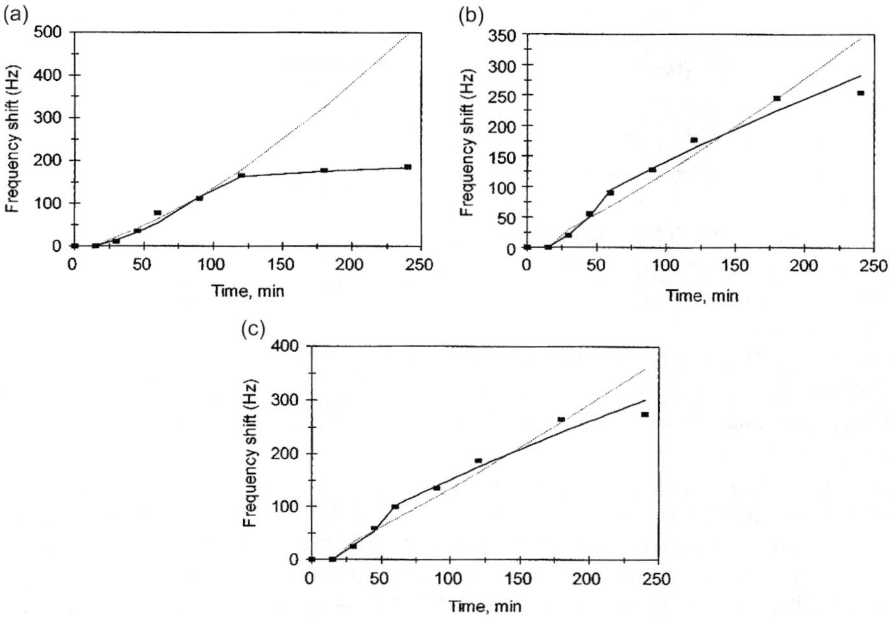

Figure 3.6 Binding of different concentrations (in mg/ml) of the anti-TB antibody on the SBS copolymer surface coated on a quartz crystal (He *et al.*, 2002): (a) 0.25, (b) 0.5, (c) 2.5. (When only a solid line (—) is used, then a single-fractal analysis applies. When a dashed (--) and a solid line (—) is used, then the dashed line represents a single-fractal analysis, and the solid line represents a dual-fractal analysis.)

Figure 3.6c shows the binding of 2.50 mg/ml antibody (anti-TB) to the SBS copolymer coated on the quartz crystal surface. Once again, a dual-fractal analysis is required to adequately describe the binding kinetics. The values of (a) the binding rate coefficient, k, and the fractal dimension, D_f for a single-fractal analysis, and (b) the binding rate coefficients, k_1 and k_2, and the fractal dimensions, D_{f1} and D_{f2} for a dual-fractal analysis are given in Table 3.4a.

Table 3.4a

Fractal dimensions and binding rate coefficients for the binding of different concentrations of antibody (Rabbit IgG against *M. tuberculosis (M. TB))* in solution (in mg/ml) to the polymer surface coated on a quartz crystal (He *et al.*, 2002)

Antibody concentration in solution (ng/ml)	k	k_1	k_2	D_f	D_{f1}	D_{f2}
0.25	0.1507 ± 0.333	0.0290 ± 0.009	71.92 ± 0.337	0.0436 ± 1.271	~0	2.6525 ± 0.019
0.5	0.5637 ± 0.1845	0.0120 ± 0.00014	3.668 ± 0.368	0.6594 ± 0.308	~0	1.414 ± 0.174
2.5	0.7928 ± 0.212	0.1433 ± 0.032	4.223 ± 0.386	0.7684 ± 0.266	~0	1.4418 ± 0.1592

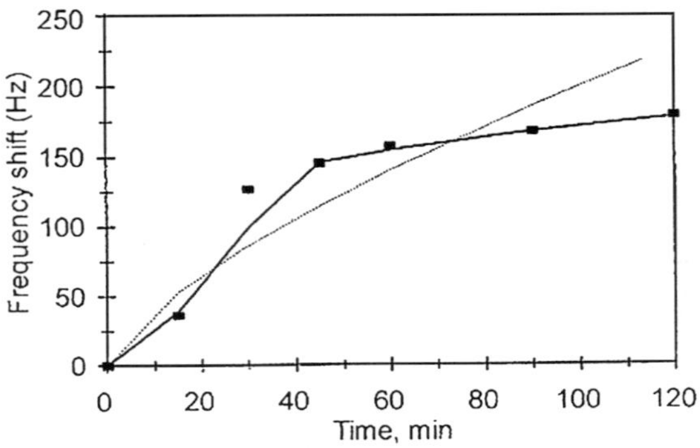

Figure 3.7 Binding of human *M. TB* ($H_{37}R_v$) antigen (10,000 cells/ml) in solution to the anti-tuberculosis (*M. TB*) antibody immobilized on the SBS copolymer coated on the quartz crystal surface (He *et al.*, 2002).

An increase in the antibody concentration in solution by a factor of five from 0.5 to 2.5 mg/ml leads to (a) an increase in the fractal dimension, D_{f2} by factor of 1.02 and to (b) an increase in the binding rate coefficient, k_2 by factor of 1.15. Once again, changes in the fractal dimension, D_{f2} (or the degree of heterogeneity) on the SBS copolymer coated quartz crystal surface, and in the binding rate coefficient, k_2 are in the same direction.

Figure 3.7 shows the binding of the antigen (human *M. TB* (H37RV); 10,000 cells/ml) in solution to the anti-tuberculosis (*M. TB*) antibody immobilized on the SBS copolymer coated on the quartz crystal surface (He *et al.*, 2002). Table 3.4b shows the values of (a) the binding rate coefficient, k, and the fractal dimension, D_f for a single-fractal analysis, and (b) the binding rate coefficients, k_1 and k_2, and the fractal dimensions, D_{f1} and D_{f2} for a dual-fractal analysis. Only a single experiment is available, and is presented.

Figure 3.8 and Tables 3.4a and 3.4b show the increase in the binding rate coefficient, k_2 with an increase in the fractal dimension, D_{f2}. Note that the data for (a) the binding of different concentrations of antibody in solution (0.25–2.5 mg/ml) to the polymer surface coated on a quartz crystal, and (b) the binding of 10,000 cells/ml in solution to antibody immobilized on a piezoelectric immunosensor surface are presented together. This was due to the scarcity of the experimental data available. For the data presented in Figure 3.8 and Tables 3.4a and 3.4b, the binding rate coefficient, k_2 is given by:

$$k_2 = (0.7313 \pm 0.0240)D_{f2}^{4.725 \pm 0.0528} \quad (3.3)$$

Table 3.4b

Fractal dimensions and binding rate coefficients for the binding of 10,000 cells (human *M. TB* ($H_{37}R_V$)) in solution to antibody immobilized on a piezoelectric immunosensor surface (He *et al.*, 2002)

k	k_1	k_2	D_f	D_{f1}	D_{f2}
8.0573 ± 3.149	1.0854 ± 0.382	68.064 ± 0.822	1.607 ± 0.389	0.3476 ± 0.768	2.5986 ± 0.032

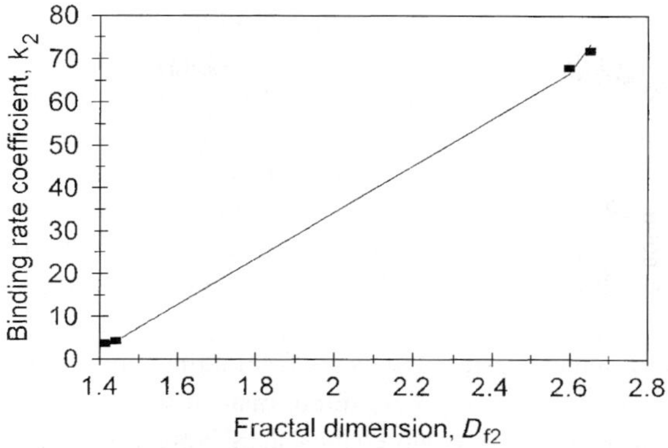

Figure 3.8 Increase in the binding rate coefficient, k_2 with an increase in the fractal dimension, D_{f2}.

The fit is quite good in spite of the fact that data from two different sets, as indicated earlier, have been used to obtain the plot shown in Figure 3.8. The availability of more data points would provide for a more reliable fit. The binding rate coefficient, k_2 is very sensitive to the degree of heterogeneity (or the fractal dimension, D_{f2}) on the TSM surface as noted by the higher than fourth order (equal to 4.725) of dependence of the binding rate coefficient, k_2 on D_{f2}.

3.4 CONCLUSIONS

A fractal analysis is presented for the binding and dissociation kinetics exhibited by pathogens such as *F. tularensis*, *Y. pestis*, *B. anthracis*, VEE virus, Vaccinia virus, and *E. coli* using a CANARY (cellular analysis and notification of antigens risks and yields) biosensor (Rider *et al.*, 2003). The binding of *M. tuberculosis* to a TSM immunosensor (He *et al.*, 2002) is also presented. The binding and dissociation kinetics may be adequately described by either a single-, dual-, or triple-fractal analysis. A double-fractal analysis was used only if a single-fractal analysis did not provide an adequate fit. This was determined by a regression analysis provided by Sigmaplot (1993).

Due to the dilute nature of the analyte(s) present in the environment, a triple-fractal analysis was used, when necessary, to adequately describe the binding kinetics. This was noted at the lower end of the analyte concentration spectrum analyzed for the VEE virus (5000 pfu, dual-fractal analysis; 500 pfu, triple-fractal analysis), and for *B. anthracis* (10,000 cfu, dual-fractal analysis; 1000 cfu, triple-fractal analysis). Only two data sets are presented here. However, if this trend is observed for the detection of other pathogens, then this makes the detection of these pathogens at their lower end of the concentration spectrum more and more challenging. This is further exacerbated by the fact that the dilute nature of the analyte also leads to variation in the original experimental data of Rider *et al.* (2003); though this is not presented in this analysis, since only discrete points at regular intervals were taken to model the kinetics.

The change from the binding rate coefficients, k_1 to k_2 to k_3, indicates a change in the binding mechanism on the biosensor surface. Also, the change in the fractal dimension, D_{f1} to D_{f2} to D_{f3}, indicates a change in the degree of heterogeneity on the surface.

For example, for the VEE virus at 500 pfu concentration, in about 25 sec (duration of the binding phase) there is a double change in the binding mechanism (from k_1 to k_2; and from k_2 to k_3). At 5000 pfu VEE concentration in about 15 sec (duration of the binding phase), there is a single change in the binding mechanism (from k_1 to k_2).

A dual-fractal analysis is required to adequately model (a) the binding of different concentrations of the anti-TB antibody to the SBS copolymer coated on the quartz crystal surface, and (b) the binding of the antigen (human *M. TB* ($H_{37}R_V$; 10,000 cells/ml)) in solution to the anti-tuberculosis (*M. TB*) antibody immobilized on the SBS copolymer coated on the quartz crystal surface.

Finally, in a general sense, fractal models are fascinating. Newer avenues are required to analyze and help detect pathogens at very dilute concentration levels. The analysis of the studies of the boundaries (scale) over which the fractal behavior occurs should prove useful. The real interesting test of the fractal model would be if it can make a prediction that turns out to be correct. This would prove to be extremely valuable, especially in the detection of pathogens. For example, if the fractal analysis enhances (or predicts) a biosensor performance parameter(s) (such as stability, sensitivity, response time, etc.) as an experimental variable is changed, then the value of the analysis will be substantially increased. Any increase in time that is made available to help in the evacuation process (for example, by making better biosensors) after the establishment of a pathogenic threat is invaluable.

REFERENCES

Alwing, A, Biowarfare defense sensors: applications and research, keynote address, *The First IEEE International Conference on Sensors, IEEE Sensors 2002*, Orlando, Florida, June 11–14, 2002.

DARPA (Defense Advanced Research Projects Agency), Special Projects Office, SIMBAD program, http://www.darpa.mil/spo/programs/simbad.htm, 2004.

Fitch, JP, E Raber and DE Imbro, Technology challenges in responding to biological or chemical attacks in the civilian sector. *Science*, 2003, **302**, 1350–1354.

Havlin, S, Molecular diffusion and reactions. *The Fractal Approach to Heterogeneous Chemistry: Surfaces, Colloids, Polymers* (ed. D Avnir), Wiley, New York, 1989, pp. 251–269.

He, F, L Zhang, J Zhao, B Hu and J Lei, A TSM immunosensor for detection of *M. tuberculosis* with a new membrane material. *Sensors and Actuators B*, 2002, **85**, 284–290.

Lee CK and SL Lee, Multi-fractal scaling analysis of reactions over fractal surfaces. *Surface Science*, 1995, **325**, 294–310.

Ligler, FS, Optical biosensors for the detection of biological warfare agents, Session on Chemical Methods for Biological Detection, *Illicit Substance Detection: Chemical/Biological Agents*, Mount Holyoke College, South Hadley, MA, June 24–29, 2001.

Ramakrishnan, A and A Sadana, A single-fractal analysis of cellular analyte–receptor binding kinetics using biosensors. *BioSystems*, 2001, **59**, 35–51.

Rider, TH, MS Petrovic, FE Nargi, JD Harper, ED Schwoebel, RH Mathews, DJ Blanchard, LT Bortolin, AM Young, J Chen and MA Hollis, A cell-based sensor for rapid identification of pathogens. *Science*, 2003, **301**, 213–215. Science Online, www.sciencemag.org/cgi/content/full/031/5630/213/DC1.

Rowe, CA, *et al.*, *Analytical Chemistry*, 1999, **71**, 3846.

Sadana, A, A fractal analysis for the evaluation of hybridization kinetics in biosensors. *Journal of Colloid and Interface Science*, 2001, **151**(1), 166–177.

Sadana A and A Ramakrishnan, A kinetic study of analyte-receptor binding and dissociation to biosensor applications: a fractal analysis for cholera toxin and peptide-protein interactions, *Sensors & Actuators B*, 2002, **85**, 61–72.

Salamone, S, Bio-IT World, Nanotechnology electrifies biological agent detection, April 18, http://www.bio-itworld.Com/news/041803-report2370.html, 2003.

Sauerbrey, GZ, *Zeitschrift fur Physics*, 1996, **155**, 206.

Schneider, B, Optical Engineering, SPIE, special focus section, April 2002.

Sigmaplot (1993). Scientific Graphic Software. User's Manual, San Rafael, CA; Jandel Scientific.

Song, X, J Shi and B Swanson, Flow-cytometry-based biosensor for detection of multi-valent proteins. *Analytical Chemistry*, 2000, **284**, 35–41.

Stowers, M, On-line bioaerosol characterization using time-of-flight mass spectrometry and laser-induced fluorescence, Session on Physical Methods for Biological Detection, *Illicit Substance Detection, Chemical/Biological Agents*, D Franz, TA Kovacs, co-chairs, Mount Holyoke College, South Hadley, MA, June 24–29, 2001.

Swanson, B, Structure-function biosensors for detection of pathogens, *Illicit Substance Detection, Chemical/Biological Agents*, D Franz, TA Kovacs, co-chairs, Mount Holyoke College, South Hadley, MA, June 24–29, 2001.

Sveshnikov, P, Fast and sensitive immunoassays for pathogen detection, Session on Rapid Detection of Infections by Bioterrorism Agents, *Illicit Substance Detection, Chemical/Biological Agents*, D Franz, TA Kovacs, co-chairs, Mount Holyoke College, South Hadley, MA, June 24–29, 2001.

– 4 –

Heat Shock Protein Fractal Binding and Dissociation Kinetics

4.1 INTRODUCTION

Biosensors are finding increasing applications in a wide variety of areas. A major application is in the biomedical area, and more recently there is increasing application of biosensors to detect chemical and biological threats to civilian and military personnel. Heat shock proteins (Hsp) have biomedical implications, and the levels of these heat shock proteins change when the cell responds to different physiological stresses (Schleiker *et al.*, 2002; Yura and Nakahigashi, 1999). Bukau and Horwich (1998); Wickner *et al.* (1999) indicate that the DnaK–DnaJ–GrpE and GroEl–GroEs chaperone systems are among the most important ones in *Escherichia coli* and other bacteria. Steel *et al.* (2004) have recently analyzed the coordinated activation of the two heat shock proteins, Kar2p and Lhs1p. These Hsps are involved in protein biogenesis in yeast endoplasmic reticulum. These authors further indicate that these two chaperones functionally overlap. Vostiar *et al.* (2003) indicate that the heat shock response in cells has been studied by different methods that include two-dimensional electrophoresis with matrix-assisted laser desorption time-of-flight mass spectrometry (Jurgen *et al.*, 2000), and DNA-microarray technology (Oh and Liao, 2000). Vostiar *et al.* (2003) have used the surface plasmon resonance (SPR) biosensor to monitor the concentration of the chaperone DnaK in a bacterial cell lysate. SPR was used to monitor off-line the intracellular DnaK level in *E. coli* subjected to heat stress.

There are two factors that need to be considered in the analysis of sensorgrams using the SPR manufactured by BIACORE (2002), and perhaps also by other manufacturers. This involves the inclusion of diffusion effects that will, more often than not, be present in these types of systems. This is in spite of assurances by the manufacturer(s) that these effects are either not present or are minimized if the SPR (or other biosensor) is operated properly. It would be prudent to include the presence of these diffusional effects right from the start. Furthermore, heterogeneity of the diffusion of the analyte may also be involved. As the receptors are immobilized on the surface, there is bound to be a degree of heterogeneity present on the SPR (or other) biosensor surface, and its effects on the association (or binding) rate coefficients are very significant (Ramakrishnan and Sadana 2000; 2001) and need to be included in a kinetic analysis.

The heterogeneous distribution of receptors on the sensing surface is theoretically included in a kinetic model. This becomes more significant if the degree of heterogeneity of the receptors on the surface affects the binding rate coefficient to a large degree. This is apparently true in most cases. One possible way of accounting for the presence of heterogeneity that exists on the surface is by using fractals. A characteristic feature of fractals is the self-similarity at different levels of scale. Fractals exhibit self-similarity. Fractals are particularly useful for this type of analysis because they help characterize the heterogeneity that exists on the surface by a lumped parameter, the fractal dimension.

In this chapter, we provide an analysis of the sensorgrams for the binding of chaperone DnaK in recombinant *E. coli* to anti-DnaK immobilized on a SPR biosensor chip surface (Vostiar *et al.*, 2003). No kinetic analysis was provided in the original paper. In this analysis, and as mentioned above, we include the involvement of heterogeneity on the receptor surface and the presence of diffusional limitations on analyte (DnaK)–receptor (anti-DnaK) binding kinetics. We assume that the surface heterogeneity is responsible, along with our factors, for the complex nature of interactions to be evaluated and presented. Particular attention is paid to a case where a very dilute analyte concentration is involved. This has implications not only for the present case of binding kinetics (and detection) of Hsp, but also, presumably for the detection of chemical and biological pathogens that are often present in very dilute solutions.

4.2 THEORY

Havlin (1989) has reviewed and analyzed the diffusion of reactants towards fractal surfaces. The details of the theory and the equations involved for the binding and the dissociation phases for analyte–receptor binding are available (Ramakrishnan and Sadana 2000). The details are not repeated here; except that just the equations are given to permit an easier reading. These equations have been applied to other biosensor systems (Ramakrishnan and Sadana 2000; 2001). Here we will attempt to apply these equations to the binding of chaperone DnaK in recombinant *E. coli* to anti-DnaK immobilized on a SPR biosensor chip surface (Vostiar *et al.*, 2003). We recognize that these systems are very dilute. To accommodate this very dilute nature of these systems, a triple-fractal analysis may be used if necessary. For most applications, a single- or a dual-fractal analysis is often adequate to describe the binding kinetics. Peculiarities in the values of the binding rate coefficients, as well as in the values of the fractal dimensions with regard to the dilute analyte systems being analyzed, will be carefully noted, if applicable.

4.2.1 Single-fractal analysis

Binding comprises two phases, an association phase and a dissociation phase. Here we will analyze only the association phase. Some authors prefer this. We prefer binding as just binding. Both views are, however, presented in this chapter.

Association rate coefficient

Havlin (1989) indicates that the diffusion of a particle (analyte [Ag]) from a homogeneous solution to a solid surface (e.g. receptor [Ab]-coated surface) on which it reacts to form a product (analyte–receptor complex; (Ab·Ag)) is given by:

$$(\text{Ab}\cdot\text{Ag}) \approx \begin{cases} t^{(3-D_{\text{f,bind}})/2} = t^p & t < t_c \\ t^{1/2} & t > t_c \end{cases} \tag{4.1a}$$

Here $D_{\text{f,bind}}$ or D_f (used later on in the manuscript) is the fractal dimension of the surface during the binding step. t_c is the cross-over value. Havlin (1989) indicates that the cross-over value may be determined by $r_c^2 \sim t_c$. Above the characteristic length, r_c, the self-similarity of the surface is lost and the surface may be considered homogeneous. Above time t_c, the surface may be considered homogeneous, since the self-similarity property disappears, and 'regular' diffusion is now present. For a homogeneous surface where D_f is equal to 2, and when only diffusional limitations are present, $p = 1/2$ as it should be. Another way of looking at the $p = 1/2$ case (where $D_{\text{f,bind}}$ is equal to 2) is that the analyte in solution views the fractal object, in our case, the receptor-coated biosensor surface, from a 'large distance'. In essence, in the association process, the diffusion of the analyte from the solution to the receptor surface creates a depletion layer of width $(Đt)^{1/2}$ where Đ is the diffusion constant. This gives rise to the fractal power law, (analyte·receptor) $\sim t^{(3-D_{\text{f,bind}})/2}$. For the present analysis, t_c is arbitrarily chosen and we assume that the value of the t_c is not reached. One may consider the approach as an intermediate 'heuristic' approach that may be used in the future to develop an autonomous (and not time-dependent) model for diffusion-controlled kinetics.

4.2.2 Dual-fractal analysis

Sometimes, the association curve exhibits complexities and two parameters (k, D_f) are not sufficient to adequately describe the binding kinetics. This is further corroborated by low values of r^2 factor (goodness-of-fit). In that case, one resorts to a dual-fractal analysis (four parameters; k_1, k_2, D_{f1}, and D_{f2}) to adequately describe the binding kinetics. The single-fractal analysis presented above is thus extended to include two fractal dimensions. At present, the time $(t = t_1)$ at which the 'first' fractal dimension 'changes' to the 'second' fractal dimension is arbitrary and empirical. For the most part, it is dictated by the data analyzed and experience gained by handling a single-fractal analysis. A smoother curve is obtained in the 'transition' region, if care is taken to select the correct number of points for the two regions. In this case, the product (antibody-antigen; or analyte–receptor complex, Ab·Ag or analyte·receptor) is given by:

$$(\text{Ab}\cdot\text{Ag}) \approx \begin{cases} t^{(3-D_{\text{f1,bind}})/2} = t^{p_1} & t < t_1 \\ t^{(3-D_{\text{f2,bind}})/2} = t^{p_2} & t_1 < t < t_2 \\ t^{1/2} & t > t_c \end{cases} \tag{4.1b}$$

In some cases, as mentioned above, a triple-fractal analysis with six parameters (k_1, k_2, k_3, D_{f1}, D_{f2}, and D_{f3}) may be required to adequately model the binding kinetics. This is when the binding curve exhibits convolutions and complexities in its shape due to perhaps to the very dilute nature of the analyte (in some of the cases to be presented) or for some other reasons.

4.3 RESULTS

The fractal analysis will be applied to the binding of chaperone DnaK in recombinant *E. coli* to anti-DnaK immobilized on a SPR biosensor chip surface (Vostiar *et al.*, 2003).

Alternate expressions for fitting the data are available that include saturation, first-order reaction, and no diffusion limitations, but these expressions are apparently deficient in describing the heterogeneity that inherently exists on the surface. One might justifiably argue that appropriate modeling may be achieved by using a Langmuirian or other approach. The Langmuirian approach may be used to model the data presented if one assumes the presence of discrete classes of sites (for example, double exponential analysis as compared with a single-fractal analysis). Lee and Lee (1995) indicate that the fractal approach has been applied to surface science, for example, adsorption and reaction processes. These authors emphasize that the fractal approach provides a convenient means to represent the different structures and morphology at the reaction surface. These authors also emphasize using the fractal approach to develop optimal structures and as a predictive approach. Another advantage of the fractal technique is that the analyte–receptor association is a complex reaction, and the fractal analysis via the fractal dimension and the rate coefficient provide a useful lumped parameter(s) analysis of the diffusion-limited reaction occurring on a heterogeneous surface.

In a classical situation, to demonstrate fractality, one should make a log–log plot, and one should definitely have a large amount of data. It may be useful to compare the fit to some other forms, such as exponential, or one involving saturation, etc. At present, we do not present any independent proof or physical evidence of fractals in the examples presented. It is a convenient means (since it is a lumped parameter) to make the degree of heterogeneity that exists on the surface more quantitative. Thus, there is some arbitrariness in the fractal model to be presented. The fractal approach provides additional information about interactions that may not be obtained by a conventional analysis of biosensor data.

There is no nonselective adsorption of the analyte. Nonselective adsorption would skew the results obtained very significantly. In these types of systems, it is imperative to minimize this nonselective adsorption. We also do recognize that, in some cases, this nonselective adsorption may not be a significant component of the adsorbed material and that this rate of association, which is of a temporal nature, would depend on surface availability. Vostiar *et al.* (2003) have been careful to eliminate or minimize nonspecific association. Their cross-reactivity test verified that their reaction was very specific and demonstrated that non-specific effects would not significantly influence the detection of DnaK in complex culture samples. The nonspecific adsorption effects were also minimized by subtracting a response reference channel with an irrelevant antibody from the response channel with immobilized anti-DnaK.

Figure 4.1a shows the binding of 30 mM DnaK in solution to anti-DnaK immobilized on a SPR sensor chip surface (Vostiar *et al.*, 2003). A single-fractal analysis is inadequate to describe the binding kinetics. A dual-fractal analysis improves the fit. But, the R squared value is rather low. It is equal to 0.722 for the first phase of binding, and equal to 0.865 for the

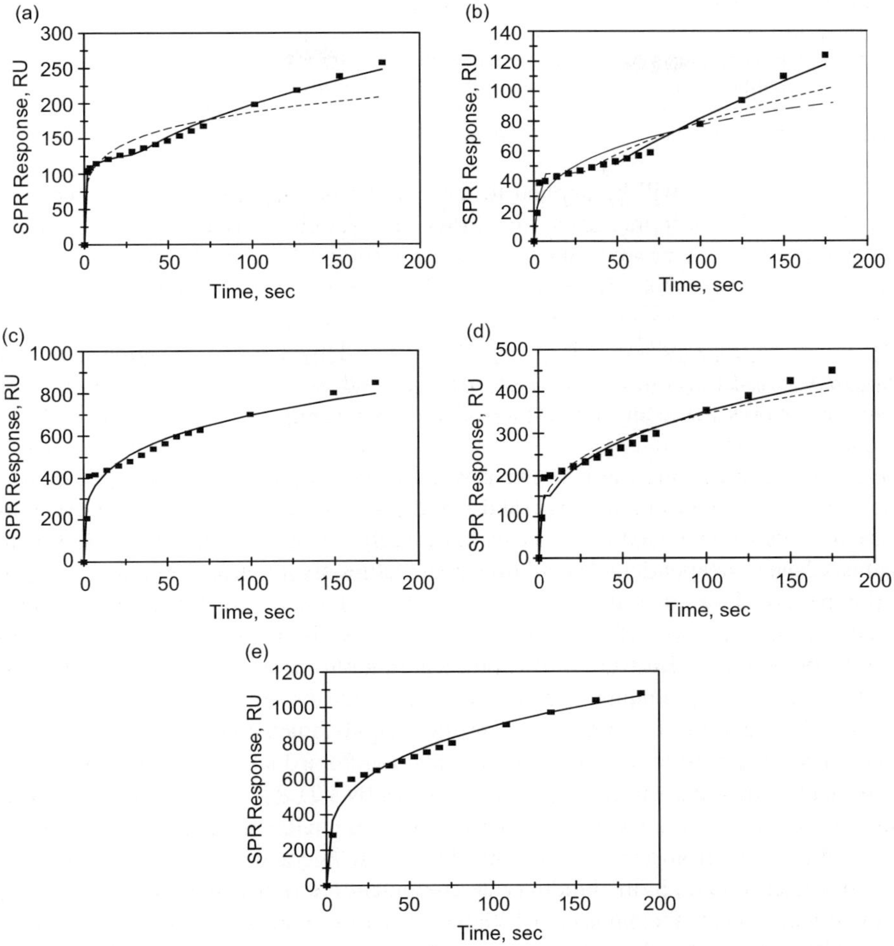

Figure 4.1 Binding rate curves for different concentrations (in mM) of chaperone DnaK in *Escherichia coli* cell culture lysates to anti-DnaK monoclonal antibody immobilized on an optical SPR biosensor (Vostiar *et al.*, 2003): (a) 30 (dual-fractal analysis), (b) 30 (triple-fractal analysis), (c) 60 (dual-fractal analysis), (d) 120 (dual-fractal analysis), (e) 240 (single-fractal analysis), (f) 320 (single-fractal analysis). The degree of fractal analysis that provides the best fit is shown in brackets. As indicated in the text as one goes towards lower DnaK concentrations, a higher degree of fractal analysis is required to adequately describe the binding kinetics. In all of the cases presented the solid line provides the best fit.

second phase of binding. This may also be seen in Figure 4.1a. Due to the complexities exhibited by the binding curve at this dilute concentration, a triple fractal analysis is used. The fit is now very adequate. The R squared values are now 0.722, 0.986, and 0.969 for the first, second, and third phases, respectively. A single- or a dual-fractal analysis is usually sufficient to adequately describe the binding kinetics of quite a few analyte–receptor binding reactions. In this case, due to the very dilute nature of the analyte in solution, a triple-fractal analysis is required to adequately describe the binding kinetics. The binding rate

Table 4.1a

Binding rate coefficients for chaperone DnaK in cell culture lysates to anti-DnaK monoclonal antibody immobilized on an optical SPR biosensor (Vostiar *et al.*, 2003)

DnaK concentration (mM)	k	k_1	k_2	k_3
30[a]	18.195 ± 3.79	14.727 ± 5.45	11.151 ± 1.61	na
30[b]	18.195 ± 3.79	14.727 ± 5.45	31.255 ± 0.034	4.008 ± 0.263
60	77.827 ± 9.89	99.714 ± 0.265	34.12 + 35.49	na
120	102.67 ± 14.29	75.71 ± 27.06	81.57 ± 4.90	na
240	227.02 ± 29.43	na	na	na
320	260.81 ± 29.58	na	na	na

[a] a dual-fractal analysis used here. Hence D_{f1} and D_{f2} values given.
[b] a triple-fractal analysis used here. Hence D_{f1}, D_{f2}, and D_{f3} values given.

coefficient values are presented in two rows in Table 4.1a. The first row gives the binding rate coefficient values for a single (k) and a dual-fractal (k_1 and k_2) analysis. The second row gives the binding rate coefficient values for a single (k) and a triple-fractal (k_1, k_2, and k_3) analysis. The corresponding fractal dimension values are given in Table 4.1b.

Figure 4.1b shows the binding of 60 mM DnaK in solution to anti-DnaK immobilized on an SPR sensor chip surface. A dual-fractal analysis is required to adequately describe the binding kinetics. The values of (a) the binding rate coefficient, k, for a single-fractal analysis, and (b) the binding rate coefficients, k_1 and k_2, for a dual-fractal analysis are given in Table 4.1a. The values of (a) the fractal dimension, D_f for a single-fractal analysis, and (b) the fractal dimensions, D_{f1} and D_{f2}, for a dual-fractal analysis are given in Table 4.1b. The fractal dimension value decreases by a 21.8% from a value of D_{f1} equal to 2.8544 to D_{f2} equal to 2.2320. This leads to a decrease in the binding rate coefficient value by a factor of 2.92 from a value of k_1 equal to 99.71 to k_2 equal to 34.12. The changes in the fractal dimension and in the binding rate coefficient are in the same direction.

Figure 4.1c shows the binding of 120 mM DnaK in solution to anti-DnaK immobilized on a SPR sensor chip surface. A dual-fractal analysis is required to adequately describe the

Table 4.1b

Binding rate coefficients for chaperone DnaK in cell culture lysates to anti-DnaK monoclonal antibody immobilized on an optical SPR biosensor (Vostiar *et al.*, 2003)

DnaK concentration (mM)	D_f	D_{f1}	D_{f2}	D_{f3}
30[a]	2.3714 ± 0.074	1.8554 ± 0.709	2.1418 ± 0.120	na
30[b]	2.3714 ± 0.074	1.8554 ± 0.709	2.7538 ± 0.017	1.6910 ± 0.087
60	2.6166 ± 0.047	2.8544 ± 0.006	2.2320 ± 0.041	na
120	2.4706 ± 0.051	1.8872 ± 0.689	2.3644 ± 0.044	na
240	2.5132 ± 0.049	na	na	na
320	2.4562 ± 0.051	na	na	na

[a] a dual-fractal analysis used here. Hence D_{f1} and D_{f2} values given.
[b] a triple-fractal analysis used here. Hence D_{f1}, D_{f2}, and D_{f3} values given.

binding kinetics. The values of (a) the binding rate coefficient, k, for a single-fractal analysis, and (b) the binding rate coefficients, k_1 and k_2, for a dual-fractal analysis are given in Table 4.1a. The values of (a) the fractal dimension, D_f, for a single-fractal analysis, and (b) the fractal dimensions, D_{f1} and D_{f2}, for a dual-fractal analysis are given in Table 4.1b. The fractal dimension value increases by 25.3% from a value of D_{f1} equal to 1.8872 to D_{f2} equal to 2.3644. This leads to an increase in the binding rate coefficient value by 7.7% from a value of k_1 equal to 75.71 to k_2 equal to 81.57. Once again, the changes in the fractal dimension and in the binding rate coefficient are in the same direction.

Figure 4.1d shows the binding of 240 mM DnaK in solution to anti-DnaK immobilized on an SPR biosensor chip surface. A single-fractal analysis is adequate to describe the binding kinetics. The values of the binding rate coefficient, k, and the fractal dimension for binding, D_f, are given in Table 4.1a and 4.1b, respectively. It is of interest to note that at the lower analyte concentration of 120 mM DnaK and below, a dual- or a triple-fractal analysis is required to describe the binding kinetics. However, at 240 mM DnaK concentration, a single-fractal analysis is adequate to describe the binding kinetics. Apparently, in this case, as one goes to lower analyte (DnaK) concentrations, a higher degree of fractal analysis is required to describe the binding kinetics. Note that at 30 mM DnaK concentration in solution, a triple-fractal analysis is required to describe the binding kinetics.

Figure 4.1e shows the binding of 320 mM DnaK in solution to anti-DnaK immobilized on an SPR biosensor chip surface. A single-fractal analysis is, once again, adequate to describe the binding kinetics. The values of the binding rate coefficient, k, and the fractal dimension for binding, D_f, are given in Table 4.1a and 4.1b, respectively. An increase in the analyte (DnaK) concentration in solution from 240 to 320 mM by 33% leads to (a) a decrease in the fractal dimension, D_f, value by 2.3% from a D_f value of 2.5132 to 2.4562, and to (b) an increase in the binding rate coefficient, k, value by 14.9% from a k value of 227.02 to 260.81. In this case, the changes in the binding rate coefficient, k, and in the fractal dimension, D_f, are in opposite directions.

Figure 4.2 shows how the fractal dimension, D_f, and the binding rate coefficient, k, change chronologically for 30 mM DnaK concentration in solution. Initially, in the first phase of binding the fractal dimension, D_{f1} is equal to 1.8554, and the binding rate

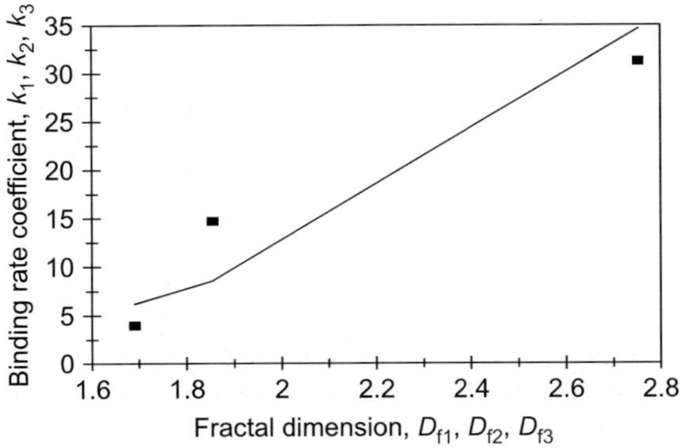

Figure 4.2 The chronological change of the binding rate coefficient, k, and the fractal dimension, D_f with time, t, for 30 mM DnaK in *Escherichia coli* cell culture lysates for a triple-fractal analysis (Vostiar *et al.*, 2003).

coefficient, k_1, is equal to 14.727. Then, in the second phase of binding, the fractal dimension value D_{f2} is equal to 2.7538, and the binding rate coefficient, k_2, value is 31.255. A 48.4% increase in the fractal dimension value from D_{f1} equal to 1.8554 to D_{f2} equal to 2.7538, leads to an increase in the binding rate coefficient by factor of 2.12 from k_1 equal to 14.727 to k_2 equal to 31.255.

In the third phase of binding, the fractal dimension, D_{f3} is equal to 1.6910, and the binding rate coefficient, k_3, is equal to 4.008. A 38.6% decrease in the fractal dimension from D_{f2} equal to 2.7538 to D_{f3} equal to 1.6910, leads to a decrease in the binding rate coefficient by factor of 7.80 from k_2 equal to 31.255 to k_3 equal to 4.008. Once again, the changes in the fractal dimension and in the binding rate coefficient are in the same direction.

For the 30, 60, and 120 mM DnaK concentration in solution, and for a dual-fractal analysis, Figure 4.3a shows that the binding rate coefficient, k_1, increases as the DnaK concentration in solution increases. In this 30–120 mM DnaK concentration range, the binding rate coefficient, k_1, is given by:

$$k_1 = (0.0880 \pm 0.0094)\ [\text{DnaK}]^{1.4354 \pm 0.1028} \qquad (4.2a)$$

The fit is quite good. More data points are required to more firmly establish the fit of the curve. The binding rate coefficient is quite sensitive to the DnaK concentration in solution as noted by the greater than first order dependence on DnaK concentration. The noninteger dependence on the DnaK concentration in solution lends support to the fractal nature of the system.

For the 30, 60, and 120 mM DnaK concentration in solution, and for a dual-fractal analysis, Figure 4.3b shows that the binding rate coefficient, k_2, increases as the fractal dimension, D_{f2} increases. In this 30–120 mM DnaK concentration range, the binding rate coefficient, k_2, is given by:

$$k_2 = (3.13\text{E}-06 \pm 0.23\text{E}-06) D_{f2}^{19.806 \pm 3.334} \qquad (4.2b)$$

Once again, the fit is quite good. More data points are required to more firmly establish the fit of the curve. The binding rate coefficient, k_2, is very sensitive to the degree of

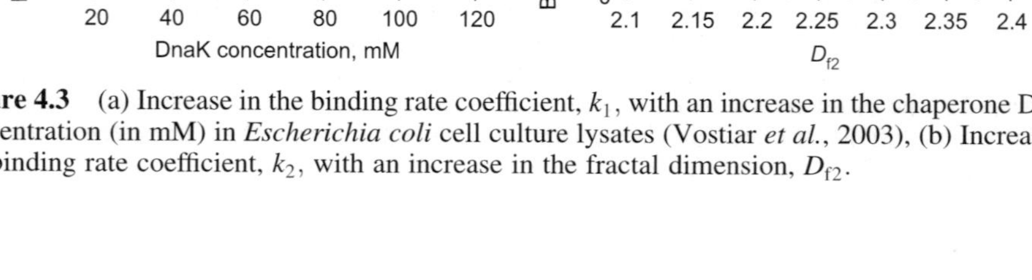

Figure 4.3 (a) Increase in the binding rate coefficient, k_1, with an increase in the chaperone DnaK concentration (in mM) in *Escherichia coli* cell culture lysates (Vostiar *et al.*, 2003), (b) Increase in the binding rate coefficient, k_2, with an increase in the fractal dimension, D_{f2}.

heterogeneity that exists on the biosensor surface as noted by the very high value of the dependence of k_2 on D_{f2}.

Figure 4.4a shows the binding of chaperone DnaK in heat-stressed cell (*E. coli*) lysates to anti-DnaK monoclonal antibody using ELISA (Vostiar *et al.*, 2003). A single-fractal analysis is adequate to describe the binding kinetics. The binding rate coefficient, k, value and the fractal dimension, D_f, are given in Table 4.2.

Figure 4.4b shows the binding of the chaperone DnaK in unstressed (*E. coli*) lysates to anti-DnaK using ELISA. As expected, the level of DnaK is lower in unstressed (*E. coli*) than that in the stressed *E. coli* culture. However, in this case, a dual-fractal analysis is required to adequately describe the binding kinetics in unstressed *E. coli*. As noted previously for the binding of DnaK in solution to anti-DnaK immobilized on an SPR biosensor chip surface, a decrease in the DnaK (analyte) concentration in solution (unstressed compared to stressed *E. coli* culture) leads to a higher degree of fractal analysis to adequately describe the binding kinetics.

The dilute nature of DnaK in unstressed *E. coli* culture leads to a significant amount of variation in the experimental data obtained by Vostiar *et al.* (2003). This variation is also reflected in the error obtained in the estimated values of the binding rate coefficients, k_1 and k_2. For k_1 only the positive error is presented (see Table 4.2), since a negative binding rate coefficient has no physical significance.

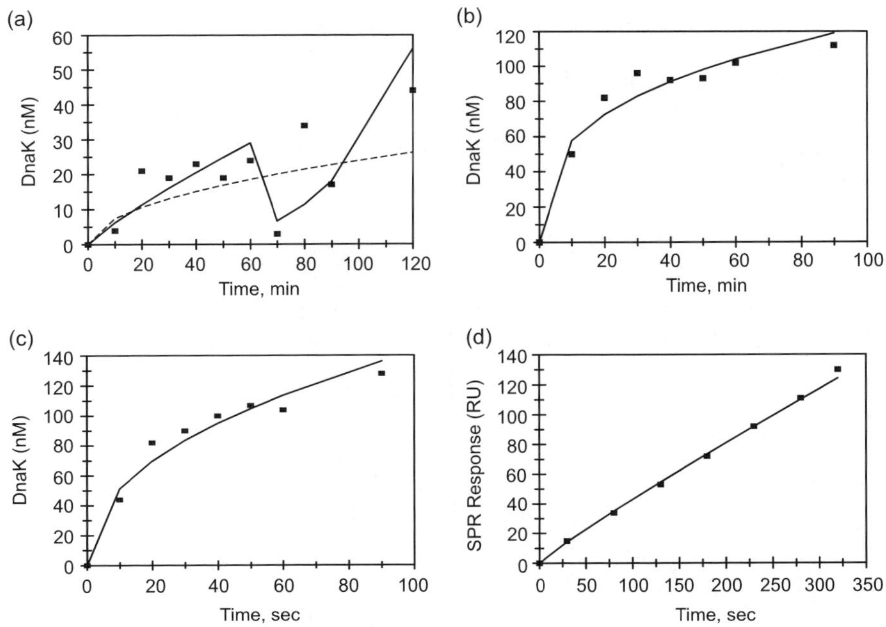

Figure 4.4 Binding rate curves for chaperone DnaK in *Escherichia coli* cell culture lysates to anti-DnaK immobilized on an optical SPR biosensor or for ELISA [6] under stressed and unstressed conditions [6]: (a) ELISA stressed culture (single-fractal analysis), (b) ELISA unstressed culture (dual-fractal analysis), (c) SPR (stressed culture) (single-fractal analysis), (d) SPR (injection cycle of cell lysate). The degree of fractal analysis that provides the best fit is shown in brackets. In all of the cases presented the solid line provides the best fit.

Table 4.2

Binding rate coefficients and fractal dimensions for chaperone DnaK in cell culture (*E. coli*) lysates to anti-DnaK monoclonal antibody immobilized on an optical SPR biosensor or for ELISA (Vostiar *et al.*, 2003)

Technique	k	k_1	k_2	D_f	D_{f1}	D_{f2}
ELISA (stressed culture)	26.975 ± 3.239	na	na	2.34 ± 0.126	na	na
ELISA (unstressed culture)	2.331 ± 2.980	$3.3 \times 10^{-7} + 5.5 \times 10^{-7}$	0.8626 ± 0.460	1.9868 ± 0.725	1.282 ± 0.725	0
SPR (stressed culture)	18.484 ± 2.319	na	na	2.1126 ± 0.131	na	na
SPR (injection cycle of cell lysate)	0.6544 ± 0.023	na	na	1.819 ± 0.035	na	na

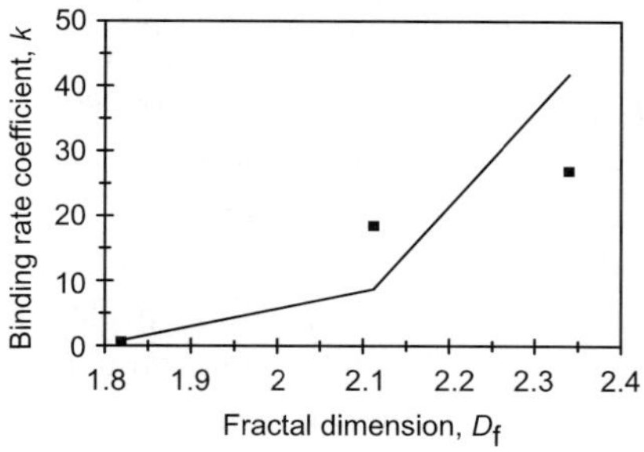

Figure 4.5 Increase in the binding rate coefficient, k, with an increase in the fractal dimension, D_f for the binding of chaperone DnaK in *Escherichia coli* cell culture lysates to anti-DnaK monoclonal antibody immobilized on an optical SPR biosensor or for ELISA [6].

Figure 4.4c shows the binding of the chaperone DnaK in heat-stressed cell (*E. coli*) lysates using an SPR biosensor. A single-fractal analysis is required to adequately describe the binding kinetics. The binding rate coefficient, k, value and the fractal dimension, D_f, are given in Table 4.2.

It is of interest to compare the binding rate coefficient, k, and fractal dimension, D_f, values obtained for the binding of heat-stressed DnaK in *E. coli* by the ELISA and the SPR biosensor analysis methods. Vostiar *et al.* (2003) indicate that the limit of detection (in mM) was 0.05 and 1.0 for the ELISA and the SPR methods, respectively. The present fractal analysis method is consistent with the Vostiar *et al.* (2003) analysis in that the binding coefficient, k, value is higher by 31.5% for the ELISA ($k = 26.975$), when compared with the SPR ($k = 18.484$). Note that the corresponding fractal dimension, D_f value for the ELISA is also higher by 9.7%. The fractal dimension, D_f, for ELISA and SPR methods are 2.34 and 2.1126, respectively.

Figure 4.4d shows the binding of chaperone DnaK in solution to anti-DnaK during the injection cycle of cell lysate using a SPR biosensor. A single-fractal analysis is adequate to describe the binding kinetics. The values of the binding rate coefficient, k, and the fractal dimension, D_f, are given in Table 4.2.

For the data presented in Table 4.2 and for a single-fractal analysis, Figure 4.5 shows that the binding rate coefficient, k, increases as the degree of heterogeneity on the biosensor surface or the fractal dimension, D_f, increases. The binding rate coefficient, k, is given by:

$$k = (0.000092 + 0.00023)D_f^{15.32 \pm 5.13} \tag{4.3}$$

The fit is quite good. More data points are required to provide a better fit. The binding rate coefficient is very sensitive to the degree of heterogeneity that exists on the biosensor surface as noted by the very high value of the dependence of the binding rate coefficient, k, on the fractal dimension, D_f.

4.4 CONCLUSIONS

A fractal analysis is presented for the binding of chaperone DnaK in unstressed and stressed *E. coli* cell lysate to anti-DnaK immobilized on a SPR biosensor chip surface or by ELISA (Vostiar *et al.*, 2003). The binding kinetics is described by either a single-, dual-, or triple-fractal analysis. In general, a higher degree of fractal analysis is required to describe the binding kinetics as one lowers the analyte (DnaK) concentration in *E. coli* cell lysates. As expected, the levels of chaperone DnaK expressed in stressed *E. coli* cell lysate is higher than that expressed in unstressed *E. coli* cell lysate. There is a change in the binding mechanism as one goes from a single- to a dual- and to a triple-fractal analysis.

For 30 mM DnaK *E. coli* cell lysate concentration a triple-fractal analysis is required to describe the binding kinetics. As the fractal dimension changes from D_{f1} to D_{f2} to D_{f3} there is a change in the binding rate coefficient from k_1 to k_2 to k_3 in the same direction. In general, the binding rate coefficient increases as the degree of heterogeneity or the fractal dimension, D_f, increases. For a dual-fractal analysis, the binding rate coefficient, k_2, is very sensitive to the degree of heterogeneity or fractal dimension, D_{f2}, that exists on the surface. The order of dependence of k_2 on D_{f2} is 19.806. For the binding of chaperone DnaK in *E. coli* cell lysates to anti-DnaK monoclonal antibody immobilized on an optical SPR biosensor or for ELISA when a first-order fractal analysis applies, the binding rate coefficient, k, is very sensitive to the degree of heterogeneity or the fractal dimension, D_f, as noted by the very high order of dependence (equal to 15.32).

The binding rate coefficient, k_1, increases as the DnaK concentration increases from 30 to 120 mM. The binding rate coefficient, k_1, exhibits a 1.4354 order of dependence on the DnaK *E. coli* cell lysate concentration.

It is of interest to note that for the unstressed *E. coli* cell culture analysis of DnaK by ELISA, there is a significant amount of variation in the DnaK concentration with time. It is this dilute nature of the DnaK concentration in solution that requires a higher degree of fractal analysis (dual- or triple-) to adequately model the binding kinetics.

The analysis provides values of the binding rate coefficient and the degree of heterogeneity on the biosensor surface. Different binding mechanisms are apparently prevalent at different *E. coli* lysate DnaK concentrations due to the fact that different (higher) degrees of fractal analysis are required to adequately model the binding kinetics (especially at the lower end of the DnaK concentrations). The analysis provides fresh physical insights into the binding of *E. coli* chaperone DnaK to anti-DnaK under unstressed and heat-stressed conditions.

REFERENCES

BIACORE AB, BIAEvaluation, 3.2 software, Uppsala, Sweden, 2002.

Bukau, B and AL Horwich, The Hsp70 and Hsp60 chaperone machines. *Cell*, 1998, **92**, 351–366.

Havlin, S, Molecular diffusion and reaction, in *The Fractal Approach to Heterogeneous Chemistry: Surfaces, Colloids, Polymers*, (ed. D. Avnir), Wiley, New York, 1989, pp. 251–269.

Jurgen, B, HY Link, S Riemschneider, C Scharf, P Neubauer, R Schmid, M Hecker and T Schweder, Monitoring of genes that respond to overproduction of an insoluble recombinant protein in *Escherichia coli* glucose-limited fed-batch fermentations. *Biotechnology Bioengineering*, 2000, **70**, 217–224.

Lee, CK and SL Lee, Multi-fractal scaling analysis of reactions over fractal surfaces. *Surface Science*, 1995, **325**, 294.

Oh, M-K and JC Liao, DNA microarray detection of metabolic responses to protein overproduction in *Escherichia coli. Metabolic Engineering*, 2000, **2**, 201–209.

Ramakrishnan, A and A Sadana, Analyte–receptor binding and dissociation kinetics for biosensor applications: a fractal analysis. *Biosensors and Bioelectronics*, 2000, **15**(11–12), 651.
Ramakrishnan, A and A Sadana, A single-fractal analysis of cellular analyte–receptor binding kinetics utilizing biosensors. *Biosystems*, 2001, **59**(1), 35.
Schlieker, C, B Bukau and A Mogk, Prevention and reversion of protein aggregation by molecular chaperones in the *E. coli* cytosol: implications for the applicability in biotechnology. *Journal of Biotechnology*, 2002, **96**, 13–21.
Steel, GJ, DM Fullerton, JR Tyson and CJ Stirling, Coordinated activation of Hsp70 chaperones. *Science*, 2004, **303**, 98–101.
Vostiar, J, J Tkac and C-F Mandenius, Monitoring of the heat-shock response in *Escherichia coli* using an optical biosensor. *Analytical Biochemistry*, 2003, **322**, 156–163.
Wickner, S, MR Maurizi and S Gottesman, Posttranslational quality control: folding, refolding, and degrading protein. *Science*, 1999, **286**, 1888–1893.
Yura, T and K Nakahigashi, Regulation of the heat-shock response. *Current Opinion in Microbiology*, 1999, **2**, 153–158.

– 5 –

Fractal Analysis of Binding and Dissociation Interactions of Prions on Biosensor Surfaces

5.1 INTRODUCTION

Prion represents a distinct class of infectious agents that are proteinaceous and devoid of nucleic acid (Caughey, 2000). The prion is an abnormal form of a host-encoded protein that can interact with its normal counterpart and cause it to become abnormal. This adaptation of the word 'prion' involves a broadening of the definition from a focus on the proteinaceous infectious agent of transmissible spongiform encephalopathy (TSE) diseases to include protein-based elements. These diseases are termed transmissible since they can be transmitted across species barriers, from cattle (cow, lamb) to humans. Studies that provide insight into the mechanisms involved are of importance. Caughey and Kocisko (2003) emphasize that nucleic acids may be a partner in crime for prion proteins that trigger a cascade of protein misfolding in the brain. This is thought to be the sole transmissible cause of some brain-destroying diseases.

Hileman (2004) recently indicates that the control of bovine spongiform encephalopathy (BSE) presents a regulatory challenge. The human form of BSE is variant Creutzfeldt–Jakob disease (vCJD). This is caused by eating BSE-infected beef and exposure to prions. This author emphasizes that (a) the human incubation period for vCJD is unknown, and (b) the minimum dose of prions (misfolded prion proteins) needed to cause vCJD is also unknown. Hileman (2004) indicates that a better understanding of the science involved in prion diseases would lead to a better and a more balanced regulatory approach to help control these types of diseases.

Mobley *et al.* (2004) very recently indicate the recent interest in understanding another prion-related disease— Alzheimer's disease (AD). Selkoe (2001) indicates that with an increase in life expectancy, AD has become the most prevalent form of late-life mental failure in humans. Here too, amyloid plaques (extracellular plaques) lead to this neurodegenerative disease wherein there is a progressive memory loss, decline in language function, disordered cognitive function, decline in motor function, and eventual death.

A proposed mechanism for prion aggregation is shown in Figure 5.1 (Kelley, 1998). PrP^C (Normal Protease sensitive form) is formed from the unfolded protein via

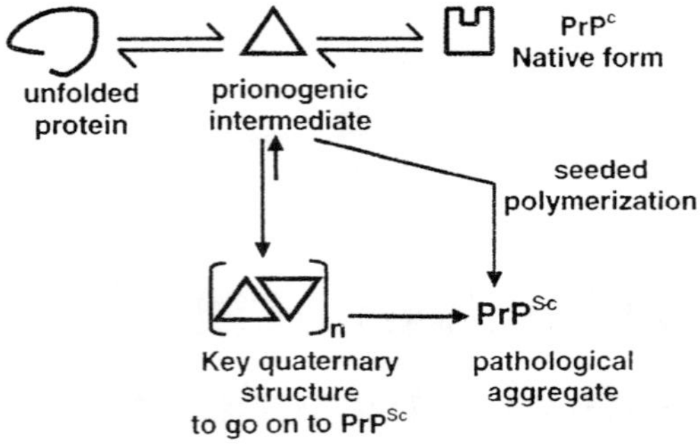

Figure 5.1 A proposed mechanism for prion aggregation (Kelley, 1998).

a prionogenic intermediate. This misfolded form PrP^{Sc} (protein resistance form) or scrapie may form by:

(a) Propagation of the prionogenic intermediate to form a template that terminates to form the PrP^{Sc} fibril.
(b) The PrP^{Sc} formed catalyzes the conversion of the prionogenic intermediate as well as the normal form of PrP^{C}. Since scarpie acts as a 'seed' that catalyzes the conversion to the misfolded form, the conversion is termed as 'seeded polymerization'.

Pruisner (1998) indicates that an understanding of the biochemical and biophysical properties of PrP is essential since it is associated with the fatal neurodegenerative disorders such as Creutzfeldt–Jakob disease in humans and BSE in animals. The mammalian PrP divides into:

(a) A C-terminal folded domain of ~100 residues.
(b) An N-terminal flexibly disordered region of ~100 residues also.

The C-terminal-folded domain of ~ 100 residues undergoes a conformational change in the generation of PrP^{Sc}. The N-terminal region includes five octapeptide repeats in the middle. Flechsig *et al.* (2000) emphasize that though the repeats are not essential for mediating pathogenisis, they do modulate the extent of these events and also of disease presentation.

PrP^{C} which is the native form, acts as a self-chaperone to form PrP^{C}, is called an auto-chaperone. This is similar to autocatalysis in chemical reactions (Liautard, 1999). Chaperones are compounds that help proteins fold properly and to their native and active form. Since PrP^{Sc} also converts the normal form of the PrP^{C} to the misfolded form, there exists a kinetic competition between these two reactions. Figure 5.2 shows the Liautard (1999) kinetic competition between the normal folding mechanism via prionogenic intermediate, auto-chaperone folding, and seeded polymerization misfolding. Liautard (1999) suggests that the key step to prevent aggregate formation is to block the formation of PrP^{C} via the prionogenic intermediate step. The model is also called auto-chaperone misfolding invasion. Jeyashekar (2003) has recently reviewed protein folding and misfolding mechanisms and their pathological implications.

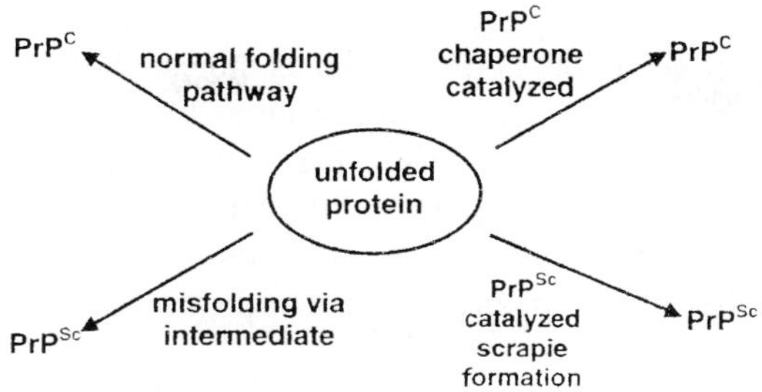

Figure 5.2 Kinetic competition between the normal folding mechanism via prionogenic intermediate, auto-chaperone folding, and seeded polymerization misfolding (Liautard, 1999).

Hetz *et al.* (2003) indicate that prions stress out the endoplasmic reticulum (ER). The neurotoxic form, PrP^{Sc} results from a conformational change in the normal form of the prion protein, PrP. These authors noted that the application of PrP^{Sc} to N2A neuroblastoma cells resulted in an increase in intracellular calcium concentration. The calcium was released from the ER. Caspase-12 is one of the proteases that mediates programmed cell death. Caspase-12 resides in the ER and was activated on the application of PrP^{Sc} to the cells. Hetz *et al.* (2003) indicate that the ER stress response may play a critical role in the toxicity of prions.

The mechanism of mitochondrial damage in neural cells by β-amyloid in Alzheimer's disease is unclear (Science and Technology Concentrates, 2004). Lustbader *et al.* (2004) postulate that β-amyloid forms a complex with β-amyloid-building alcohol dehydrogenase (ABAD) in the mitochondria of neuron cells. The enzyme is distorted by the β-amyloid, and then cannot bind with nicotinamide adenine dinucleotide. This is a cofactor that ABAD needs for its activity. Eventually, this leads to mitochondrial dysfunction and cell death.

Mallucci *et al.* (2003) indicate that the mechanisms involved in prion toxicity are unclear. They emphasize that therapies preventing the accumulation of PrP^{Sc} (the disease-associated form of the prion protein, PrP as indicated above) apparently do not significantly prolong survival in mice. This is for mice in which prion has infected the central nervous system. They emphasize that arresting the continued conversion of PrP^{C} to PrP^{Sc} within neurons either limits or prevents neurotoxicity even during scrapie infection. Furthermore, the authors suggest that depleting neuronal PrP in prion infection not only prevents disease but also reverses spongiosis. Their arresting procedure of neuronal conversion of PrP^{C} to PrP^{Sc} apparently prevents progression from preclinical central nervous system (CNS) prion infection to clinically manifest forms of the disease.

Finally, Mallucci *et al.* (2003) suggest targeting PrP^{C} as a therapeutic intervention in prion disease. This, they indicate could prevent the progression to clinical disease even for individuals with prions or with pathogenic PRNP mutations, and who are in a presymptomatic state. PRNP is the gene that encodes PrP^{C} and has been identified in hamster by Charles Weissman and independently in mice by Bruce Chesboro (Aguzzi and Heppner, 2000).

A therapeutic strategy has been suggested to prevent or help minimize aggregate formation that lead to a whole family of amyloid diseases (Dobson, 2001). In principle, masking the hydrophobic surfaces on intermediates and monomers prevents aggregation.

This would reduce the rate of aggregate deposition. A system of artificial chaperones (compounds that help proteins fold correctly and as indicated earlier) such as detergents and cyclodextrin have been successfully used to minimize aggregation in the renaturation of the denatured lysozyme. Osmolytes and other stabilizing ligands may decrease the amount of aggregation. This probably reflects differential stabilization of the native state. Osmolytes have been shown to inhibit prion formation in the scrapie system. Figure 5.3 shows that this is consistent with the involvement of protein conformation changes in the formation of prions (Fink, 1998).

Rovner (2004) recently indicates that immunotherapy has been used to treat Alzheimer's disease. The treatment is based on the amyloid cascade theory wherein plaques form first. This then leads to the formation of neurofibrillary tangles. Antibodies to the plaques of amyloid-β-peptide (Aβ) are being used in the immune therapy method to get rid of the Aβ from the brain.

Couzin (2004) has recently commented on the prion debate for Alzheimer's. This author indicates that controversy will continue in this area until the gold-standard experiment is performed. Infect normal mice, not transgenic ones, with prion proteins. This would help validate the proposal that these proteins alone are responsible for mad cow disease and similar illnesses.

Mattson (2004) has recently reviewed the pathways that lead to and away from Alzheimer's disease. The author emphasizes that though drugs may temporarily alleviate memory loss, there are no medical treatments that can presently either reverse or stop this neurodegenerative process.

Knight (2003) very recently indicates that a 'quantum leap' in the sensitivity of diagnostic procedures or assays are required to prevent a future outbreak of prion diseases. He emphasizes that prion diseases, to TSEs can be confirmed only after death. Even then, they are often missed. A sensitive test is required that is at least three orders of magnitude (1000 times) greater than present-day tests. Prion diseases may also be considered as a potential bioterror agent.

Hopkin (2004) has recently mentioned the increase in apprehension with the passing of blood stocks in which the prions are undetected. He indicates that certain genetic types of individuals may incubate variant Creutzfeldt–Jakob disease (vCJD), and pass it on whilst appearing healthy.

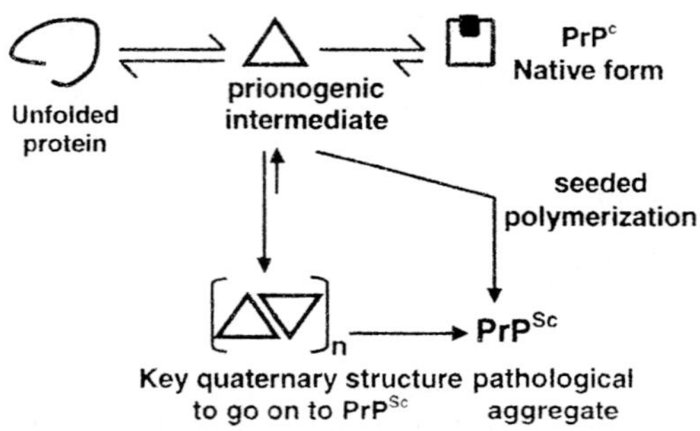

Figure 5.3 Protein conformational changes in the formation of prions (Fink, 1998).

Considering the importance to detect the presence of prions in cattle before they are consumed, it is important to quantitatively estimate the amount of scrapie prions present for effective treatment. In this chapter a conformation dependent immunoassay (CDI) developed by Dr. Pruisner's group at the University of California at San Francisco (UCSF) is presented, and the data obtained by this assay procedure is analyzed using fractals (Pruisner, 2002). CDI is an effective assay for prion detection. A biosensor can be developed based on this assay to detect scrapie prions. The effect of heterogeneity of the receptor surface on binding and dissociation kinetics is analyzed by fractal modeling. The fractal modeling is applied to the sensorgram data obtained from experiments performed in Dr. Pruisner's laboratory at UCSF.

5.2 THEORY

UCSF-led researchers (Safar *et al.*, 2002) have developed a highly sensitive automated test that they report significantly improves the accuracy and speed of detecting the various forms of infectious agent, which causes a set of neurodegenerative diseases in cattle, sheep, deer, and elk. It is called CDI. The test is able to detect much smaller levels of the infectious prion protein that can be seen with the current standard immunological procedures. As in the previous chapters, fractal analysis is used to analyze the binding and dissociation of an analyte in solution to an appropriate receptor immobilized on a surface plasmon resonance (SPR) biosensor surface.

Havlin (1989) has reviewed and analyzed the diffusion of reactants towards fractal surfaces. The details of the theory and the equations involved for the binding and the dissociation phases for analyte–receptor are available (Ramakrishnan and Sadana, 2001; Sadana, 2001). The analyte in solution is the fluorescently lit antibody (Fab)-Phuman, and the receptor on the SPR biosensor surface is the Bo 96-115. Bo 96-115 is a type of prion strain causing BSE in cows, and can be transmitted to humans upon consumption. Here we will attempt to apply these equations to the antibody (FAB)-Phuman-Bo 96-115 interaction (Pruisner, 2002).

The basic idea is to fold everything into the lumped parameter, the fractal dimension, D_f. This lumped parameter takes care of the chemical, orientational, and geometrical details. It is also assumed that the kinetic rate depends only on the geometric (static) parameter, D_f. It would also be useful to indicate the low and high limits in angstroms of the fractal character to be presented.

5.2.1 Single-fractal analysis

Binding rate coefficient

Havlin (1989) indicates that the diffusion of a particle (analyte) from a homogeneous solution to a solid surface (e.g. receptor-coated surface) on which it reacts to form

a product (analyte–receptor) complex is given by:

$$(\text{Analyte}\cdot\text{Receptor}) \sim \begin{cases} t^{(3-D_{\text{f,bind}})/2} = t^p & (t < t_c) \\ t^{1/2} & (t > t_c) \end{cases} \tag{5.1a}$$

where the analyte–receptor represents the association complex formed on the surface. Once again, the analyte in solution is the fluorescently lit antibody (Fab)-Phuman, and the receptor on the SPR biosensor surface is Bo 96-115. Here $p = -b$, and D_f is the fractal dimension of the surface. Havlin (1989) states that the crossover value may be determined by $r_c^2 \sim t_c$. Above the characteristic length, r_c, the self-similarity of the surface is lost and the surface may be considered homogeneous. Eq. 5.1a indicates that the concentration of the product [analyte–receptor] on a solid fractal surface scales at short and intermediate times as analyte–receptor $\sim t^p$ with the coefficient $p = (3 - D_f)/2$ at short time scales and $p = 1/2$ at intermediate time scales. Note that D_f, $D_{f,assoc}$, and $D_{f,bind}$ are used interchangeably. This equation is associated with the short-term diffusional properties of a random walk on a fractal surface. Note that, in perfectly stirred kinetics on a regular (nonfractal) structure (or surface), k_1 is a constant, that is, it is independent of time. In other words, the limit of regular structures (or surfaces) and the absence of diffusion-limited kinetics leads to k_1 being independent of time. In all other situations, one would expect a scaling behavior given by $k_1 \sim k' t^{-b}$ with $-b = p < 0$. Also, the appearance of the coefficient, p different from $p = 0$ is the consequence of two different phenomena, that is, the heterogeneity (fractality) of the surface and the imperfect mixing (diffusion-limited) condition. Finally, for a homogeneous surface where D_f is equal to 2, and when only diffusional limitations are present, $p = 1/2$ as it should be. Another way of looking at the $p = 1/2$ case (where $D_{f,bind}$ is equal to two) is that the analyte in solution views the fractal object, in our case the receptor-coated biosensor surface, from a 'large distance'. In essence, in the association process, the diffusion of the analyte from the solution to the receptor surface creates a depletion layer of width $(Ɖt)^{1/2}$ where $Ɖ$ is the diffusion constant. This gives rise to the fractal power law, $(\text{Analyte}\cdot\text{Receptor}) \sim t^{(3-D_{\text{f,bind}})/2}$.

The values of the parameters k (association rate coefficient), p, and D_f in eq. 5.1a may be obtained for analyte–receptor binding kinetics data. This may be done by a regression analysis using, for example, Quattro Pro (1997) along with eq. 5.1a where $(\text{analyte}\cdot\text{receptor}) = kt^p$. The fractal dimension may be obtained from the parameter, p. Higher values of the fractal dimension would indicate higher degrees of 'disorder' or heterogeneity or inhomogeneity on the surface.

Another way of looking at the diffusive process is that it inherently involves fluctuations at the molecular level that may be described by a random walk (Weiss, 1994). This author indicates that the kinetics of transport on disordered (or heterogeneous) media itself needs to be described by a random-walk model. When both of these are present, that is, the diffusion phenomena as well as a fractal surface, then one needs to analyze the interplay of both of these fluctuations. In essence, the disorder on the surface (or a higher fractal dimension, D_f) tends to slow down the motion of a particle (analyte in our case) moving in such a medium. Basically, according to Weiss (1994) the particle (random walker analyte) is trapped in regions in space, as it oscillates for a long time before resuming its motion.

Havlin (1989) indicates that the crossover value may be determined by $r_c^2 \sim t_c$. Above the characteristic length, r_c, the self-similarity of the surface of the surface is lost. Above t_c, the surface may be considered homogeneous, and 'regular' diffusion is now present. One may consider the analysis to be presented as an intermediate 'heuristic' approach in that in the future one may also be able to develop an autonomous (and not time-dependent) model of diffusion-limited kinetics in disordered media. No evidence of fractality is presented.

Dissociation rate coefficient

The diffusion of the dissociated particle (receptor or analyte) from the solid surface (e.g. analyte–receptor complex coated surface) into solution may be given, as a first approximation by:

$$\begin{aligned}(\text{Analyte}\cdot\text{Receptor}) &\sim -t^{(3-D_{\mathrm{f,diss}})/2}, \qquad t > t_{\mathrm{diss}} \\ &= -k_{\mathrm{diss}}t^{(3-D_{\mathrm{f,diss}})/2}\end{aligned} \tag{5.1b}$$

Here $D_{\mathrm{f,diss}}$ is the fractal dimension of the surface for the dissociation step. t_{diss} represents the start of the dissociation step. This corresponds to the highest concentration of the analyte–receptor complex on the surface. Henceforth, its concentration only decreases. $D_{\mathrm{f,assoc}}$ may or may not be equal to $D_{\mathrm{f,diss}}$. k_d and k_{diss}, and $D_{\mathrm{f,d}}$ and $D_{\mathrm{f,diss}}$ are used interchangeably in the chapter.

5.2.2 Dual-fractal analysis

Binding rate coefficient

The single-fractal analysis we have just presented is extended to include two fractal dimensions. At present, the time ($t = t_1$) at which the first fractal dimension 'changes' to the second fractal dimension is arbitrary and empirical. For the most part it is dictated by the data analyzed and the experience gained by handling a single-fractal analysis. The r^2 (regression coefficient) value obtained is also used to determine if a single-fractal analysis is sufficient, or one needs to use a dual-fractal analysis to provide an adequate fit. Only if the r^2 value is less than 0.97 for a single-fractal analysis, do we use a dual-fractal model.

In this case, the analyte–receptor complex is given by:

$$(\text{Analyte}\cdot\text{Receptor}) \sim \begin{cases} t^{(3-D_{\mathrm{f1,bind}})/2} = t^{p_1} & (t < t_1) \\ t^{(3-D_{\mathrm{f2,bind}})/2} = t^{p_2} & (t_1 < t < t_2 = t_c) \\ t^{1/2} & (t > t_c) \end{cases} \tag{5.1c}$$

In analyte–receptor association or binding the analyte–receptor associates with the active site on the surface and the product is released. In this sense the catalytic surface exhibits an unchanging fractal surface to the reactant in the absence of fouling and other complications. In the case of analyte–receptor association the biosensor surface exhibits a changing fractal surface to the analyte in solution. This occurs since as each association (binding) reaction takes place, smaller and smaller amounts of 'association' sites or receptors are available on the biosensor surface to which the analyte may bind.

Furthermore, as the reaction proceeds, there is an increasing degree of heterogeneity on the biosensor surface for some reaction systems. This is manifested by two degrees of heterogeneity or two fractal dimensions on the biosensor surface. When dilute analyte systems are involved (for example in the detection of pathogens; see for example Chapter 3), then a triple-fractal analysis is required to adequately model the binding kinetics. Then, in this case three fractal dimensions are required to describe the degree of heterogeneity on the CM5 sensor chip in the binding phase. In the theoretical limit one might envisage a temporal fractal dimension wherein there is a continuous change in the degree of heterogeneity on the surface; though of course, such situations would be very rare, if at all. The dissociation phase may also be represented by one or two fractal dimensions, depending on whether a single- or a dual-fractal analysis is required to describe the dissociation kinetics. This may also be noted by the complexity of the curve exhibited in the dissociation phase. The fractal dimension(s) in the dissociation phase may or may not be equal to the fractal dimension(s) in the binding phase.

Note that, at present, the dual-fractal analysis does not have a basis at the molecular level. This represents two different levels of heterogeneity on the biosensor surface. But, in some examples presented, a single-fractal analysis is clearly inadequate to model the data. Only in these cases does one resort to a dual-fractal analysis.

Dissociation rate coefficient

In this case the dissociation rate coefficient is given by:

$$(\text{Analyte}\cdot\text{Receptor}) \sim \begin{cases} -t^{(3-D_{\text{f1,diss}})/2} & (t_{\text{diss}} < t < t_{\text{d1}}) \\ -t^{(3-D_{\text{f2,diss}})/2} & (t_{\text{d1}} < t < t_{\text{d2}}) \end{cases} \tag{5.1d}$$

Here $D_{\text{f,diss}}$ is the fractal dimension of the surface for the dissociation step. t_{diss} represents the start of the dissociation step. This corresponds to the highest concentration of the analyte·receptor complex on the surface. Henceforth, its concentration only decreases. As indicated above, $D_{\text{f,bind}}$ or $D_{\text{f,assoc}}$ may or may not be equal to $D_{\text{f,diss}}$.

5.3 RESULTS

The CDI developed in Dr. Pruisner's laboratory involves revealing the region of PrP^C that is exposed to the normal PrP^C but is buried in infectious PrP^{Sc}. This is done on using high-affinity, newly generated antibodies that identify PrP^{Sc} through the distinct shape of the molecule, independent of proteolytic treatments. This permits the detection of potentially large concentrations of protease sensitive PrP^{Sc} molecules (Jeyashekar, 2003).

Jeyashekar (2003) indicates that two steps are involved in the CDI procedure:

(a) Expose a tissue containing the infectious prion protein in its natural state to the antibody and measure its reactivity.
(b) Unfold the protein by chemical means so that the hidden region is exposed. The antibody's immunoreactivity to this denatured region, as measured by its degree

of binding to the molecule is much higher than it is to the diseased protein in its native state. The ratio of denatured to native infectious protein indicates the amount of PrP^{Sc}.

Figure 5.4a shows the binding and dissociation of 62.5 mM Phuman Fab antibody in solution to Bo 96-115 immobilized on a CM5 sensor chip of an SPR biosensor (Pruisner, 2002). A single-fractal analysis is adequate to describe the binding as well as the dissociation kinetics. The values of the fractal dimensions for the binding and the dissociation phases are presented in Table 5.1b. The values of the binding and dissociation rate coefficient(s), and the fractal dimensions for the binding and the dissociation phase presented in Table 5.1 were obtained using Corel Quattro Pro 8.0 (1989). The binding and dissociation rate coefficient values obtained are within 95% confidence limits. For example, for the binding of 62.5 mM Phuman Fab antibody in solution to Bo 96-115 immobilized on a CM5 sensor chip surface, the binding rate coefficient, k is given by 6.099 ± 0.563. The 95% confidence limit indicates that the k value lies between 5.53 and 6.662. This indicates that the value is precise and significant.

Affinity values are of interest to practicing biosensorists. Define affinity as $K = k/k_d$, the ratio of the binding and dissociation rate coefficients. Then, the affinity, K is equal to 8.80 for a single-fractal analysis. The ratio of fractal dimension, D_f/D_{fd} is 0.876.

Figure 5.4b shows the binding and dissociation of 125 mM Phuman antibody in solution to Bo 96-115 immobilized on a CM5 sensor chip (Pruisner, 2002). A dual-fractal analysis is required to adequately describe the binding and dissociation kinetics.

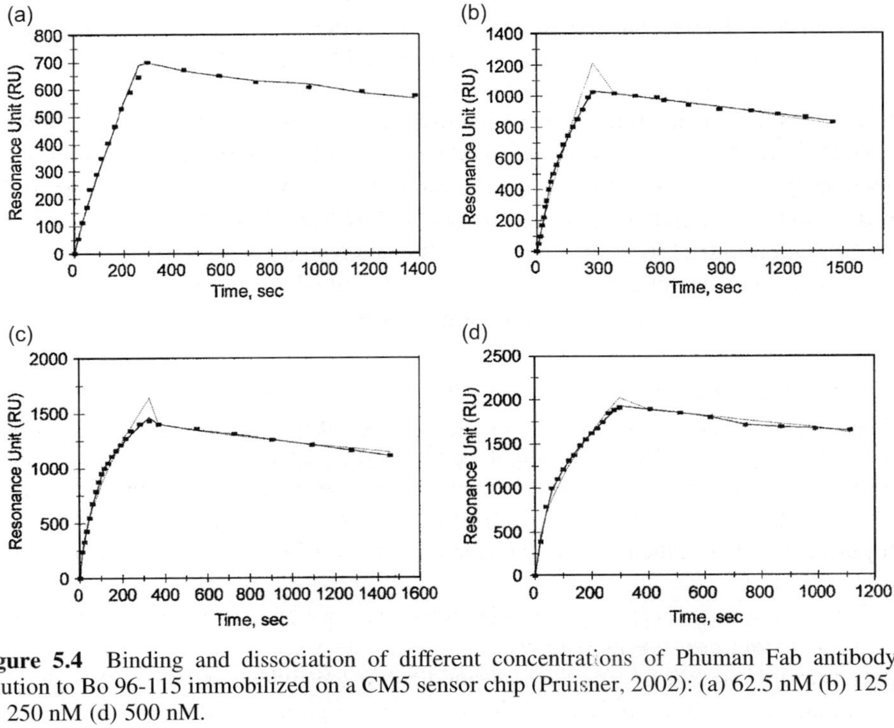

Figure 5.4 Binding and dissociation of different concentrations of Phuman Fab antibody in solution to Bo 96-115 immobilized on a CM5 sensor chip (Pruisner, 2002): (a) 62.5 nM (b) 125 nM (c) 250 nM (d) 500 nM.

Table 5.1a

Binding and dissociation rate coefficients for Phuman Fab antibody in solution to Bo 96-115 immobilized on a CM5 sensor chip (Pruisner, 2002)

Phuman Fab antibody concentration, mM	k	k_1	k_2	k_d	k_{d1}	k_{d2}
62.5	6.099 ± 0.563	na	na	0.6928 ± 0.076	na	na
125	10.43 ± 1.599	5.0697 ± 0.3839	35.537 ± 0.493	0.3593 ± 0.005	0.0148 ± 0.002	0.100 ± 0.003
250	65.43 ± 4.94	48.784 ± 2.090	177.57 ± 2.575	1.612 ± 0.248	2.374 ± 0.392	0.4498 ± 0.001
500	101.9 ± 9.50	42.149 ± 5.049	168.61 ± 1.63	0.1096 ± 0.028	0.0193 ± 0.002	10.283 ± 0.098

The values of (a) the binding rate coefficient, k for a single-fractal analysis, and (b) the binding rate coefficients, k_1 and k_2 for a dual-fractal analysis, (c) the dissociation rate coefficient, k_d for a single-fractal analysis, and (d) the dissociation rate coefficients, k_{d1} and k_{d2} for a dual-fractal analysis are given in Table 5.1a. The values of (a) the fractal dimension, D_f for binding using a single-fractal analysis, (b) the fractal dimensions, D_{f1} and D_{f2} for a dual-fractal analysis, (c) the fractal dimension, D_{fd} for dissociation using a single-fractal analysis, and (d) the fractal dimensions, D_{fd1} and D_{fd2} for dissociation using a dual-fractal analysis are given in Table 5.1b.

Note that an increase in the Phuman Fab antibody concentration in solution from 62.5 to 125 mM leads to a change in the mechanism for binding as well as for dissociation. At the lower 62.5 mM Phuman antibody concentration in solution a single-fractal analysis is adequate to describe the binding and dissociation kinetics. At the higher 125 mM Phuman Fab antibody concentration in solution a dual-fractal analysis is required to adequately describe the binding and dissociation kinetics.

Also, an increase in the fractal dimension in the binding phase by a factor of 2.04 from a value of D_{f1} equal to 0.8832 to D_{f2} equal to 1.802 leads to an increase in the binding rate coefficient by a factor of 7.0 from a value of k_1 equal to 5.0697 to k_2 equal to 35.537. Similarly, in the dissociation phase, an increase in the fractal dimension by a factor of 3.99 from D_{fd1} equal to 0.2156 to D_{fd2} equal to 0.861 leads to an increase in the dissociation rate coefficient by a factor of 6.76 from a value of k_{d1} equal to 0.0148 to value of k_{d2} equal to 0.10. Apparently, an increase in the degree of heterogeneity (increase in the fractal

Table 5.1b

Fractal dimensions for the binding and dissociation phase for Phuman Fab antibody in solution to Bo 96-115 immobilized on a CM5 sensor chip (Pruisner, 2002)

Phuman Fab antibody concentration, mM	D_f	D_{f1}	D_{f2}	D_{fd}	D_{fd1}	D_{fd2}
62.5	1.3022 ± 0.059	na	na	1.487 ± 0.126	na	na
125	1.3100 ± 0.071	0.8832 ± 0.0719	1.802 ± 0.0226	0.5422 ± 0.121	0.2156 ± 0.191	0.861 ± 0.169
250	1.887 ± 0.0386	1.727 ± 0.038	2.271 ± 0.029	1.522 ± 0.103	1.6854 ± 0.154	1.128 ± 0.012
500	1.950 ± 0.061	1.472 ± 0.217	2.145 ± 0.136	0.6558 ± 0.234	0.01 + 0.179	2.033 ± 0.038

dimension value) on the CM5 sensor chip surface leads to an increase in the binding and in the dissociation rate coefficients. .The binding and dissociation rate coefficients are quite sensitive to the fractal dimension or the degree of heterogeneity present on the biosensor surface in both the binding and in the dissociation phases.

In this case since a dual-fractal analysis is required to describe the binding and dissociation kinetics, one may define two values of the affinity. K_1 which is equal to k_1/k_{d1} and K_2 which is equal to k_2/k_{d2}. Then, K_1 is equal to 342.5, and K_2 is equal to 355.7. The affinity value, K_2 is higher than that of K_1 by 3.75%.

Figure 5.4c shows the binding and dissociation of 125 mM Phuman antibody in solution to Bo 96-115 immobilized on a CM5 sensor chip (Pruisner, 2002). A dual-fractal analysis is required to adequately describe the binding and dissociation kinetics. The values of (a) the binding rate coefficient, k for a single-fractal analysis, and (b) the binding rate coefficients, k_1 and k_2 for a dual-fractal analysis, (c) the dissociation rate coefficient, k_d for a single-fractal analysis, and (d) the dissociation rate coefficients, k_{d1} and k_{d2} for a dual-fractal analysis are given in Table 5.1a. The values of (a) the fractal dimension, D_f for binding using a single-fractal analysis, (b) the fractal dimensions, D_{f1} and D_{f2} for a dual-fractal analysis, (c) the fractal dimension, D_{fd} for dissociation using a single-fractal analysis, and (d) the fractal dimensions, D_{fd1} and D_{fd2} for dissociation using a dual-fractal analysis are given in Table 5.1b.

Note once again that an increase in the fractal dimension in the binding phase by a factor of 1.31 from a value of D_{f1} equal to 1.727 to D_{f2} equal to 2.271 leads to an increase in the binding rate coefficient by a factor of 3.64 from a value of k_1 equal to 48.784 to k_2 equal to 172.57. In the dissociation phase, the fractal dimension decreases by 66.92% from D_{fd1} equal to 1.6854 to D_{fd2} equal to 1.128. This leads to a decrease in the dissociation rate coefficient by a factor of 5.28 from a value of k_{d1} equal to 2.374 to value of k_{d2} equal to 0.4498. Apparently, in this case, a decrease in the degree of heterogeneity on the CM5 sensor chip surface leads to a decrease in the in the dissociation rate coefficient. Note, however, that changes in the degree of heterogeneity on the CM5 sensor chip surface and in the dissociation rate coefficient are in the same direction for either the 125 or the 250 nM Phuman Fab concentration in solution.

The affinity value, K_1 is equal to 20.55, and the affinity value K_2 is equal to 394.8. The corresponding ratio of fractal dimension values D_{f1}/D_{fd1} and D_{f2}/D_{fd2} are 1.025 and 2.013, respectively. An increase in the ratio of the fractal dimension values by a factor 1.96 from a D_{f1}/D_{fd1} equal to 1.025 to D_{f2}/D_{fd2} equal to 2.013 leads to an increase in the affinity value by a factor of 19.2 from a value of K_1 equal to 20.55 to K_2 equal to 394.8.

Figure 5.4d shows the binding and dissociation of 500 nM Phuman antibody in solution to Bo 96-115 immobilized on a CM5 sensor chip. A dual-fractal analysis is required to adequately describe the binding and dissociation kinetics. The values of (a) the binding rate coefficient, k_1 for a single-fractal analysis, (b) the binding rate coefficient, k_1 and k_2 for a dual-fractal analysis, (c) the dissociation rate coefficient, k_d for a single-fractal analysis, and (d) the dissociation rate coefficients, k_{d1} and k_{d2} for a dual-fractal analysis are given in Table 5.1a. The values of the fractal dimension, D_f for binding using a single-fractal analysis, (b) the fractal dimensions, D_{f1} and D_{f2} for binding for a dual-fractal analysis, (c) the fractal dimensions, D_{fd} for dissociation using a single-fractal analysis, and (d) the fractal dimensions, D_{fd1} and D_{fd2} for a dual-fractal analysis for dissociation are given in Table 5.1b.

An increase in the fractal dimension in the binding phase by a factor of 1.46 from a value of D_{f1} equal to 1.472 to D_{f2} equal to 2.271 leads to an increase in the binding rate coefficient by a factor of 4.0 from a value of k_1 equal to 42.15 to k_2 equal to 168.6. Similarly, in the dissociation phase, an increase in the fractal dimension value by a factor of 203.3 from D_{fd1} equal to 0.01 to D_{fd2} equal to 2.033 leads an increase in the dissociation rate coefficient by a factor of 532.8 from a value of k_{d1} equal to 0.0193 to k_{d2} equal to 10.283. Once again, changes in the degree of heterogeneity (or the fractal dimension value) on the CM5 sensor chip surface in either the binding or the dissociation phase lead to changes in the binding and in the dissociation rate coefficients in the same direction.

The affinity, K_1 value is equal to 2184 and the affinity, K_2 value is equal to 16.4. The corresponding ratio of fractal dimension values D_{f1}/D_{fd1} and D_{f2}/D_{fd2} are 147.2 and 1.055, respectively. Changes in the affinity value are in the same direction as changes in the corresponding ratio of fractal dimension values.

Figure 5.5a and Table 5.1 show for a dual-fractal analysis the increase in the binding rate coefficient, k_1 with an increase in the fractal dimension, D_{f1}. For the 125–500 nM Phuman Fab antibody concentration in solution, the binding rate coefficient, k_1 is given by:

$$k_1 = (38.36 \pm 3.0)D_{f1}^{3.564 \pm 0.620} \tag{5.2a}$$

The fit is quite good. Only three data points are available. The availability of more data points will lead to a more reliable fit. The binding rate coefficient, k_1 is sensitive to the degree of heterogeneity or the fractal dimension, D_{f1} on the CM5 sensor chip surface as noted by the greater than third-order dependence (equal to 3.564) of k_1 on D_{f1}.

Figure 5.5b and Table 5.1 show for a dual-fractal analysis the increase in the binding rate coefficient, k_2 with an increase in the fractal dimension, D_{f2}. For the 125–500 nM Phuman Fab antibody concentration in solution, the binding rate coefficient, k_2 is given by:

$$k_2 = (0.4730 + 0.6199)D_{f2}^{7.42 \pm 1.59} \tag{5.2b}$$

The fit is quite good. Only three data points are available. The availability of more data points will lead to a more reliable fit. The error is quite large; thus only the positive error is given. The binding rate coefficient, k_2 is very sensitive to the degree of heterogeneity or the fractal dimension, D_{f2} on the CM5 sensor chip surface as noted by the greater than seventh-order dependence (equal to 7.42) of k_2 on D_{f2}.

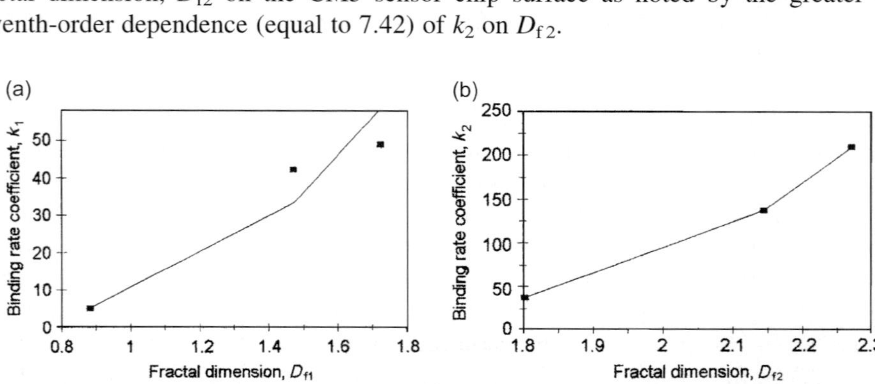

Figure 5.5 (a) Increase in the binding rate coefficient, k_1 with an increase in the fractal dimension, D_{f1}. (b) Increase in the binding rate coefficient, k_2 with an increase in the fractal dimension, D_{f2}.

Figure 5.6a and Table 5.1 show for a dual-fractal analysis the increase in the fractal dimension in the dissociation phase, D_{fd2} with an increase in the Phuman Fab antibody concentration in solution. For the 125–500 mM Phuman Fab antibody concentration in solution, the fractal dimension, D_{fd2} is given by:

$$D_{fd2} = (0.041 \pm 0.006)[\text{Phuman Fab antibody}]^{0.620 \pm 0.133} \tag{5.3a}$$

The fit is quite good. Only three data points are available. The availability of more data points would lead to a more reliable fit. The fractal dimension in the second phase of dissociation, D_{fd2} is only mildly sensitive to the Phuman Fab antibody concentration in solution as noted by the 0.620 order of dependence.

Figure 5.6b and Table 5.1 show for a dual-fractal analysis the increase in the dissociation rate coefficient, k_{d2} with an increase in the Phuman Fab antibody concentration in solution. For the 125–500 mM Phuman Fab antibody concentration in solution, the dissociation rate coefficient, k_{d2} is given by:

$$k_{d2} = (7.6 \pm 7.4)[\text{Phuman Fab antibody}]^{3.34 \pm 0.679} \tag{5.3b}$$

The fit is quite good. Only three data points are available. The availability of more data points would lead to a more reliable fit. There is some scatter in the data, and this is reflected in the error of the estimate for the dissociation rate coefficient. The dissociation rate coefficient, k_{d2} is very sensitive to the Phuman Fab antibody concentration in solution

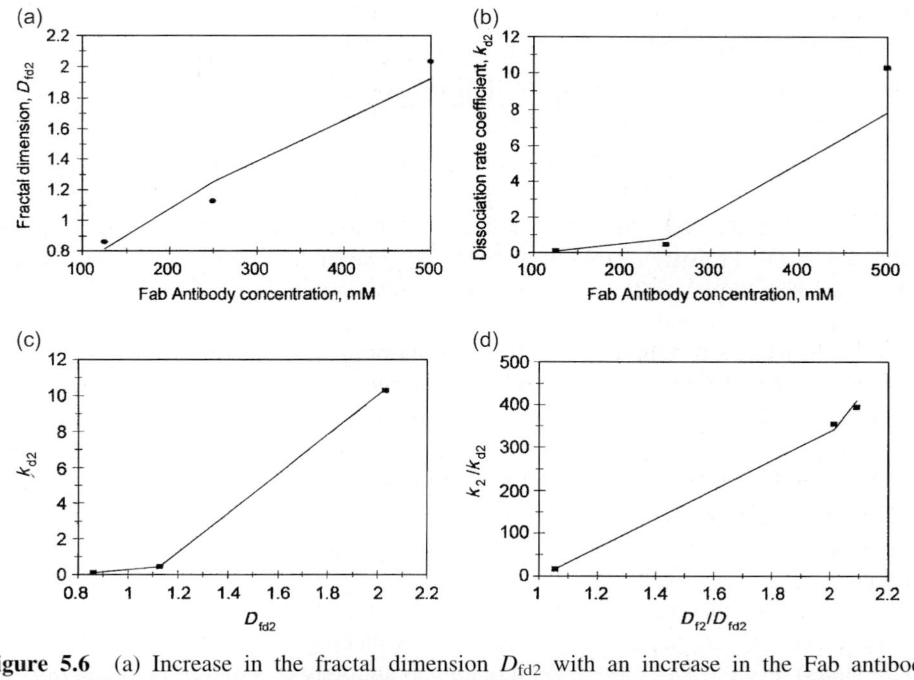

Figure 5.6 (a) Increase in the fractal dimension D_{fd2} with an increase in the Fab antibody concentration in solution. (b) Increase in the dissociation rate coefficient, k_{d2} with an increase in the Fab antibody concentration in solution. (c) Increase in the dissociation rate coefficient, k_{d2} with an increase in the fractal dimension D_{fd2}. (d) Increase in the ratio of k_2/k_{d2} with an increase in the ratio, D_{f2}/D_{fd2}.

as noted by the greater than third-order (equal to 3.34) dependence of k_{d2} on the Phuman Fab antibody concentration in solution.

Figure 5.6c and Table 5.1 show for a dual-fractal analysis the increase in the dissociation rate coefficient, k_{d2} with an increase in the fractal dimension in the second phase, D_{fd2}. For the 125–500 mM Phuman Fab antibody concentration in solution, the dissociation rate coefficient, k_{d2} is given by:

$$k_{d2} = (0.2290 \pm 0.0078) D_{fd2}^{5.372 \pm 0.0537} \quad (5.4a)$$

The fit is good. Only three data points are available. The availability of more data points would lead to a more reliable fit. The dissociation rate coefficient, k_{d2} is very sensitive to the degree of heterogeneity on the CM5 sensor chip surface as noted by the greater than fifth-order (equal to 5.372) of dependence of k_{d2} on D_{fd2}.

Figure 5.6d shows for a dual-fractal analysis the increase in the affinity, $K_2(= k_2/k_{d2})$ with an increase in the fractal dimension ratio, D_{f2}/D_{fd2}. For the 125–500 mM Phuman Fab antibody concentration in solution, the affinity, K_2 is given by:

$$K_2 = (0.228 \pm 0.009)(D_{f2}/D_{fd2})^{5.379 \pm 0.061} \quad (5.4b)$$

The fit is good. Only three data points are available. The availability of more data points would lead to a more reliable fit. The affinity, K_2 is very sensitive to the fractal dimension ratio as noted by the higher than fifth-order dependence on the fractal dimension ratio, (D_{f2}/D_{fd2}).

5.4 CONCLUSIONS

A fractal analysis is presented for the binding and dissociation of the analyte, fluorescently lit Phuman Fab antibody in solution to the receptor, Bo-96-115 immobilized on an SPR biosensor surface (Pruisner, 2002). This is for the CDI developed in Dr. Pruisner's laboratory at UCSF. The binding and dissociation kinetics may be described by a single- and a dual-fractal analysis. A dual-fractal analysis is only used when a single-fractal analysis does not provide an adequate fit. This was done using Corel Quattro Pro 8.0 (1989), and only when the regression coefficient for the sum of least squares was less than 0.97. The fractal dimension values provide a quantitative indication of the degree of heterogeneity present on the CM5 sensor chip surface. Binding and dissociation rate coefficient values, and affinity values are provided. The fractal dimension for the binding and the dissociation phase, D_f and D_{fd}, respectively, is not a typical independent variable, such as analyte concentration, that may be directly manipulated. It is estimated from eqs. (5.1a–d), and one may consider it as a derived variable.

An increase in the fractal dimension value or the degree of heterogeneity on the surface leads, in general, to an increase in the binding and in the dissociation rate coefficient(s). For the 125–500 mM Phuman Fab antibody concentration in solution, the binding rate coefficients, k_1 and k_2 are very sensitive to the fractal dimension or the degree of heterogeneity on the CM5 sensor chip surface as noted by the 3.564 and 7.420 order of dependence exhibited, respectively. Predictive equations are also developed for the dissociation rate coefficient, k_{d2} and the fractal dimension, D_{fd2} as a function of the Phuman Fab antibody concentration in solution. The dissociation rate coefficient, k_{d2} is also very sensitive to the fractal dimension, D_{fd2} or the degree of heterogeneity on the CM5 sensor chip surface as noted by the 5.372 order of dependence exhibited. Affinity defined as $K = k/k_d$, $K_1 = k_1/k_{d1}$, and $K_2 = k_2/k_{d2}$ also exhibits an increase with the respective ratio(s) of the fractal dimensions obtained in the binding and in the dissociation phases.

The predictive relationships developed for the binding rate coefficients, k_1 and k_2 as a function of the fractal dimensions, D_{f1} and D_{f2}, respectively, are of considerable value because they directly link the binding rate coefficient to the degree of heterogeneity that exists on the CM5 sensor chip surface, and provides a means by which the binding rate coefficient may be manipulated by changing the degree of heterogeneity on the sensor chip surface. Note that a change in the degree of heterogeneity on the CM5 sensor chip surface would, in general, lead to changes in both the binding and in the dissociation rate coefficient. Thus, this may require a little thought and manipulation. The binding and the dissociation rate coefficients are rather sensitive to their respective fractal dimensions or the degree of heterogeneity that exists on the biosensor surface. This may be noted by the high orders of dependence, as indicated above. It is suggested that the fractal surface (roughness) leads to turbulence, which enhances mixing, decreases diffusional limitations, and leads to an increase in the binding (and dissociation) rate coefficient(s) (Martin *et al.*, 1991).

More such studies are required to determine whether the binding and dissociation rate coefficient(s) are sensitive to their respective fractal dimensions or the degree of heterogeneity that exists on the CM5 sensor chip surface. If this is correct, then experimentalists may find it worth their effort to pay a little more attention to the nature of the surface, and how it may be manipulated to control the relevant parameters and biosensor performance in desired directions. This is particularly true for the conformation dependent immunoassay (CDI) developed for prion detection. Any improvement, even very small in the early detection of prions by the conformation dependent assay is worth the effort. The earlier one is able to detect the presence of prions, the earlier one may institute a protocol that better helps manage the onset and progress of these serious prion-related diseases.

PrP^{Sc} and PrP^{Sc}-aggregated states are termed PrP_{res} as they are partially resistant to destruction by proteases. Delault *et al.* (2003) were able to use RNA to amplify PrP_{res}. This will significantly assist in improving the sensitivity and reliability of diagnostic tests. It is absolutely necessary to be able to detect very small amounts of PrP_{res} in both the food supply and in the early diagnosis of diseases such as bovine spongiform encephalopathy.

The fractal analysis provided in this chapter also aids in the this direction by re-analyzing data on the conformation dependent assay (CDI) developed by Dr. Pruisner's laboratory (Pruisner, 2002) by providing fresh physical insights into the Phuman Fab antibody interaction with Bo 96-115. Hopefully, this will lead to a more sensitive, rapid, and reliable diagnostic procedure for the detection of the abnormal form(s) of prions. In fact, CDI is being developed by InPro Biotechnology in South San Francisco, CA (Hileman, 2004). Hileman (2004) further indicates that other rapid tests include the Idexx laboratories test, the Abott laboratories test, and two tests made by Prionics in Zurch, Switzerland. No comparison was apparently made between the CDI procedure being developed by InPro Biotechnology and other rapid tests such as the Idexx laboratories test, the Abott laboratories test, and the two tests made by Prionics.

Finally, Hon and Zagorski (2004) indicate that there are 24 proteins that produce amyloid fibrils associated with human disease (Westermark *et al.*, 2002). These include the Aβ of Alzheimer's disease, prion of TSEs, transthyretin of familial amyloidosis, huntington of Huntington's disease, α-synuyclein of Parkinson's disease, and amylin of maturity-onset disease. Hon and Zagorski (2004) emphasize that there is a lack of molecular mechanisms of amyloidosis. Particularly, lacking are high-resolution structural data, such as atomic level structural information. These authors emphasize that it would be useful to know which amino acids are interacting during the aggregation processes, and if fibrils adopt unique, three-dimensional folded structures.

REFERENCES

Aguzzi, B and FL Heppner, Pathogenisis of prion diseases: a progress report. *Cell Death and Differentiation*, 2000, **7**, 889–902.

Caughey, B, Transmissible spongiform encephalopathies, amylodoses, and yeast prions. Common threads? *Nature Medicine*, 2000, **6**, 751–754.

Caughey, B and DA Kocisko, A nucleic-acid accomplice. *Nature*, 2003, **425**, 673–674.

Corel Quattro Pro, 8.0., Corel Corporation, Ottawa, Canada, 1989.

Couzin, J, An end to the prion debate? Don't count on it. *Science*, 2004, **305**, 589.

Delault, NR, RW Lucassen and S Supattapone, RNA molecules stimulated prion protein conversion. *Nature*, 2003, **425**, 717–720.

Dobson, CM, The structural basis of protein folding and its links with human disease. *Philosophical Transactions of the Royal Society of London, Series B, Biological Sciences*, 2001, **356**, 133–145.

Fink, AL, Protein aggregation: folding aggregates, inclusion bodies, and amyloid. *Folding & Design*, 1998, **3**, R9–R23.

Flechsig, E, D Shmerling, I Hegyi, AJ Raeber, M Fischer, A Cozzio, C von Mering, A Aguzzi and C Weissman, Prion protein devoid of the octapeptide region restores susceptibility to scarpie in PrP knockout mice. *Neuron*, 2000, **27**, 399–408.

Havlin, S, Molecular diffusion and reactions. *The Fractal Approach to Heterogeneous Chemistry: Surfaces, Colloids, Polymers*. (ed. D Avnir), Wiley, New York, 1989.

Hetz C, M Rossdakis-Carneiro, K Maundrell, J Castilla and C Soto, Caspase-12 and endoplasmic reticulum stress mediate neurotoxicity of pathological prion protein. *EMBO*, 2003, **22**, 5345.

Hileman, B, Mad cow disease. Regulatory changes stemming from discovery of one diseased cow create new conflicts exacerbated by gaps in scientific understanding. *Chemical & Engineering News*, 2004, **82**(22), 21–25.

Hon, L and MG Zagorski, Sorting out the driving forces for parallel and antiparallel alignment in the Aβ peptide fibril structure. *Biophysical Journal*, 2004, **86**, 1–2.

Hopkin, M, Fears grow as blood stocks pass on proteins undetected. *Nature*, 2004, **430**, 712.

Jeyshekar, NS, Protein Folding and Misfolding Mechanisms–Pathological Implications, Detection and Modeling, Master of Science Thesis, The University of Mississippi, University, MS, USA, August 2003.

Kelley, JW, The alternative conformations of amyloidogenic proteins and their multi-step assembly pathways. *Current Opinion in Structural Biology*, 1998, **8**, 101–106.

Knight, J, Academy calls for improved tests to beat prion disease. *Nature*, 2003, **206**, 216 www.nap.edu.

Liautard, JP, Analytical background and discussion of the chaperone model of prion diseases. *Biotheoretica*, 1999, **47**, 219–238.

Lustbader, JW, M Cirilli, C Lin, HW Xu, K Takuma, N Wang, C Casperson, X Chen, S Pollak, M Chaney, F Trinchese, S Liu, F Gunn-Moore, LF Lue, DG Walker, P Kuppusamy, ZL Zewier,

D Arancio, D Stern, SS Yan and H Wu, ABAD directly links AB to mitochondrial toxicity in Alzheimer's disease. *Science*, 2004, **304**, 448–452.

Mallucci, G, A Dickinson, J Linehan, P-C Klohn, S Brandner and J Collinge, Depleting neuronal PrP in prion infection prevents disease and reverses spongiosis. *Science*, 2003, **302**, 871–874.

Martin, SJ, VE Granstaff and GC Frye, Effect of surface roughness on the response of thickness-shear mode resonators in liquids. *Analytical Chemistry*, 1991, **65**, 2910–2922.

Mattson, MP, Pathways towards and away from Alzheimer's disease. *Nature*, 2004, **430**, 631–639.

Mobley, DL, DL Cox, RP Singh, MW Maddox and MW Longo, Modeling amyloid β-peptide insertion into lipid bilayers. *Biophysical Journal*, 2004, **86**, 3585–3597.

Pruisner, SB, Prions. *Proceedings of the National Academy of Sciences USA*, 1998, **95**, 13363–13383.

Pruisner, SB, Personal communication to Mr. Nigil Jeyshekar, Masters student, University of Mississippi, Chemical Engineering Department, 2002.

Ramakrishnan, A and A Sadana, A fractal analysis for cellular analyte–receptor binding kinetics: biosensor applications. *Automedica*, 2001, **20**, 313–340.

Rovner, S, Antibodies fight Alzheimer's. Immune therapy clears both plaques and tangles of mice. *Chemical & Engineering News*, 2004, **82**(32) http://www.cen-online.org.

Sadana, A, A fractal analysis approach for the evaluation of hybridization kinetics in biosensors. *Journal of Colloid & Interface Science*, 2001, **234**, 9–18.

Safar, JG, M Scott, J Monoghan, C Deering, S Didorenko, J Vergana, H Ball, G Legname, E Leclerc, L Solforosi, H Serban, D Groth, DR Burton, SB Pruisner and A Williamson, Measuring prions causing bovine spongiform encephalopathy or chronic wasting disease by immunoassays and transgenic mice. *Nature Biotechnology*, 2002, **748**, 1–4.

Science and Technology Concentrates, *Chemical and Engineering News*, April 19, 2004, pp. 35–45.

Selkoe, DJ, Alzheimer's disease: genes, proteins, and therapy. *Physiological Reviews*, 2001, **82**, 741–766.

Weiss, GH, *Fractals in Science*. Springer, Berlin, 1994.

Westermark, P, MD Benson, JN Buxbaum, AS Cohen, B Frangione, S Ikeda, CL Masters, G Merlini, MJ Sarava and JD Sipe, Amyloid fibril protein nomenclature. *Amyloid*, 2002, **9**, 197–200.

– 6 –

Fractal Analysis of Binding and Dissociation of Analytes Related to Human Health on Biosensor Surfaces

6.1 INTRODUCTION

A promising approach in the understanding of biomolecular interactions is the development of biosensors. These biosensors are finding increasing application in the area of biotechnology, physics, chemistry, medicine, aviation, oceanography, environmental control, safe guarding of civilian infrastructures, protection of military and civilian personnel, and food product applications. Some of the not too distant future sensor applications include biomedical health monitoring, monitoring of terrestrial and aquatic environments, surveillance technology, and crisis management systems (Sensors and Sensor Networks (Sensors), 2003: NSF program Solicitation). These future applications may or may not involve biological molecules as detectors; thus they are classified as sensors. In the not too distant past the mad cow incident in North Western United States underscored the importance of protecting the food products fit for human consumption, and the need to identify the possible presence of pathogens in these products. The detection of biomedical analytes for the maintenance of human health is not only an important area of investigation to help provide a safe and healthy environment for humans, but is also one of the areas where the economics for biosensor usage exhibits the greatest potential for further development. Caelen *et al.* (2004) have very recently developed a surface plasmon resonance (SPR) biosensor method for the detection of riboflavin (Rf) in milk-based products. These authors indicate that (a) riboflavin is an essential protein in human nutrition, and (b) the human body obtains a quarter of the average Rf intake from milk and milk-based products.

Pei *et al.* (2004) very recently indicate that Diabetes is amongst the most prevalent and costly diseases in the world. In the year 2004, these authors estimated approximately 17 million people in the United States have diabetes. This is roughly 6.2% of the population. Yonzon *et al.* (2004) further indicate that there are 16 million prediabetics in the United States. The American Diabetic Association (2003) indicates that the economic estimated annual cost of diabetes is $132 billion. There have recently been news reports that indicate that diabetes is reaching epidemic proportions. As expected, a considerable amount of research has been done and effort been spent in detecting glucose levels, which are critical in this disease (Henry, 1998; Thundat *et al.*, 1994, 1995; Ramanathan *et al.*, 2001).

Caelen *et al.* (2004) used the Biacore Q system (Biacore AB, Uppsala, Sweden) and a carboxymethylated dextran sensor chip (CM5, Biacore AB), but did not provide a kinetic analysis and values of the binding rate coefficients and affinities for detecting riboflavin. Lofas *et al.* (1990) indicate that the SPR technique permits a kinetic analysis, and the software that comes with it does provide values of the binding rate coefficients and the affinities (if the dissociation phase, if present, is also analyzed). The SPR software that provides the kinetic analysis does not incorporate the presence of diffusional limitations and the degree of heterogeneity present on the SPR chip surface. The SPR manufacturers (Biacore AB) claim that diffusional limitations may be minimized if the SPR biosensor is run properly. Also, for the analysis of low-molecular weight compounds, such as vitamins, as is analyzed by Caelen *et al.* (2004) for riboflavin, an indirect assay is required (Caselunghe and Lindeberg, 2000; Indyk *et al.*, 2000; Grace and Stenberg, 2002; Haines *et al.*, 2001).

In this chapter, we re-analyze using fractal analysis the diffusion-limited binding data of biomedical analytes that are useful for human health such as riboflavin (Caelen *et al.*, 2004), glucose (Pei *et al.*, 2004), acetylcholine and choline (Mitchell, 2004), and Pb^{++} (Liu and Lu, 2003). Fractal analysis has been used previously to analyze the diffusion-limited analyte–receptor reactions occurring on heterogeneous biosensor surfaces (Butala *et al.*, 2003a,b; Sadana, 2003). Caelen *et al.*, (2004) indicate that natural riboflavin does not bind to the chip surface since it does not contain the required chemical groups. Thus, it needs to be modified by introducing reactive ester groups at the $N-3$ position of the molecule (Merril and McCormick, 1988). Values of the binding rate coefficient, k and the fractal dimension, D_f, are provided. The fractal dimension, D_f, is a quantitative measure of the degree of heterogeneity on the surface. An increase in the value of the fractal dimension on the surface indicates an increase in the degree of heterogeneity on the sensor chip surface.

6.2 THEORY

Havlin (1989) has reviewed and analyzed the diffusion of reactants towards fractal surfaces. The details of the theory and the equations involved for the binding and the dissociation phases for analyte–receptor binding are available (Butala *et al.*, 2003b). The details are not repeated here; except that just the equations are given to permit an easier reading. These equations have been applied to other analyte–receptor reactions occurring on biosensor surfaces (Sadana, 2003; Butala *et al.*, 2003b). Here, we will attempt to apply these equations to the binding of biomedical analytes of interest such as riboflavin (Caelen *et al.* 2004), glucose (Pei *et al.* 2004), acetylcholine and choline (Mitchell 2004), and Pb^{++} (Liu and Lu, 2003). Other reactions are also analyzed. For most applications, a single- or a dual-fractal analysis is often adequate to describe the binding kinetics.

6.2.1 Single-fractal analysis

Binding rate coefficient

Havlin (1989) indicates that the diffusion of a particle (analyte [Ag]) from a homogeneous solution to a solid surface (e.g. receptor [Ab]-coated surface) on which it reacts to form a product (analyte–receptor complex; (Ab·Ag)) is given by:

$$(\text{Ab}\cdot\text{Ag}) \approx \begin{cases} t^{(3-D_{\text{f,bind}})/2} = t^{\text{p}} & t < t_{\text{c}} \\ t^{1/2} & t > t_{\text{c}} \end{cases} \tag{6.1a}$$

Here $D_{\text{f,bind}}$ or D_{f} (used later on in the chapter) is the fractal dimension of the surface during the binding step. t_{c} is the crossover value. Havlin (1989) indicates that the crossover value may be determined by $r_{\text{c}}^2 \sim t_{\text{c}}$. Above the characteristic length, r_{c}, the self-similarity of the surface is lost and the surface may be considered homogeneous. Above time, t_{c}, the surface may be considered homogeneous, since the self-similarity property disappears, and 'regular' diffusion is now present. For a homogeneous surface where D_{f} is equal to two, and when only diffusional limitations are present, $p = \frac{1}{2}$ as it should be. Another way of looking at the $p = \frac{1}{2}$ case (where $D_{\text{f,bind}}$ is equal to two) is that the analyte in solution views the fractal object, in our case, the receptor-coated biosensor surface, from a 'large distance.' In essence, in the binding process, the diffusion of the analyte from the solution to the receptor surface creates a depletion layer of width $(Đt)^{\frac{1}{2}}$ where $Đ$ is the diffusion constant. This gives rise to the fractal power law, (Analyte·Receptor) $\sim t^{(3-D_{\text{f,bind}})/2}$. For the present analysis, t_{c} is arbitrarily chosen and we assume that the value of the t_{c} is not reached. One may consider the approach as an intermediate 'heuristic' approach that may be used in the future to develop an autonomous (and not time-dependent) model for diffusion-controlled kinetics.

6.2.2 Dual-fractal analysis

Binding rate coefficient

Sometimes, the binding curve exhibits complexities and two parameters (k, D_{f}) are not sufficient to adequately describe the binding kinetics. This is further corroborated by low values of r^2 factor (goodness-of-fit). In that case, one resorts to a dual-fractal analysis (four parameters; k_1, k_2, D_{f1}, and D_{f2}) to adequately describe the binding kinetics. The single-fractal analysis presented above is thus extended to include two fractal dimensions. At present, the time $(t = t_1)$ at which the 'first' fractal dimension 'changes' to the 'second' fractal dimension is arbitrary and empirical. For the most part, it is dictated by the data analyzed and experience gained by handling a single-fractal analysis. A smoother curve is obtained in the 'transition' region, if care is taken to select the correct number of points for the two regions. In this case, the product (antibody–antigen; or analyte–receptor complex, Ab·Ag) is given by:

$$(\text{Ab}\cdot\text{Ag}) \approx \begin{cases} t^{(3-D_{\text{f1,bind}})/2 = t^{p1}} & t < t_1 \\ t^{(3-D_{\text{f2,bind}})/2 = t^{p2}} & t_1 < t < t_2 \\ t^{1/2} & t > t_c \end{cases} \tag{6.1b}$$

In some cases, a triple-fractal analysis with six parameters (k_1, k_2, k_3, D_{f1}, D_{f2}, and D_{f3}) may be required to adequately model the binding kinetics. This is when the binding curve exhibits convolutions and complexities in its shape due to, perhaps to the very dilute nature of the analyte (as for example, in the detection of pathogens in the environment) or for some other reasons.

6.3 RESULTS

The fractal analysis will be applied to the binding of biomedical analytes of interest such as riboflavin (Caelen *et al.*, 2004), glucose (Pei *et al.*, 2004; Leegsma-Vogt *et al.*, 2004; Hsieh *et al.*, 2004), acetylcholine and choline (Mitchell, 2004), and Pb^{++} (Liu and Lu, 2003). Other reactions are also analyzed.

Alternate expressions for fitting the data are available that include saturation, first-order reaction, and no diffusion limitations, but these expressions are apparently deficient in describing the heterogeneity that inherently exists on the surface. One might justifiably argue that appropriate modeling may be achieved by using a Langmuirian or other approach. The Langmuirian approach may be used to model the data presented if one assumes the presence of discrete classes of sites (for example, double exponential analysis as compared with a single-fractal analysis). Lee and Lee (1995) indicate that the fractal approach has been applied to surface science, for example, adsorption and reaction processes. These authors emphasize that the fractal approach provides a convenient means to represent the different structures and morphology at the reaction surface. These authors also emphasize using the fractal approach to develop optimal structures and as a predictive approach. Another advantage of the fractal technique is that the analyte–receptor association is a complex reaction, and the fractal analysis via the fractal dimension and the rate coefficient provide a useful lumped parameter(s) analysis of the diffusion-limited reaction occurring on a heterogeneous surface.

In a classical situation, to demonstrate fractality, one should make a log–log plot, and one should definitely have a large amount of data. It may be useful to compare the fit to some other forms, such as exponential, or one involving saturation, etc. At present, we do not present any independent proof or physical evidence of fractals in the examples presented. It is a convenient means (since it is a lumped parameter) to make the degree of heterogeneity that exists on the surface more quantitative. Thus, there is some arbitrariness in the fractal model to be presented. The fractal approach provides additional information about interactions that may not be obtained by a conventional analysis of biosensor data.

There is no nonselective adsorption of the analyte. Nonselective adsorption would skew the results obtained very significantly. In these types of systems, it is imperative to minimize this nonselective adsorption. We also do recognize that, in some cases, this nonselective adsorption may not be a significant component of the adsorbed material and that this rate of association, which is of a temporal nature, would depend on surface availability.

Figure 6.1a shows the binding rate curve for the direct binding assay of 0.1 μg/ml Rf binding protein (RBP) in solution to Rf derivative immobilized on a SPR biosensor chip surface. A dual-fractal analysis is required to adequately describe the binding kinetics. The values of (a) the binding rate coefficient, k, and the fractal dimension, D_f, for a single-fractal analysis, and (b) the binding rate coefficient, k_1 and k_2, and the fractal dimension, D_{f1} and D_{f2}, for a dual-fractal analysis are given in Table 6.1. The values of the binding rate coefficient(s), and the fractal dimension for binding presented in Table 6.1 were obtained from a regression analysis using Corel Quattro Pro 8.0 (1997) to model the data using eqs. (6.1a and b) wherein $(\text{Ab·Ag}) = kt^{(3-D_f)/2}$ for a single-fractal analysis and $(\text{Ab·Ag}) = k_1 t^{(3-D_{f1})/2}$ for time, $t < t_1$ and $(\text{Ab·Ag}) = k_1 t^{(3-D_{f2})/2}$ for time, $t_1 < t < t_2 = t_c$, for a dual-fractal analysis. The binding rate coefficients presented in Table 6.1 are within 95% confidence limits. For example, for the binding of 0.1 μg/ml RBP concentration in solution to immobilized Rf derivative, the binding rate coefficient, k_1, for a dual-fractal analysis is equal to 96.66 ± 2.77. The 95% confidence limit indicates that the k_1 value lies between 93.89 and 99.43. This indicates that the values are precise and significant. A decrease in the fractal dimension value by 44.1% from a D_{f1} value equal to 2.763 to D_{f2} equal to 1.5456 leads to a decrease in the binding rate coefficient value by 14.8% from a k_1 value equal to 96.66 to a k_2 value equal to 82.31. Note that the changes in the fractal dimension and in the binding rate coefficient are in the same direction. A decrease in the degree of heterogeneity on the surface (decrease in the fractal dimension value) leads to a decrease in the binding rate coefficient.

Figure 6.1b shows the binding rate curve for the direct binding assay of 0.2 μg/ml RBP in solution to Rf derivative immobilized on a SPR biosensor chip surface. In this case, a single-fractal analysis is adequate to describe the binding kinetics. The values of the binding rate coefficient, k, and the fractal dimension, D_f, for a single-fractal analysis are given in Table 6.1. It is of interest to note for 0.1 and 0.2 μg/ml RBP in solution, a dual- and a single-fractal analysis, respectively, are required to adequately describe the binding kinetics. This indicates that there is a change in the binding mechanism as one goes from 0.1 to 0.2 μg/ml RBP in solution. In fact, Table 6.1 indicates that in the 0.2–10 μg/ml range RBP in solution analyzed that a single-fractal analysis is adequate to describe the binding kinetics. It is of interest to note that a more complex mechanism is required to adequately describe the binding at the lowest (0.1 μg/ml) RBP concentration in solution.

Figure 6.1c shows the binding rate curve for the direct binding assay of 0.4 μg/ml RBP in solution to Rf derivative immobilized on a SPR biosensor chip surface. Once again, and as indicated above, a single-fractal analysis is required to adequately describe the binding kinetics. The values of the binding rate coefficient, k, and the fractal dimension, D_f, for a single-fractal analysis are given in Table 6.1. Table 6.1 indicates that as one goes from 0.2 to 0.4 μg/ml RBP in solution, the fractal dimension, D_f, value increases by 5.02% from a value of 1.9746 to 2.079, and the binding rate coefficient, k, value increases by 42.6% from a value of 145.38 to 207.27. Once again, the changes in the fractal dimension, D_f, and in the binding rate coefficient, k, are in the same direction.

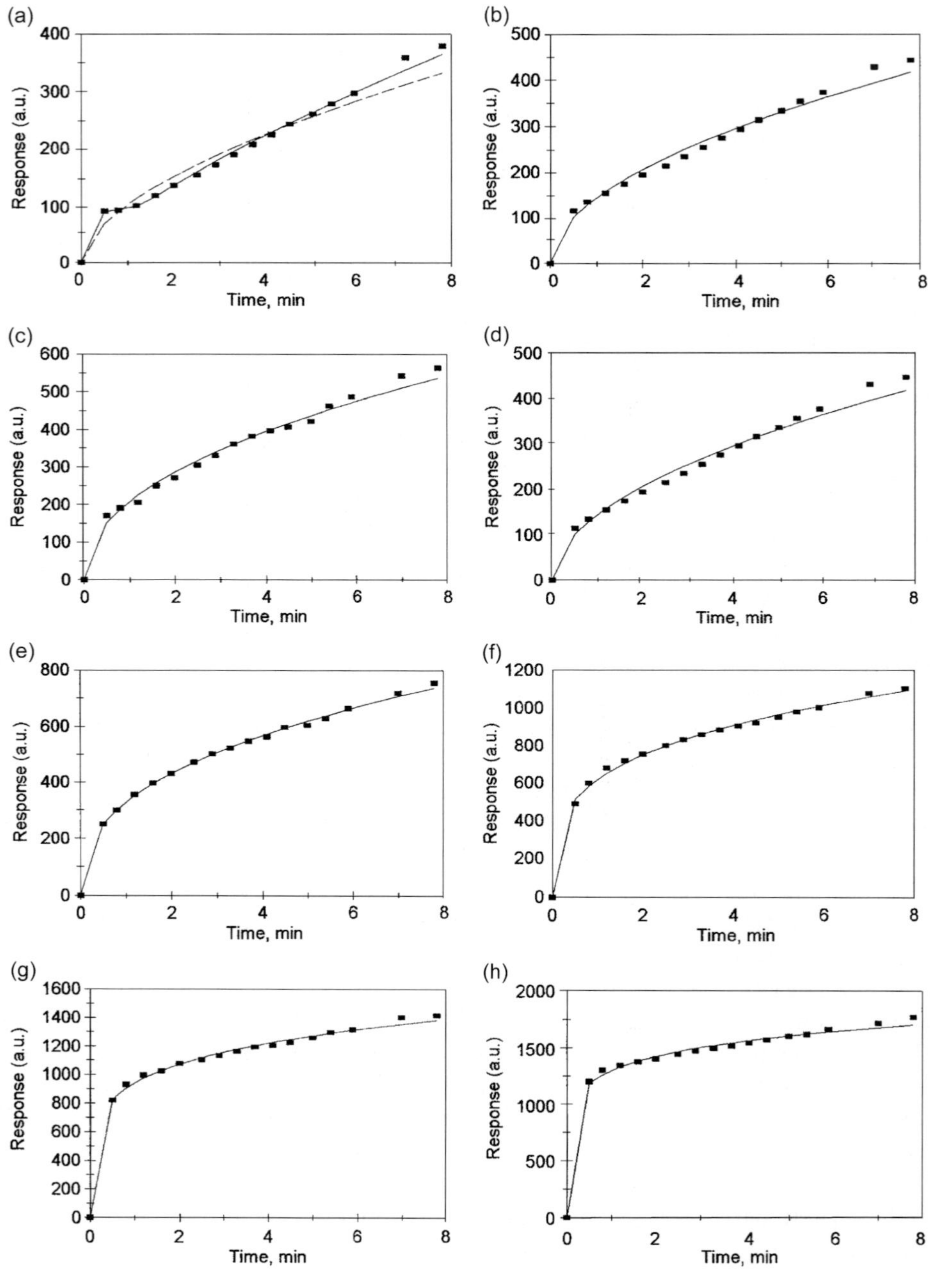

Figure 6.1 Binding of different concentrations (in μg/ml) of RBP in solution to Rf immobilized on a sensor chip surface using a direct binding assay (Caelen *et al.*, 2004): (a) 0.1, (b) 0.2, (c) 0.4, (d) 0.8, (e) 1.0, (f) 2.5, (g) 5.0, (h) 10.0. When only a solid line is used, then a single-fractal analysis applies as in Figures b–h. When both a dashed (--) line and a solid (—) line is used as in Figure a then a solid line represents the dual-fractal analysis, and the dashed line represents the single-fractal analysis. In both cases, the solid line is the best fit line.

Table 6.1

Binding rate coefficients and fractal dimensions for the direct binding assay of RBP to Rf derivative immobilized on a sensor chip surface (Caelen *et al.*, 2004)

RBP concentration (in μg/ml)/immobilized Rf derivative	k	k_1	k_2	D_f	D_{f1}	D_{f2}
0.1	101.60 ± 12.51	96.66 ± 2.77	82.31 ± 2.88	1.8418 ± 0.076	2.763 ± 0.09	1.5456 ± 0.003
0.2	145.38 ± 9.18	na	na	1.9746 ± 0.04	na	na
0.4	207.27 ± 11.24	na	na	2.079 ± 0.034	na	na
0.8	312.68 ± 10.07	na	na	2.160 ± 0.021	na	na
1.0	327.61 ± 4.04	na	na	2.215 ± 0.008	na	na
2.5	621.94 ± 4.04	na	na	2.455 ± 0.014	na	na
5.0	942.54 ± 17.10	na	na	2.631 ± 0.012	na	na
10	1299.6 ± 25.05	na	na	2.738 ± 0.012	na	na

Figure 6.1d shows the binding rate curve for the direct binding assay of 0.8 μg/ml RBP in solution to Rf derivative immobilized on a SPR biosensor chip surface. A single-fractal analysis is required to adequately describe the binding kinetics. The values of the binding rate coefficient, k, and the fractal dimension, D_f, for a single-fractal analysis are given in Table 6.1.

Figure 6.1e–h shows the binding rate curve(s) for the direct binding assay of 1.0, 2.5, 5.0, and 10.0 μg/ml RBP in solution, respectively, to Rf derivative immobilized on a SPR biosensor chip surface. The values of the binding rate coefficient, k, and the fractal dimension, D_f, are given in Table 6.1. It is of interest to note that as the RBP concentration in solution increases from 0.2 to 10.0 μg/ml, there is a gradual increase in (a) the fractal dimension, D_f, and in (b) the binding rate coefficient, k, for a single-fractal analysis.

Caelen *et al.* (2004) have also compared the binding of RBP in solution to Rf immobilized on the sensor chip with the binding of RBP that forms a complex with Rf in solution to Rf immobilized on a sensor chip. They did this at two concentrations of free RBP at 500 ng/ml and at 10 μg/ml. This permitted them to analyze their results at a low and a high response. Their results were very similar in both of these cases, indicating that the binding of RBP on the surface is not changed in the presence of these bound molecules in solution. It would be of interest to obtain values of the binding rate coefficient and the fractal dimensions in these two cases.

Figure 6.2a shows the binding of 500 ng/ml RBP + 500 ng/ml Rf in solution to Rf immobilized on a sensor chip surface. A dual-fractal analysis is required to adequately describe the binding kinetics. The values of (a) the binding rate coefficient, k, and the fractal dimension, D_f, for a single-fractal analysis, and (b) the binding rate coefficient, k_1 and k_2, and the fractal dimension, D_{f1} and D_{f2}, for a dual-fractal analysis are given in Table 6.2. An increase in the fractal dimension value by 27.5% from a D_{f1} value equal to 1.849 to D_{f2} equal to 2.357 leads to an increase in the binding rate coefficient value by 24.1% from a k_1 value equal to 96.92 to a k_2 value equal to 120.27. Note that the changes in the fractal dimension and in the binding rate coefficient are in the same direction. An increase in the degree of heterogeneity on the surface (increase in the fractal dimension value) leads to an increase in the binding rate coefficient.

Figure 6.2b shows the binding of 500 ng/ml RBP in solution to Rf immobilized on a sensor chip surface. A dual-fractal analysis is required to adequately describe the binding kinetics. The values of (a) the binding rate coefficient, k, and the fractal dimension, D_f, for a single-fractal analysis, and (b) the binding rate coefficient, k_1 and k_2, and the fractal dimension, D_{f1} and D_{f2}, for a dual-fractal analysis are given in Table 6.2. An increase in the fractal dimension value by 79.5% from a D_{f1} value equal to 1.407 to D_{f2} equal to 2.526 leads to an increase in the binding rate coefficient value by 63.2% from a k_1 value equal to 89.74 to a k_2 value equal to 146.48. Note that the changes in the fractal dimension and in the binding rate coefficient are, once again, in the same direction. An increase in the degree of heterogeneity on the surface (increase in the fractal dimension value) leads to an increase in the binding rate coefficient. One notes that the k_1 value is higher by 8% and the k_2 value is lower by 17.9% when Rf is absent is compared with when it is present in solution with RBP. The fractal dimension values on the sensor chip surface also exhibit small differences. This is to be expected since the binding rate coefficients are sensitive to the degree of heterogeneity that exists on the surface. These results are consistent with the results of Caelen *et al.* (2004) who indicated very similar behavior (in the shapes and

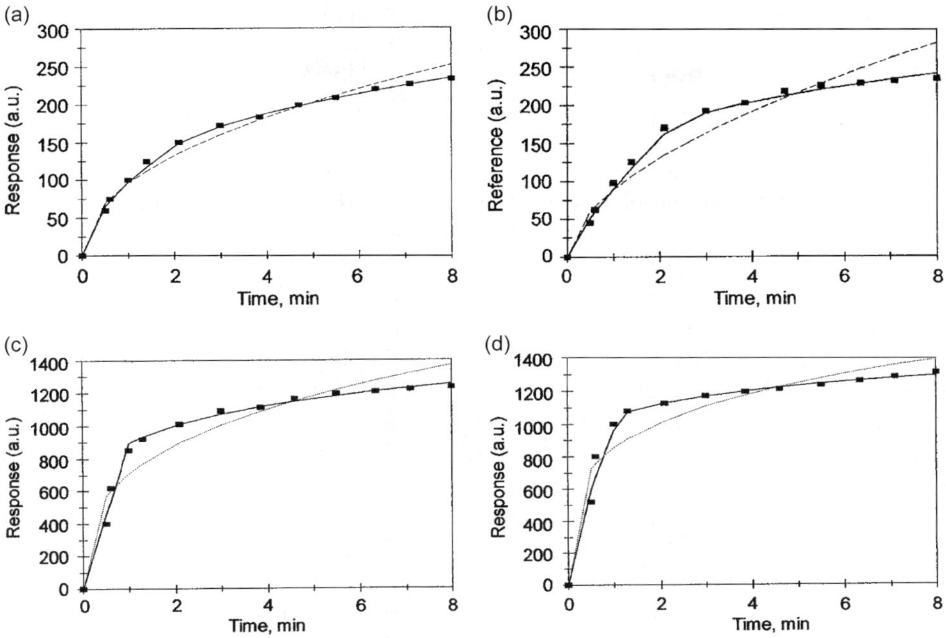

Figure 6.2 Binding of 500 ng/ml RBP (a) with 500 ng/ml Rf and (b) without Rf in solution to immobilized Rf on a sensor chip (Caelen *et al.*, 2004): binding of 10 μg/ml RBP, (c) with 500 ng/ml Rf and (d) without Rf in solution to immobilized Rf on a sensor chip (Caelen *et al.*, 2004). When both a dashed (--) line and a solid (—) line is used then a solid line represents the dual-fractal analysis, and the dashed line represents the single-fractal analysis. The solid line is the best fit line.

response values of the binding curves) when Rf was absent and when it was present in solution with RBP.

Figure 6.2c shows the binding of 10 μg/ml RBP + 500 ng/ml Rf in solution to Rf immobilized on a sensor chip surface. A dual-fractal analysis is required to adequately describe the binding kinetics. The values of (a) the binding rate coefficient, k, and the fractal dimension, D_f, for a single-fractal analysis, and (b) the binding rate coefficient, k_1 and k_2, and the fractal dimension, D_{f1} and D_{f2}, for a dual-fractal analysis are given in Table 6.2. An increase in the fractal dimension value by 91.25% from a D_{f1} value equal to 1.3986 to D_{f2} equal to 2.6738 leads to an increase in the binding rate coefficient value by 11.65% from a k_1 value equal to 801.28 to a k_2 value equal to 894.59. Note that the changes in the fractal dimension and in the binding rate coefficient are, once again, in the same direction. An increase in the degree of heterogeneity on the surface (increase in the fractal dimension value) once again leads to an increase in the binding rate coefficient.

Figure 6.2d shows the binding of 10 μg/ml RBP in solution to Rf immobilized on a sensor chip surface. A dual-fractal analysis is required to adequately describe the binding kinetics. The values of (a) the binding rate coefficient, k, and the fractal dimension, D_f, for a single-fractal analysis, and (b) the binding rate coefficient, k_1 and k_2, and the fractal dimension, D_{f1} and D_{f2}, for a dual-fractal analysis are given in

Table 6.2

Binding rate coefficients and fractal dimensions for the binding after complexation of RBP and by direct assay (Caelen *et al.*, 2004)

Analyte/receptor	k	k_1	k_2	D_f	D_{f1}	D_{f2}
500 ng/ml RBP + 500 ng/ml Rf/ immobilized Rf on sensor chip	96.362 ± 7.874	96.92 ± 6.35	120.27 ± 1.065	2.0768 ± 0.049	1.849 ± 0.081	2.357 ± 0.021
500 ng/ml RBP/ immobilized Rf on sensor chip	88.923 ± 10.52	89.74 ± 10.52	146.48 ± 3.54	1.896 ± 0.103	1.407 ± 0.141	2.526 ± 0.039
10 μg/ml RBP + 500 ng/ml Rf/immobilized Rf on sensor chip	710.51 ± 117.6	801.28 ± 139.39	894.59 ± 12.23	2.368 ± 0.094	1.3986 ± 0.418	2.6738 ± 0.016
10 μg/ml RBP/ immobilized Rf on sensor chip	855.95 ± 127.3	960.39 ± 169.22	1043.83 ± 8.09	2.5298 ± 0.086	1.6440 ± 0.423	2.7936 ± 0.009

Table 6.2. An increase in the fractal dimension value by 69.9% from a D_{f1} value equal to 1.6440 to D_{f2} equal to 2.7936 leads to an increase in the binding rate coefficient value by 8.69% from a k_1 value equal to 960.39 to a k_2 value equal to 1043.83. Note that the changes in the fractal dimension and in the binding rate coefficient are, once again, in the same direction. An increase in the degree of heterogeneity on the surface (increase in the fractal dimension value) leads to an increase in the binding rate coefficient value.

On comparing the binding rate coefficient values in both of these two cases one notes that the k_1 value is higher by 16.6% and the k_2 value is higher by 14.3% when Rf is absent and when it is present in solution with RBP. Similar behavior was noted at the higher response value (500 ng/ml RBP). The fractal dimension values on the sensor chip surface also exhibit small differences. This is to be expected since the binding rate coefficients are sensitive to the degree of heterogeneity that exists on the surface. Once again, these results are consistent with the results of Caelen *et al.* (2004) who indicated very similar behavior when Rf was absent and when it was present in solution with RBP.

It is of interest to note that the fractal dimension, D_{f2}, is higher for the 10 μg/ml RBP (with or without 500 ng/ml Rf) when compared with its corresponding value of 500 ng/ml RBP (with or without 500 ng/ml Rf) in solution. The increase in the fractal dimension with an increase in RBP concentration in solution is consistent with the result presented in Figure 6.3a (even though it is for a single-fractal analysis) which shows the increase in the fractal dimension, D_f, with an increase in the RBP concentration in solution. It must be remembered that the fractal dimension is on a log scale, and even

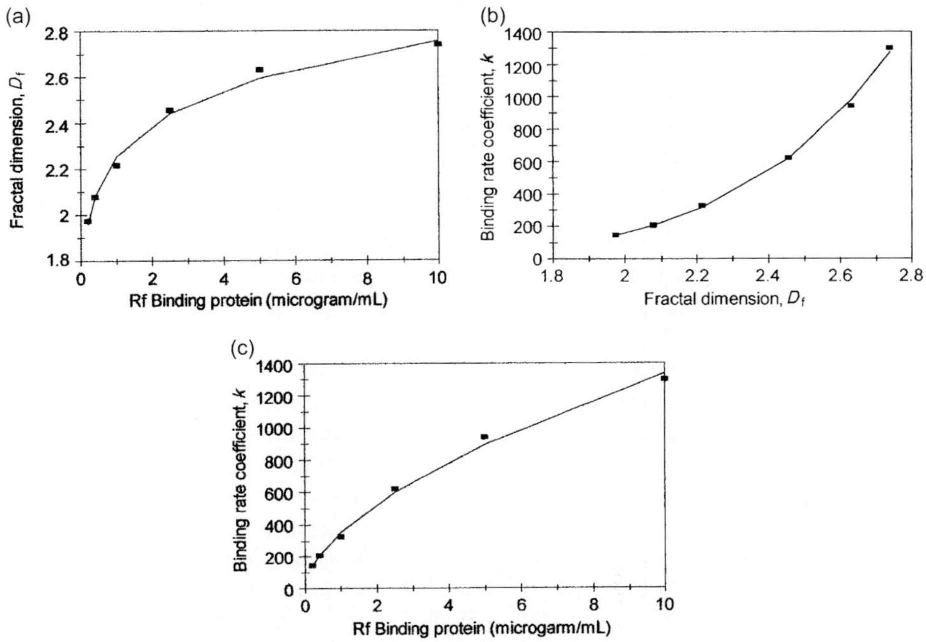

Figure 6.3 For a single-fractal analysis, an increase in: (a) the fractal dimension, D_f, with an increase in the RBP concentration, (b) the binding rate coefficient, k, with an increase in the fractal dimension, D_f, (c) the binding rate coefficient, k, with an increase in the RBP concentration.

small changes in the fractal dimension imply significant changes in the degree of heterogeneity on the biosensor chip surface. These changes in the degree of heterogeneity on the surface would lead to significant changes in, for example, the binding rate coefficient and also possibly in other relevant biosensor performance parameters.

Figure 6.3a shows for a single-fractal analysis the increase in the fractal dimension, D_f, with an increase in the Rf binding protein concentration in the range 0.2–10.0 μg/ml in solution. In this concentration range, the fractal dimension, D_f, is given by:

$$D_f = (2.255 \pm 0.029)[\mathrm{RBP}]^{0.087 \pm 0.0038} \tag{6.2a}$$

The fit is good, and the fractal dimension, D_f, exhibits a negligible order (equal to 0.087) of dependence on the RBP concentration in solution in the 0.2–10.0 μg/ml range analyzed.

Figure 6.3b shows for a single-fractal analysis the increase in the binding rate coefficient, k, with an increase in the fractal dimension, D_f, for the RBP in solution/immobilized Rf derivative interaction on the chip surface. In the 0.2–10.0 μg/ml RBP concentration range, the binding rate coefficient, k, is given by:

$$k = (1.681 \pm 0.0515) D_f^{6.583 \pm 0.103} \tag{6.2b}$$

The fit is good. The binding rate coefficient, k, is very sensitive to the fractal dimension, D_f, or the degree of heterogeneity on the sensor chip surface as noted by the high value of the order of dependence (equal to 6.583) of the binding rate coefficient, k, on D_f.

Figure 6.3c shows for a single-fractal analysis, the increase in the binding rate coefficient, k, with an increase in the RBP concentration in the range 0.2–10.0 μg/ml in solution. In this concentration range, the binding rate coefficient, k, is given by:

$$k = (354.86 \pm 20.04)[\mathrm{RBP}]^{0.576 \pm 0.0163} \tag{6.2c}$$

Once again, the fit is good, and the binding rate coefficient, k, exhibits a slightly higher than one-half order (equal to 0.576) dependence on the RBP concentration in solution in the 0.2–10.0 μg/ml range analyzed.

Mitchell (2004) has recently indicated the need for an effective method to detect acetylcholine (ACh) and choline (Ch). A rapid *in situ* determination of ACh and Ch extracellular levels would help (a) in understanding the cholinergic transmission in normal and pathological physiology, and (b) in the evaluation of pharmaceuticals that affect cholinergic activity (Lemiere *et al.*, 1999; Lenigk *et al.*, 2000). Mitchell (2004) developed a multienzyme amperometric biosensor to measure physiologically relevant levels of Ch and ACh *in vivo*. The template for the immobilization of the enzymes was poly(*m*-4,3)-phenylenediamine (pmPD) electropolymerized on a platinum iridium wire (Pt). The multienzyme layer contained choline oxidase (ChO_x) and ascorbic acid oxidase (AAO) for a choline sensor. The multienzyme layer comprised of acetylcholinesterase (AChE) and AAO for ACh/Ch sensor. This multienzyme layer was immobilized with bovine serum albumin by cross-linking with glutaraldehyde.

Mitchell (2004) indicates that the Ch and ACh sensors are based on the following reactions:

Ch sensor:

$$Ch + O_2 \overset{CHO_x}{\longrightarrow} \text{betaine aldehyde} + 2H_2O_2 \tag{6.3a}$$

ACh/Ch sensor:

$$ACh + H_2O \overset{AChE}{\longrightarrow} \text{acetate} + Ch$$

$$Ch + O_2 \overset{ChO_x}{\longrightarrow} \text{Betaine aldehyde} + 2H_2O_2 \tag{6.3b}$$

Indirect detection of enzyme substrates involves measurement of H_2O_2 production. Often, the detection mechanism involves oxidation of H_2O_2. Mitchell (2004) emphasizes that (a) the simplicity of fabrication and (b) the easy availability of sensor components necessary for fabrication is a significant advantage.

Figure 6.4a shows the binding and dissociation of exogeneous Ch to the Ch sensor. The Ch was applied by pressure ejection (10 mM, 2 s, 20 psi). Mitchell (2004) indicates that the Ch was detected immediately by the Ch sensor. Here nC is the charge. A single-fractal analysis is required to model the binding kinetics. A dual-fractal analysis is required to model the dissociation kinetics. The values of (a) the binding rate coefficient, k for a single-fractal analysis, (b) the dissociation rate coefficient, k_d for a single-fractal analysis, and the (c) the dissociation rate coefficients, k_{d1} and k_{d2} for a dual-fractal analysis, are given in Table 6.3a. Similarly, the values of (a) the fractal dimension for the binding phase, D_f for a single-fractal analysis, (b) the fractal dimension for the dissociation phase, D_{fd} for a single-fractal analysis, and the (c) the fractal dimensions for the dissociation phase for a dual-fractal analysis, D_{fd1} and D_{fd2} for a dual-fractal analysis, are given in Table 6.3b.

Figure 6.4b shows the binding and dissociation of exogeneous ACh to the ACh sensor. Mitchell (2004) indicates that the pressure ejection (10 mM, 2 s, 20 psi) was detected as Ch at a later time following hydrolysis of ACh by extracellular AChE. A dual-fractal

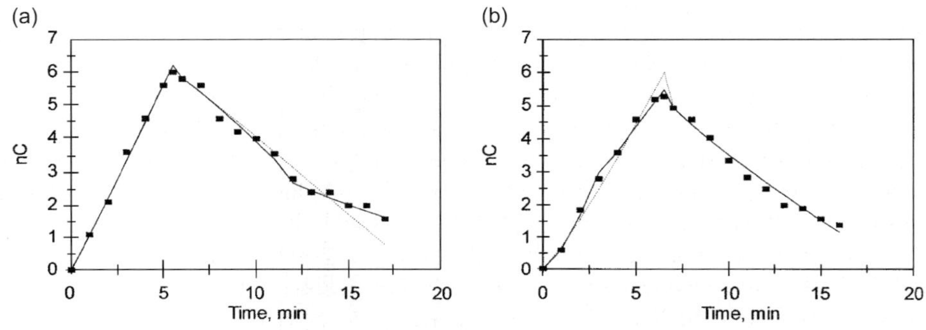

Figure 6.4 (a) Binding and dissociation of endogenous choline to a choline sensor, (b) binding and dissociation of acetylcholine to an acetylcholine sensor (Mitchell, 2004).

Table 6.3a

Rate coefficients for the binding and dissociation phase for K^{+}-evoked increases in endogenous acetylcholine (ACh) and choline (Ch) (Mitchell, 2004)

Sensor type	k	k_1	k_2	k_d	k_{d1}	k_{d2}
(a) Ch sensor	1.0959 ± 0.016	na	na	0.4037 ± 0.088	0.3865 ± 0.0931	1.3613 ± 0.0661
ACh sensor	0.7079 ± 0.1025	0.6226 ± 0.0825	1.1028 ± 0.0404	0.6024 ± 0.0699	0.5872 ± 0.0788	1.2228 ± 0.0389
(b) Ch sensor	0.0791 ± 0.0069	na	na	0.00840 ± 0.00024	0.006649 ± 0.00213	0.01089 ± 0.001
ACh/Ch sensor	0.1569 ± 0.0187	0.0986 ± 0.0072	0.3375 ± 0.0097	0.00888 ± 0.0017	0.00694 ± 0.00122	0.0621 ± 0.0014
Differential measurement for ACh	2.6732 ± 0.1723	na	na	0.1128 ± 0.0067	0.02495 ± 0.017	2.9596 ± 0.0971

Table 6.3b

Fractal dimensions for the binding and dissociation phase for K^+-evoked increases in endogenous acetylcholine (ACh) and choline (Ch) (Mitchell, 2004)

Sensor type	D_f	D_{f1}	D_{f2}	D_{fd}	D_{fd1}	D_{fd2}
(a) Ch sensor	0.9642 ± 0.0690	na	na	0.9082 ± 0.1282	0.7554 ± 0.1804	2.0480 ± 0.2502
ACh sensor	0.7104 ± 0.1642	$0.1448 + 0.3166$	1.2822 ± 0.1140	1.2912 ± 0.0752	1.1912 ± 0.1274	1.9834 ± 0.1312
(b) Ch sensor	1.4710 ± 0.0874	na	na	0.9634 ± 0.1530	0.9082 ± 0.2406	1.1072 ± 0.00684
ACh/Ch sensor	1.6564 ± 0.1032	1.2890 ± 0.1104	2.0754 ± 0.1494	1.8738 ± 0.0736	0.6874 ± 0.1052	1.671 ± 0.0622
Differential measurement for ACh	1.8816 ± 0.0748	na	na	0.9482 ± 0.2970	$0.0228 + 0.5774$	2.3056 ± 0.0626

analysis was required to model the binding and the dissociation kinetics. The values of (a) the binding rate coefficient, k for a single-fractal analysis, (b) the binding rate coefficients, k_1 and k_2 for a dual-fractal analysis, (c) the dissociation rate coefficient, k_d for a single-fractal analysis, and the (d) dissociation rate coefficients, k_{d1} and k_{d2} for a dual-fractal analysis, are given in Table 6.3a. Similarly, the values of (a) the fractal dimension for the binding phase, D_f for a single-fractal analysis, (b) the fractal dimensions for the binding phase, D_{f1} and D_{f2} for a dual-fractal analysis, (c) the fractal dimension for the dissociation phase, D_{fd} for a single-fractal analysis, and the (d) fractal dimensions for the dissociation phase, D_{fd1} and D_{fd2} for a dual-fractal analysis are given in Table 6.3b.

The binding mechanisms for the Ch sensor and the ACh sensor are different since a single-fractal analysis is adequate to model the binding kinetics observed with a Ch sensor whereas a dual-fractal analysis is required for the ACh sensor.

Mitchell (2004) also analyzed the stimulated increase of endogenous ACh and Ch brain extracellular fluid (ECF) by using the depolarizing agent, K^+. The author indicates that 200 nl of K^+ was administered by a microsyringe located approximately 400 μm away from the ACh and Ch sensors.

Figure 6.5a shows the binding and dissociation of exogeneous Ch to the Ch sensor after stimulation by K^+. As observed for the un-stimulated case (Figure 6.4a) the binding kinetics is adequately described by a single-fractal analysis. A dual-fractal analysis is required to adequately describe the dissociation kinetics. The values of (a) the binding

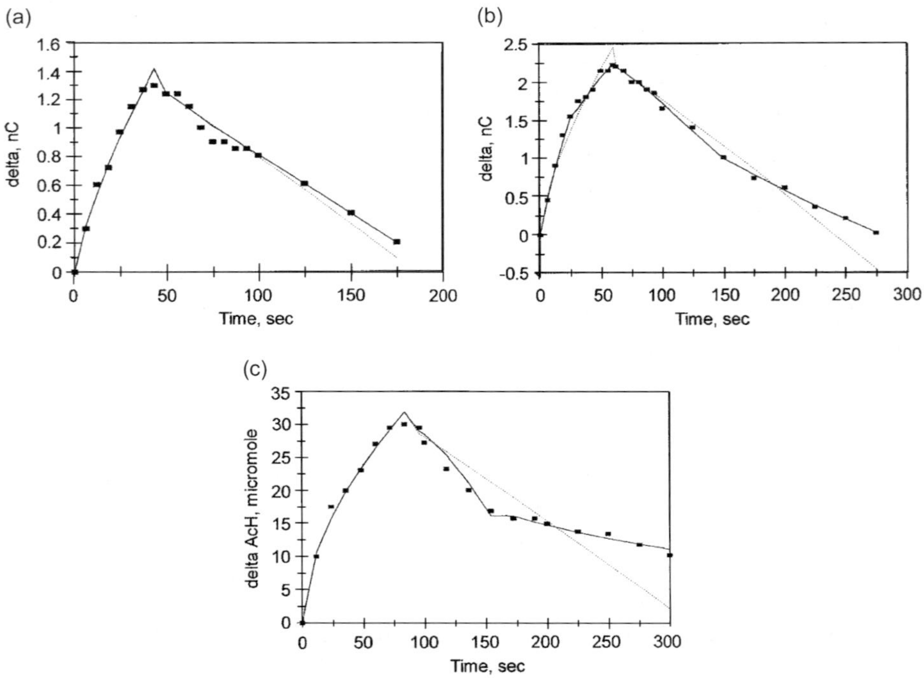

Figure 6.5 (a) Binding and dissociation of exogeneous Ch to a Ch sensor after K^+ stimulation, (b) binding and dissociation of exogeneous ACh to a ACh sensor after K^+ stimulation. (c) Binding and dissociation during differential measurement of ACh (Mitchell, 2004).

rate coefficient, k for a single-fractal analysis, (b) the dissociation rate coefficient, k_d for a single-fractal analysis, and the (c) the dissociation rate coefficients, k_{d1} and k_{d2} for a dual-fractal analysis, are given in Table 6.3a. Similarly, the values of (a) the fractal dimension for the binding phase, D_f for a single-fractal analysis, (b) the fractal dimension for the dissociation phase, D_{fd} for a single-fractal analysis, and the (c) the fractal dimensions for the dissociation phase for a dual-fractal analysis, D_{fd1} and D_{fd2} for a dual-fractal analysis, are given in Table 6.3b.

Figure 6.5b shows the binding and dissociation of exogeneous ACh to the ACh sensor after stimulation by K^+. As observed for the un-stimulated case (Figure 6.4b) the binding and the dissociation kinetics requires a dual-fractal analysis to adequately describe them. The values of (a) the binding rate coefficient, k for a single-fractal analysis, (b) the binding rate coefficients, k_1 and k_2 for a dual-fractal analysis, (c) the dissociation rate coefficient, k_d for a single-fractal analysis, and the (d) the dissociation rate coefficients, k_{d1} and k_{d2} for a dual-fractal analysis, are given in Table 6.3a. Similarly, the values of (a) the fractal dimension for the binding phase, D_f for a single-fractal analysis, (b) the fractal dimensions for the binding phase, D_{f1} and D_{f2} for a dual-fractal analysis, (c) the fractal dimension for the dissociation phase, D_{fd} for a single-fractal analysis, and the (d) the fractal dimensions for the dissociation phase, D_{fd1} and D_{fd2} for a dual-fractal analysis are given in Table 6.3b.

It is of interest to compare the binding and dissociation rate coefficients of the un-stimulated case with the K^+-evoked increase in endogenous ACh and Ch. Note that for the ACh case, as one goes from the un-stimulated case to the K^+-evoked case, (a) the binding rate coefficient, k for a single-fractal analysis decreases by a factor of 13.85 from a value of 1.0959 to 0.0791. Similar decrease in the binding rate coefficients, k_1 and k_2 may also be noted. Similarly, for a dual-fractal analysis the dissociation rate coefficient, k_{d1} decreases by a factor of 58.12, and the dissociation rate coefficient, k_{d2} decreases by a factor of 125. The stimulation provided by K^+ affects the dissociation rate coefficient on comparing with the un-stimulated case substantially, and exerts its major influence over there rather than on the binding rate coefficient. Note that, and as mentioned above the binding rate coefficient for the K^+ stimulated case is lower than in the un-stimulated case.

Figure 6.5c shows the binding and dissociation of exogeneous ACh to the (ACh/Ch–Ch) sensor. This is a differential measurement. A single-fractal analysis is adequate to describe the binding kinetics. A dual-fractal analysis is required to adequately describe the dissociation kinetics. The values of (a) the binding rate coefficient, k for a single-fractal analysis, (b) the binding rate coefficients, k_1 and k_2 for a dual-fractal analysis, (c) the dissociation rate coefficient, k_d for a single-fractal analysis, and the (d) the dissociation rate coefficients, k_{d1} and k_{d2} for a dual-fractal analysis, are given in Table 6.3a. Similarly, the values of (a) the fractal dimension for the binding phase, D_f for a single-fractal analysis, (b) the fractal dimensions for the binding phase, D_{f1} and D_{f2} for a dual-fractal analysis, (c) the fractal dimension for the dissociation phase, D_{fd} for a single-fractal analysis, and the (d) fractal dimensions for the dissociation phase, D_{fd1} and D_{fd2} for a dual-fractal analysis are given in Table 6.3b.

Diabetes is among the most prevalent and costly diseases in the world (Pei *et al.*, 2004). The authors have recently developed a microcantilever-based glucose biosensor. They immobilized glucose oxidase (GO_x) on a microcantilever. Using their micromechanical technique they were able to detect physiologically relevant glucose levels. The concept for the detection of glucose is simple. The binding of the glucose in solution to the glucose oxidase immobilized on the microcantilever surface induces a surface stress. This surface stress leads to the bending of the microcantilever. The authors emphasize that the common interference(s) observed in other glucose detection techniques are not observed in the microcantilever technique.

Figure 6.6a and Table 6.4 show the binding of 1 nM glucose in solution to the glucose oxidase immobilized on the microcantilever (Pei *et al.*, 2004). A dual-fractal analysis is required to adequately describe the binding kinetics. The values of (a) the binding rate

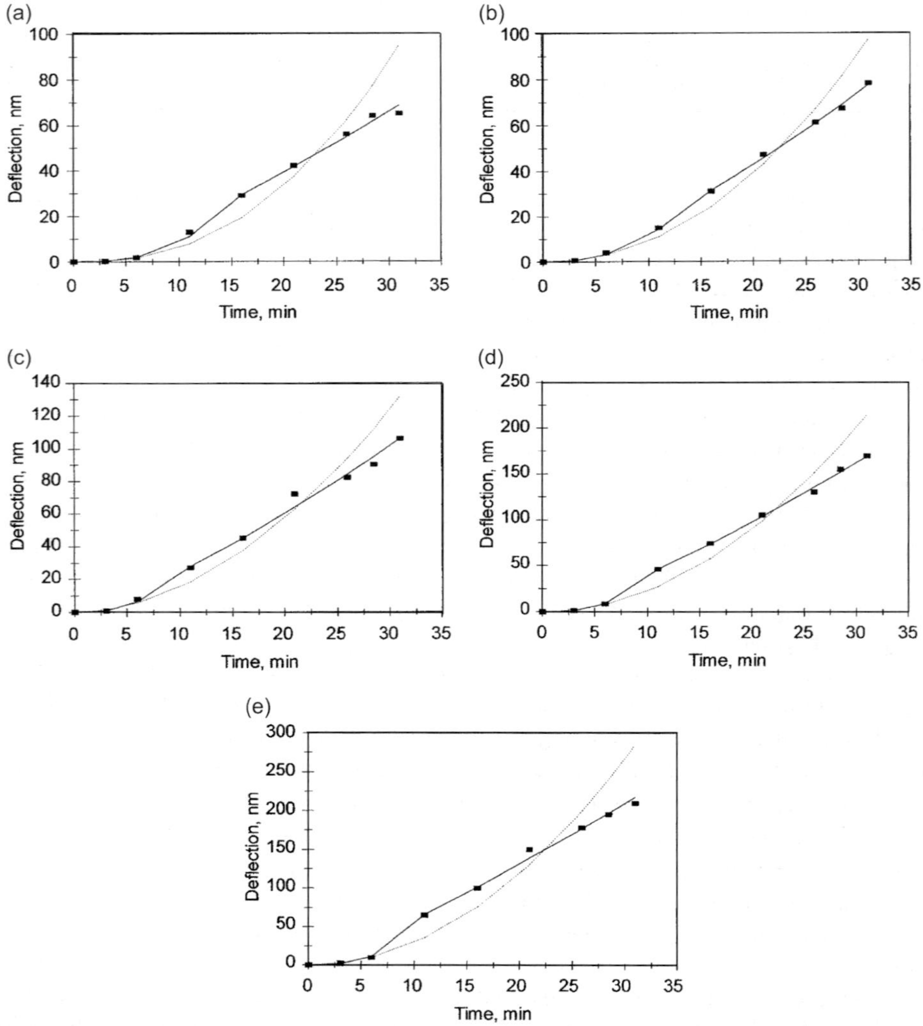

Figure 6.6 Binding of different concentrations of glucose in solution (in nM) to glucose oxidase (GO_x) immobilized on a microcantilever biosensor surface (Pei *et al.*, 2004): (a) 1, (b) 2, (c) 5, (d) 10, (e) 20. When both a dashed (--) line and a solid (—) line are used, then a solid line represents the dual-fractal analysis, and the dashed line represents the single-fractal analysis. The solid line is the best fit line.

Table 6.4

Rate coefficients along with fractal dimension values for the binding and dissociation of glucose in solution to a GO_x-functionalized cantilever (Pei *et al.*, 2004)

Glucose concentration in solution (nM)	k	k_1	k_2	D_f	D_{f1}	D_{f2}
1	0.0263 ± 0.011	0.0110 ± 0.0017	0.852 ± 0.036	~ 0	~ 0	0.445 ± 0.155
2	0.0732 ± 0.023	0.0387 ± 0.0095	0.732 ± 0.019	~ 0	~ 0	0.287 ± 0.100
5	0.195 ± 0.074	0.0672 ± 0.020	1.273 ± 0.089	~ 0	~ 0	0.426 ± 0.153
10	0.240 ± 0.085	0.082 ± 0.0023	2.282 ± 0.055	~ 0	~ 0	0.4934 ± 0.054
20	0.300 ± 0.121	0.0984 ± 0.024	4.224 ± 0.190	~ 0	~ 0	0.7054 ± 0.099

coefficient, k and the fractal dimension, D_f for a single-fractal analysis, (b) the binding rate coefficients, k_1 and k_2, and the fractal dimensions, D_{f1} and D_{f2} for a dual-fractal analysis are given in Table 6.4.

Note that the fractal dimension, D_f for a single-fractal analysis, and the fractal dimension, D_{f1} for the first binding phase for a dual-fractal analysis are approximately zero. No explanation is offered, except that the microcantilever surface appears as a Cantor-like dust to the analyte (glucose) in solution during the first phase of binding. In fact, for all of the glucose concentrations in solution used (1,2,5,10, and 20 nM), the fractal dimensions, D_f and D_{f1} are estimated to be approximately equal to zero.

Figure 6.6b and Table 6.4 show the binding of 2 nM glucose in solution to the glucose oxidase immobilized on the microcantilever (Pei *et al.*, 2004). A dual-fractal analysis is required to adequately describe the binding kinetics. The values of (a) the binding rate coefficient, k and the fractal dimension, D_f for a single-fractal analysis, (b) the binding rate coefficients, k_1 and k_2, and the fractal dimensions, D_{f1} and D_{f2} for a dual-fractal analysis are given in Table 6.4.

An increase in the glucose concentration in solution from 1 to 5 nM leads to (a) an increase in the binding rate coefficient, k_1 by a factor of 3.518 from a value of 0.0110 to 0.0387, (b) a decrease in the binding rate coefficient, k_2 by 14.1% from a value of 0.852 to 0.732, and to (c) a decrease in the fractal dimension, D_{f2} by 35.5% from a value of 0.445 to 0.287. It is of interest to note that when the results in Figure 6.6a and b are compared, the changes in the fractal dimension, D_{f1} and D_{f2} are closely followed by changes in the binding rate coefficient (k_1 and k_2), respectively. In other words, if the fractal dimension increases so does the binding rate coefficient, and *vice versa*.

Figure 6.6c and Table 6.4 show the binding of 5 nM glucose in solution to the glucose oxidase immobilized on the microcantilever (Pei *et al.*, 2004). A dual-fractal analysis is required to adequately describe the binding kinetics. The values of (a) the binding rate coefficient, k and the fractal dimension, D_f for a single-fractal analysis, and (b) the binding rate coefficients, k_1 and k_2, and the fractal dimensions, D_{f1} and D_{f2} for a dual-fractal analysis are given in Table 6.4.

An increase in the glucose concentration in solution from 1 to 5 nM in solution leads to (a) an increase in the binding rate coefficient, k_1 by a factor of 6.11 from a value of

0.0110 to 0.0672, (b) an increase in the binding rate coefficient, k_2 by 49.4% from a value of 0.852 to 1.273, and to (c) a decrease in the fractal dimension, D_{f2} by 4.3% from a value of 0.445 to 0.426.

Figure 6.6d and Table 6.4 show the binding of 10 nM glucose in solution to the glucose oxidase immobilized on the microcantilever (Pei *et al.* 2004). A dual-fractal analysis is required to adequately describe the binding kinetics. The values of (a) the binding rate coefficient, k and the fractal dimension, D_f for a single-fractal analysis, and (b) the binding rate coefficients, k_1 and k_2, and the fractal dimensions, D_{f1} and D_{f2} for a dual-fractal analysis are given in Table 6.4.

An increase in the glucose concentration in solution from 1 to 10 nM in solution leads to (a) an increase in the binding rate coefficient, k_1 by a factor of 7.45 from a value of 0.0110 to 0.082, (b) an increase in the binding rate coefficient, k_2 by a factor of 2.678 from a value of 0.852 to 2.282, and to (c) an increase in the fractal dimension, D_{f2} by 10.9% from a value of 0.445 to 0.4934.

Figure 6.6e and Table 6.4 show the binding of 20 nM glucose in solution to glucose oxidase immobilized on a microcantilever surface (Pei *et al.* 2004). A dual-fractal analysis is required to adequately describe the binding kinetics. The values of (a) the binding rate coefficient, k and the fractal dimension, D_f for a single-fractal analysis, and (b) the binding rate coefficients, k_1 and k_2, and the fractal dimensions, D_{f1} and D_{f2} for a dual-fractal analysis are given in Table 6.4.

Figure 6.7a and Table 6.4 show the increase in the binding rate coefficient, k_1 with an increase in the glucose concentration (in nM) in solution. For the 1–20 nM glucose concentration in solution, the binding rate coefficient, k_1 is given by:

$$k_1 = (0.01682 \pm 0.00815)[\text{glucose concentration, in nM}]^{0.6772 \pm 0.1640} \quad (6.4a)$$

The fit is quite good. The binding rate coefficient curve shape seems to exhibit a 'saturation' type tendency. More data points are required to provide a more reliable and better fit. The binding rate coefficient, k_1 exhibits an order of dependence between one-half and first-order (equal to 0.6772) on glucose concentration in solution. The noninteger order of dependence exhibited lends support to the fractal nature of the system.

Figure 6.7b and Table 6.4 show the increase in the binding rate coefficient, k_1 with an increase in the glucose concentration (in nM) in solution. For the 1–20 nM glucose concentration in solution, the binding rate coefficient, k_2 is given by:

$$k_2 = (0.6329 \pm 0.2048)[\text{glucose concentration, in nM}]^{0.5686 \pm 0.1164} \quad (6.4b)$$

The fit is quite good. More data points are required to provide a more reliable and better fit. The binding rate coefficient, k_2 exhibits an order of dependence between one-half and first-order (equal to 0.5686) on glucose concentration in solution. The noninteger order of dependence exhibited, once again, lends support to the fractal nature of the system. The binding rate coefficient, k_2 shows a slightly lower dependence on glucose concentration in solution than the binding rate coefficient, k_1.

Figure 6.7c and Table 6.4 show the increase in the fractal dimension, D_{f2} with an increase in the glucose concentration (in nM) in solution. For the 1–20 nM glucose

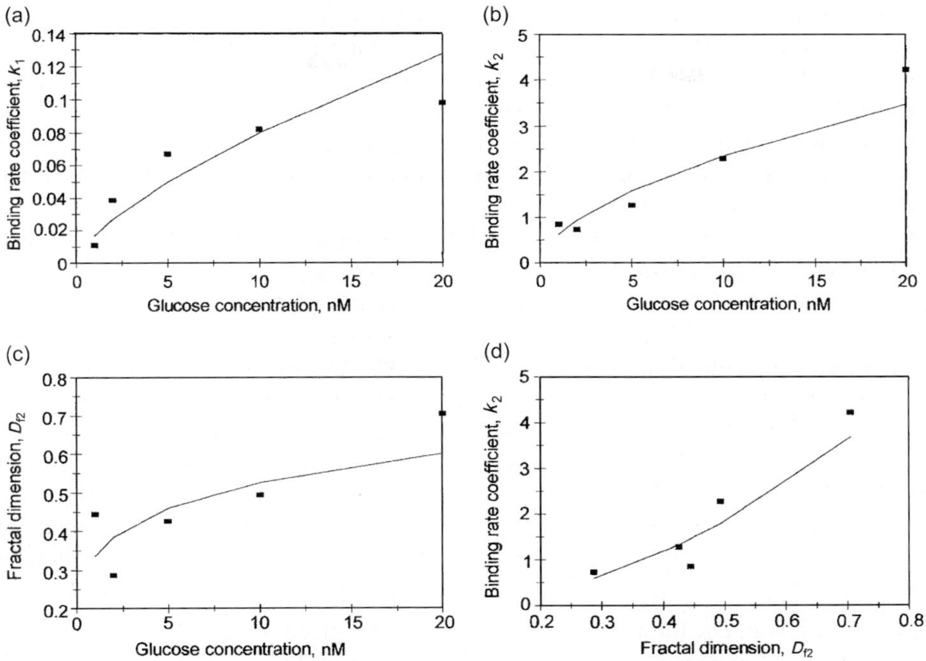

Figure 6.7 (a) Increase in the binding rate coefficient, k_1 with an increase in the glucose concentration in solution (in nM), (b) increase in the binding rate coefficient, k_2 with an increase in the glucose concentration in solution (in nM), (c) increase in the fractal dimension, D_{f2} with an increase in the glucose concentration in solution (in nM), (d) increase in the binding rate coefficient, k_2 with an increase in the degree of heterogeneity on the surface or the fractal dimension, D_{f2}.

concentration in solution, the fractal dimension, D_{f2} is given by:

$$D_{f2} = (0.3369 \pm 0.0992)[\text{glucose concentration, in nM}]^{0.1934 \pm 0.1072} \qquad (6.4c)$$

The fit is quite good. The fractal dimension, D_{f2} exhibits only a slight order of dependence (equal to 0.1934) on the glucose concentration in solution. Once again, it should be pointed out that the fractal dimension is base on a log scale, and even small changes in the fractal dimension could lead to significant changes in the degree of heterogeneity on the microcantilever surface.

Figure 6.7d and Table 6.4 show the increase in the binding rate coefficient, k_2 with an increase in the fractal dimension, D_{f2}. For the 1–20 nM glucose concentration in solution, the binding rate coefficient, k_2 is given by:

$$k_2 = (7.4866 \pm 3.3185 D_{f2}^{2.023 \pm 0.5684} \qquad (6.4d)$$

The fit is good. The fractal dimension, D_{f2} exhibits only a slightly higher than second order of dependence (equal to 2.023) on the fractal dimension, D_{f2}.

Leegsma-Vogt *et al.* (2004) have recently presented the potential of biosensor technology in clinical monitoring and experimental research. They do emphasize that for continuous *in vivo* monitoring of patients very little data is reported. For example, Rhemrev-Boom (1999) describes a biosensor device and ultrafiltration sampling for the continuous *in vivo* monitoring of glucose. Leegsma-Vogt *et al.* (2004) emphasize that biosensors may be used for the continuous online monitoring of glucose and lactate which would help facilitate therapeutic interventions when need be. The monitoring of glucose would help in managing diabetes, and the monitoring of lactate would help facilitate the optimization of sports training schedules. These authors indicate that over the years they have mainly focused on glucose and lactate biosensors. They further indicate that their biosensors have been functional for several months.

Pellerin and Magistretti (1996) indicate the importance of measuring subcutaneous levels of glucose. Figure 6.8 shows the oral glucose tolerance test (OGGT) administered by Leegsma-Vogt *et al.* (2004) with glucose and insulin measurements. Probes placed at different locations measured plasma insulin (Figure 6.8a), plasma glucose (Figure 6.8b), adipose tissue interstitial glucose (Figure 6.8c), and connective tissue interstitial glucose (Figure 6.8d).

Figure 6.8a shows the binding and dissociation of insulin in plasma. A dual-fractal analysis is required to adequately describe the binding kinetics. A single-fractal analysis is adequate to describe the dissociation kinetics. The values of (a) the binding rate coefficient, k and the fractal dimension, D_f for a single-fractal analysis, (b) the binding rate coefficients, k_1 and k_2, and the fractal dimensions, D_{f1} and D_{f2} for a dual-fractal analysis, and (c) the values of the dissociation rate coefficient, k_d and the fractal dimension in the dissociation phase, D_{fd} for a single-fractal analysis are given in Tables 6.5a and b. Note that as the fractal dimension value increases by a factor of 2.47 from D_{f1} equal to 0.6827

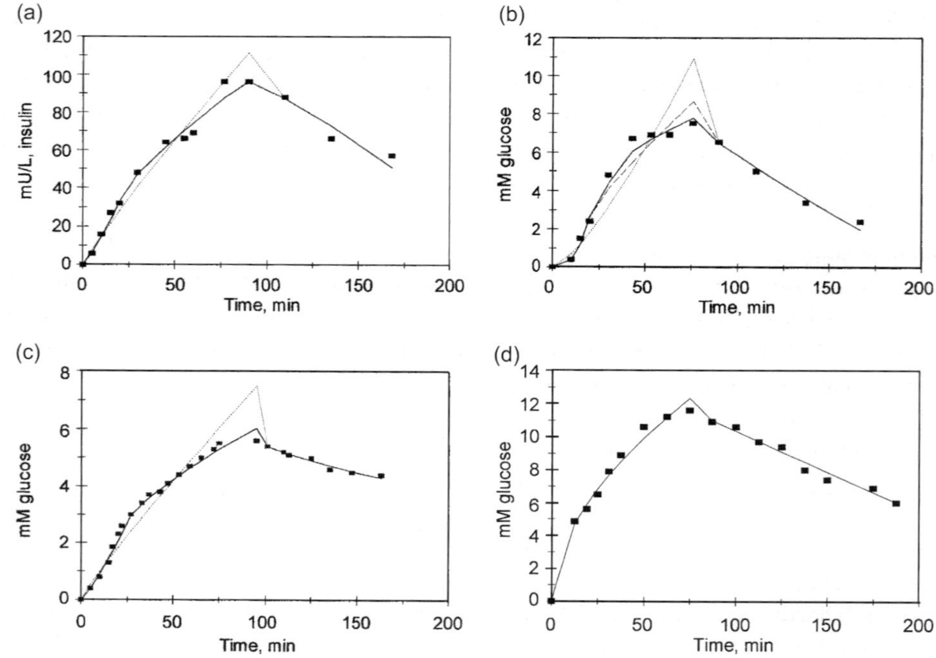

Figure 6.8 Oral glucose tolerance test (OGGT) (Leegsma-Vogt *et al.*, 2004). Binding and dissociation using the glucose biosensor for: (a) plasma insulin, (b) plasma glucose, (c) adipose tissue interstitial glucose, (d) connective tissue interstitial glucose.

Table 6.5a

Binding rate coefficients for glucose in plasma, in connective tissue, and in adipose tissue, and insulin in plasma (Leegsma-Vogt *et al.*, 2004)

Compound	Location	k	k_1	k_2	k_d
Insulin	Plasma	1.8557 ± 0.334	1.0232 ± 0.1309	5.0388 ± 0.3671	0.2436 ± 0.0875
Glucose	Plasma	0.0329 ± 0.0154	0.0101 ± 0.00025	0.0442 ± 0.0057	0.1019 ± 0.0103
Glucose	Adipose tissue interstitial	0.1246 ± 0.0242	0.0545 ± 0.0063	0.4841 ± 0.0164	0.0513 ± 0.0056
Glucose	Connective tissue interstitial	1.220 ± 0.067	na	na	0.0519 ± 0.0081

to D_{f2} equal to 1.6852, the binding rate coefficient increases by a factor of 4.92 from k_1 equal to 1.0232 to k_2 equal to 5.0388. An increase in the degree of heterogeneity on the probe surface leads to an increase in the binding rate coefficient.

Figure 6.8b shows the binding and dissociation of glucose in plasma. A dual-fractal analysis is required to adequately describe the binding kinetics. A single-fractal analysis is adequate to describe the dissociation kinetics. The values of (a) the binding rate coefficient, k and the fractal dimension, D_f for a single-fractal analysis, (b) the binding rate coefficients, k_1 and k_2, and the fractal dimensions, D_{f1} and D_{f2} for a dual-fractal analysis, and (c) the values of the dissociation rate coefficient, k_d and the fractal dimension in the dissociation phase, D_{fd} for a single-fractal analysis are given in Tables 6.5a and b. It is of interest to note that the D_{f1} value is equal to zero. This indicates that the surface exists as a Cantor-like dust. The binding rate curve exhibits an 'S' shaped curve (convex to the origin), and this results in the estimation of the fractal dimension equal to zero. Actually, the estimated value of the fractal dimension is less than zero, but since one cannot physically have a negative dimension, then the estimated value of the fractal dimension is set equal to zero. In this case too, an increase in the degree of heterogeneity on the probe surface leads to an increase in the binding rate coefficient. As the fractal dimension value increases from D_{f1} equal to zero to D_{f2} equal to 0.3022, the binding rate coefficient increases from k_1 equal to 0.0101 to k_2 equal to 0.0442.

Figure 6.8c shows the binding and dissociation of adipose tissue interstitial glucose. A dual-fractal analysis is required to adequately describe the binding kinetics. A single-fractal analysis is adequate to describe the dissociation kinetics. The values of (a) the binding rate coefficient, k and the fractal dimension, D_f for a single-fractal analysis, (b) the binding

Table 6.5b

Fractal dimensions for glucose in plasma, in connective tissue, and in adipose tissue, and insulin in plasma (Leegsma-Vogt *et al.*, 2004)

Compound	Location	D_f	D_{f1}	D_{f2}	D_{fd}
Insulin	Plasma	1.1804 ± 0.116	0.6827 ± 0.1757	1.6852 ± 0.1604	$0.602 + 0.6334$
Glucose	Plasma	0.3128 ± 0.402	0	0.3022 ± 0.450	1.2298 ± 0.1363
Glucose	Adipose tissue interstitial	1.200 ± 0.1117	0.5720 ± 0.136	1.891 ± 0.0456	1.4696 ± 0.1018
Glucose	Connective tissue interstitial	1.9284 ± 0.0664	na	na	1.0193 ± 0.1462

rate coefficients, k_1 and k_2, and the fractal dimensions, D_{f1} and D_{f2} for a dual-fractal analysis, and (c) the values of the dissociation rate coefficient, k_d and the fractal dimension in the dissociation phase, D_{fd} for a single-fractal analysis are given in Tables 6.5a and b. Note that as the fractal dimension value increases by a factor of 3.31 from D_{f1} equal to 0.5720 to D_{f2} equal to 1.891, the binding rate coefficient increases by factor of 8.88 from k_1 equal to 0.0545 to k_2 equal to 0.4841. Once again, an increase in the degree of heterogeneity on the probe surface leads to an increase in the binding rate coefficient.

On comparing the binding rate coefficient values in for glucose in plasma and in the adipose interstitial tissue one notes that both the binding rate coefficient values, k_1 and k_2 are higher in the interstitial adipose tissue than in the plasma. As expected, the corresponding values of the fractal dimensions are also higher.

Figure 6.8d shows the binding and dissociation of connective tissue interstitial glucose. A dual-fractal analysis is required to adequately describe the binding kinetics. A single-fractal analysis is adequate to describe the binding and the dissociation kinetics. The values of (a) the binding rate coefficient, k and the fractal dimension, D_f for a single-fractal analysis, and (b) the values of the dissociation rate coefficient, k_d and the fractal dimension in the dissociation phase, D_{fd} for a single-fractal analysis are given in Tables 6.5a and b.

Figure 6.9a and Tables 6.5a and b show the decrease in the dissociation rate coefficient, k_d with an increase in the fractal dimension in the dissociation phase, D_{fd}. For the data

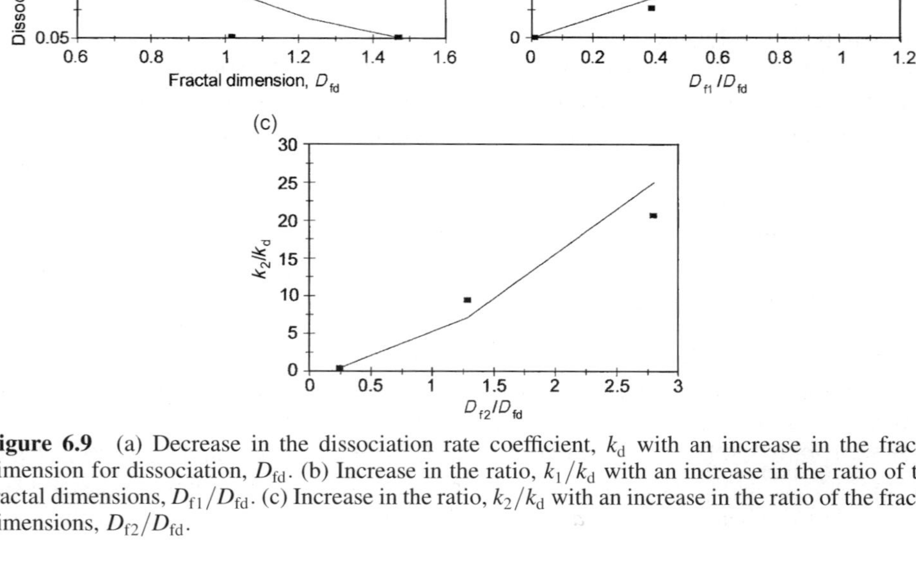

Figure 6.9 (a) Decrease in the dissociation rate coefficient, k_d with an increase in the fractal dimension for dissociation, D_{fd}. (b) Increase in the ratio, k_1/k_d with an increase in the ratio of the fractal dimensions, D_{f1}/D_{fd}. (c) Increase in the ratio, k_2/k_d with an increase in the ratio of the fractal dimensions, D_{f2}/D_{fd}.

presented in Tables 6.5a and b and for glucose and insulin present in plasma and in interstitial adipose and connective tissue, the dissociation rate coefficient, k_d is given by:

$$k_d = (0.0939 \pm 0.0614) D_{fd}^{-1.583 \pm 0.753} \tag{6.5a}$$

The fit is quite good. Only four data points are available. The availability of more data points would lead to a more reliable fit. Note that the data for glucose and insulin are plotted together. The dissociation rate coefficient, k_d exhibits close to a negative one and one-half order dependence on the degree of heterogeneity (D_{fd}) that exists on the biosensor surface.

Figure 6.9b and Tables 6.5a and b show the increase in the ratio of the binding and the dissociation rate coefficient, (k_1/k_d) with an increase in the ratio of the fractal dimensions (D_{f1}/D_{fd}). For the data presented in Tables 6.5a and b and for glucose and insulin present in plasma and for glucose in interstitial adipose tissue and in interstitial connective tissue, the ratio of the binding and the dissociation rate coefficient, (k_1/k_d) is given by:

$$(k_1/k_d) = (5.149 \pm 1.912)(D_{f1}/D_{fd})^{1.371 \pm 0.1036} \tag{6.5b}$$

The fit is quite good. Only three data points are available. The availability of more data points would lead to a more reliable fit. Note that the data for glucose and insulin are plotted together. The ratio of the binding and the dissociation rate coefficient, (k_1/k_d) exhibits an order of dependence between first and one and one-half order (equal to 1.371) on the ratio of the fractal dimensions, (D_{f1}/D_{fd}) that exists on the biosensor surface.

Figure 6.9c and Tables 6.5a and b show the increase in the ratio of the binding and the dissociation rate coefficient, (k_2/k_d) with an increase in the ratio of the fractal dimensions (D_{f2}/D_{fd}). For the data presented in Tables 6.5a and b and for glucose and insulin present in plasma and for glucose in interstitial adipose tissue and in interstitial connective tissue, the ratio of the binding and the dissociation rate coefficient, (k_2/k_d) is given by:

$$(k_2/k_d) = (4.698 \pm 2.022)(D_{f2}/D_{fd})^{1.628 \pm 0.2034} \tag{6.5c}$$

The fit is quite good. Only three data points are available. The availability of more data points would lead to a more reliable fit. Note that the data for glucose and insulin are plotted together. The ratio of the binding and the dissociation rate coefficient, (k_2/k_d) exhibits an order of dependence between one and one-half and second-order (equal to 1.628) on the ratio of the fractal dimensions, (D_{f2}/D_{fd}) that exists on the biosensor surface. This is slightly more than the order (equal to 1.371) exhibited by the ratio (k_1/k_d) on the ratio of the fractal dimensions, (D_{f1}/D_{fd}) that exists on the biosensor surface.

Hsieh *et al.* (2004) have recently detected glucose using glucose/galactose binding protein (GGBP) as the receptor immobilized on a SPR biosensor surface. These authors indicate that the detection of low-molecular weight analytes such as glucose (180 Da) by a SPR biosensor is difficult since the molecules have insufficient mass to provide a measurable change in the refractive index. These authors have used unlabeled GGBP combined with SPR for the direct detection of glucose.

These authors indicate that GGBP is a bacterial periplasmic binding protein. Upon binding of its ligand proteins, glucose or galactose GGBP exhibits a hinge-twist conformational change (Zou *et al.*, 1993; Gerstein *et al.*, 1994). This conformational

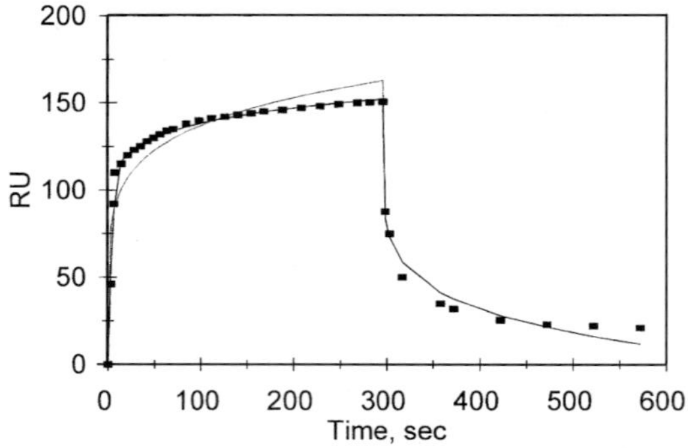

Figure 6.10 Binding and dissociation of 100 μM glucose in solution to thiol coupled E149C GGBP (glucose/galactose binding protein) immobilized on a CM5 sensor chip surface (Hsieh *et al.*, 2004).

change may be used to detect the binding of glucose (Salins *et al.*, 2001) and galactose (Zukin *et al.*, 1977).

Figure 6.10 shows the binding and dissociation of 100 μM glucose in solution to thiol coupled E149C GGBP (~10,000 RU) immobilized on a CM5 sensor chip. The GGBP was engineered to bind in the physiological range by mutation at additional sites (Hseih *et al.*, 2004). The E149 is one such mutation site. A dual-fractal analysis is required to adequately describe the binding kinetics. A single-fractal analysis is adequate to describe the dissociation kinetics. The values of (a) the binding rate coefficient, k and the fractal dimension, D_f for a single-fractal analysis, and (b) the binding rate coefficient, k_1 and k_2 for a dual-fractal analysis, and the fractal dimensions, D_{f1} and D_{f2} for a dual-fractal analysis are given in Tables 6.6a and b. Only one set of data points is available.

It is of interest to note that as the fractal dimension increases by a factor of 1.569 from a value of D_{f1} equal to 1.7992 to D_{f2} equal to 2.824, the binding rate coefficient increases by a factor of 3.31 from a value of k_1 equal to 27.851 to k_2 equal to 92.140. The ratio of the binding rate coefficient(s) to the dissociation rate coefficient(s), k_1/k_d is equal to 0.493, and k_2/k_d is equal to 1.632.

Cai *et al.* (2004) have recently developed a wireless, remote-query glucose biosensor. The sensor uses a ribbon-like, mass-sensitive magnetoelastic sensor as a transducer. The magnetoelastic sensor is initially coated with a pH-sensitive polymer. A glucose oxidase (GO_x) layer is then coated on the pH-sensitive polymer. These authors indicate that the GO_x-catalyzed oxidation of glucose produces gluconic acid. This induces the pH-responsive polymer to shrink, which decreases the polymer mass. The magnetoelastic

Table 6.6a

Binding and dissociation rate coefficients for glucose in solution to glucose/galactose binding protein (GGBP) thiol-immobilized on a CM5 sensor chip surface (Hsieh *et al.*, 2004)

k	k_1	k_2	k_d
66.178 ± 9.366	27.851 ± 7.321	92.190 ± 0.703	56.474 ± 3.522

Table 6.6b

Fractal dimension values for the binding of glucose in solution to glucose/galactose binding protein (GGBP) thiol-immobilized on a CM5 sensor chip surface (Hsieh *et al.*, 2004)

D_f	D_{f1}	D_{f2}	D_{fd}
2.6838 ± 0.0414	1.7992 ± 0.4264	2.8240 ± 0.0037	2.6802 ± 0.0268

sensor vibrations (characteristic resonance frequency) are inversely dependent on sensor mass loading, and these changes in resonance frequency (from a passive magnetoelastic transducer) may be detected on a remote basis. These authors further emphasize that compounds in clinical samples such as uric acid, acetaminophen, and ascorbic acid may interfere with accurate glucose detection.

Cai *et al.* (2004) indicate that the normal serum glucose concentration is in the 3.8–6.1 mmol/l range under physiological conditions. When the blood glucose concentration is generally higher than 9 mmol/l, then diabetic urine is present. In order that physiological conditions may be approached with their electrochemical biosensor, Cai *et al.* (2004) added 0.15 mol/l NaCl, and calibrated their biosensor in the glucose concentration range of 1–15 mmol/l.

Figure 6.11a shows the binding of 1 mmol/l glucose in solution to the sensor that is coated with 0.3 mg of GO_x, 0.03 mg of catalase, 0.75 mg of BSA, and 0.3 mg of glutaric aldehyde. A dual-fractal analysis is required to adequately describe the binding kinetics. The values of (a) the binding rate coefficient, k, and the fractal dimension, D_f for a single-fractal analysis, and (b) the binding rate coefficients, k_1 and k_2, and the fractal dimensions, D_{f1} and D_{f2} for a dual-fractal analysis are given in Table 6.7. It is of interest to note that as the fractal dimension increases by a factor of 13.023 from a value of D_{f1} equal to 0.1750 to D_{f2} equal to 2.2791, the binding rate coefficient increases by a factor of 16.56 from a value of k_1 equal to 1.2184 to k_2 equal to 20.184. An increase in the degree of heterogeneity on the electrochemical biosensor surface leads to an increase in the binding rate coefficient.

Figure 6.11b shows the binding of 4 mmol/l glucose in solution to the sensor that is coated with 0.3 mg of GO_x, 0.03 mg of catalase, 0.75 mg of BSA, and 0.3 mg of glutaric aldehyde. A dual-fractal analysis is required to adequately describe the binding kinetics. The values of (a) the binding rate coefficient, k, and the fractal dimension, D_f for a single-fractal analysis, and (b) the binding rate coefficients, k_1 and k_2, and the fractal dimensions, D_{f1} and D_{f2} for a dual-fractal analysis are given in Table 6.7. It is of interest to note that as the glucose concentration in solution increases by a factor of four from 1 to 4 mmol/l, (a) the binding rate coefficient, k_1 increases by a factor of 7.516 from a value of 1.2184 to 9.1577, and (b) the binding rate coefficient, k_2 increases by a factor of 2.155 from a value of 20.184 to 43.056.

Figure 6.11c shows the binding of 7 mmol/l glucose in solution to the sensor that is coated with 0.3 mg of GO_x, 0.03 mg of catalase, 0.75 mg of BSA, and 0.3 mg of glutaric aldehyde. A dual-fractal analysis is required to adequately describe the binding kinetics. The values of (a) the binding rate coefficient, k, and the fractal dimension, D_f for a single-fractal analysis, and (b) the binding rate coefficients, k_1 and k_2, and the fractal dimensions,

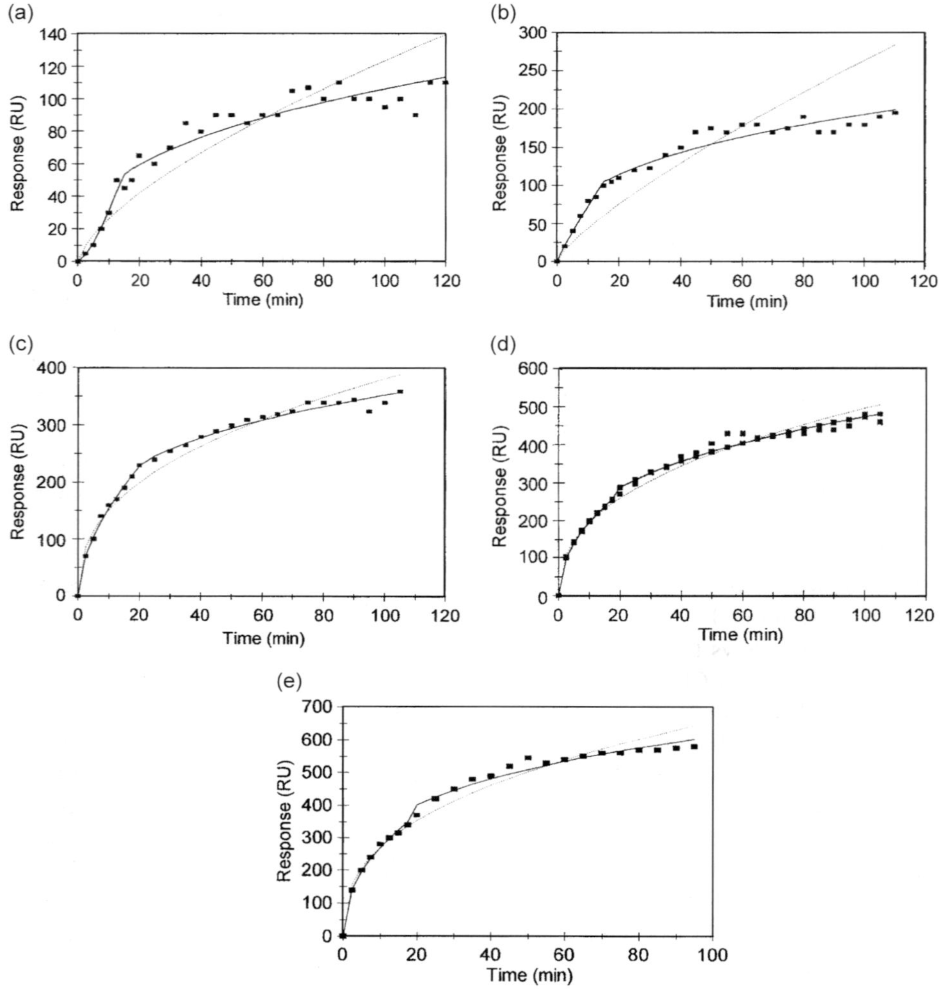

Figure 6.11 Binding of different concentrations of glucose in solution (in mmol/l) to 0.3 mg of glucose oxidase (GO_x) + 0.03 mg of catalase + 0.75 mg of BSA + 0.3 mg of glutaric dialdehyde in a coating on a magnetoelastic sensor with a pH-sensitive polymer (Cai *et al.*, 2004): (a) 1, (b) 4, (c) 7, (d) 10, (e) 15. When both a dashed (--) line and a solid (—) line is used, then a solid line represents the dual-fractal analysis, and the dashed line represents the single-fractal analysis. The solid line is the best fit line.

D_{f1} and D_{f2} for a dual-fractal analysis are given in Table 6.7. It is of interest to note that as the glucose concentration in solution increases by a factor of seven from 1 to 7 mmol/l, (a) the binding rate coefficient, k_1 increases by a factor of 34.349 from a value of 1.2184 to 41.852, and (b) the binding rate coefficient, k_2 increases by a factor of 5.286 from a value of 20.184 to 106.709.

Figure 6.11d shows the binding of 10 mmol/l glucose in solution to the sensor that is coated with 0.3 mg of GO_x, 0.03 mg of catalase, 0.75 mg of BSA, and 0.3 mg of glutaric

Table 6.7

Rate coefficients and fractal dimensions for the binding of different concentrations of glucose in solution (in mmol/l) to 0.3 mg of catalase, 0.75 mg of BSA (bovine serum albumin) and 0.3 mg of glutaric dialdehyde in a coating on a magnetoelastic sensor with a pH-sensitive polymer (Cai *et al.*, 2004)

Analyte (glucose) concentration (mmol/l)	k	k_1	k_2	D_f	D_{f1}	D_{f2}
1	5.658 ± 1.879	1.2184 ± 0.198	20.184 ± 2.049	1.661 ± 0.106	0.175 ± 0.238	2.2791 ± 0.058
4	7.461 ± 5.102	9.158 ± 0.666	43.51 ± 3.11	1.451 ± 0.167	1.1914 ± 0.095	2.3522 ± 0.05
7	60.04 ± 5.496	41.852 ± 1.634	106.71 ± 2.91	2.197 ± 0.035	1.8664 ± 0.041	2.4799 ± 0.031
10	78.66 ± 4.912	66.199 ± 1.255	114.71 ± 5.03	2.199 ± 0.024	2.051 ± 0.022	2.3845 ± 0.042
15	111.95 ± 7.64	94.92 ± 2.685	94.92 ± 2.685	2.232 ± 0.028	2.094 ± 0.033	2.4797 ± 0.039

aldehyde. A dual-fractal analysis is required to adequately describe the binding kinetics. The values of (a) the binding rate coefficient, k, and the fractal dimension, D_f for a single-fractal analysis, and (b) the binding rate coefficients, k_1 and k_2, and the fractal dimensions, D_{f1} and D_{f2} for a dual-fractal analysis are given in Table 6.7. It is of interest to note that as the glucose concentration in solution increases by a factor of 10 from 1 to 10 mmol/l, (a) the binding rate coefficient, k_1 increases by a factor of 54.33 from a value of 1.2184 to 66.199, and (b) the binding rate coefficient, k_2 increases by a factor of 5.683 from a value of 20.184 to 114.714. Once again, and as indicated above, an increase in the glucose concentration in solution leads to an increase in the values of the binding rate coefficients, k_1 and k_2.

Figure 6.11e shows the binding of 15 mmol/l glucose in solution to the sensor that is coated with 0.3 mg of GO_x, 0.03 mg of catalase, 0.75 mg of BSA, and 0.3 mg of glutaric aldehyde. A dual-fractal analysis is required to adequately describe the binding kinetics. The values of (a) the binding rate coefficient, k, and the fractal dimension, D_f for a single-fractal analysis, and (b) the binding rate coefficients, k_1 and k_2, and the fractal dimensions, D_{f1} and D_{f2} for a dual-fractal analysis are given in Table 6.7. It is of interest to note that as the glucose concentration in solution increases by a factor of 15 from 1 to 15 mmol/l, (a) the binding rate coefficient, k_1 increases by a factor of 77.9 from a value of 1.2184 to 94.918, and (b) the binding rate coefficient, k_2 increases by a factor of 9.13 from a value of 20.184 to 184.213.

Figure 6.12a shows the increase in the binding rate coefficient, k_1 with an increase in the glucose concentration in solution in the range 1–15 mmol/l. In this concentration range, the binding rate coefficient, k_1 is given by:

$$k_1 = (1.1764 \pm 0.3813)[\text{glucose}]^{1.6910 \pm 0.1335} \quad (6.6a)$$

The fit is very good. The binding rate coefficient, k_1 exhibits an order of dependence between one and one-half and second order (equal to 1.6910) on the glucose concentration in solution. The noninteger order of dependence exhibited by the binding rate coefficient, k_1 on the glucose concentration in solution lends support to the fractal nature of the system.

Figure 6.12b shows the increase in the binding rate coefficient, k_2 with an increase in the glucose concentration in solution in the range 1–15 mmol/l. In this concentration range, the binding rate coefficient, k_2 is given by:

$$k_2 = (18.283 \pm 4.074)[\text{glucose}]^{0.8241 \pm 0.0956} \quad (6.6b)$$

The fit is good. The binding rate coefficient, k_2 exhibits an order of dependence less than first order (equal to 0.8241) on the glucose concentration in solution. The noninteger order of dependence exhibited by the binding rate coefficient, k_2 on the glucose concentration in solution, once again, lends support to the fractal nature of the system.

It is of interest to note that the order of dependence exhibited by k_2 (equal to 0.8241) is less than half of that exhibited by k_1 (equal to 1.6910) on the glucose concentration in solution. In general, the values of k_2 are higher than those of k_1 for any particular analyte–receptor reaction occurring on biosensor surfaces, and this is reflected in the constant values for k_1 (equal to 1.1764) and k_2 (equal to 18.283), respectively.

Figure 6.12c shows the increase in the ratio of the binding rate coefficients, k_1/k_2 with an increase in the ratio of the fractal dimensions, D_{f1}/D_{f2}. For the data given in Table 6.7,

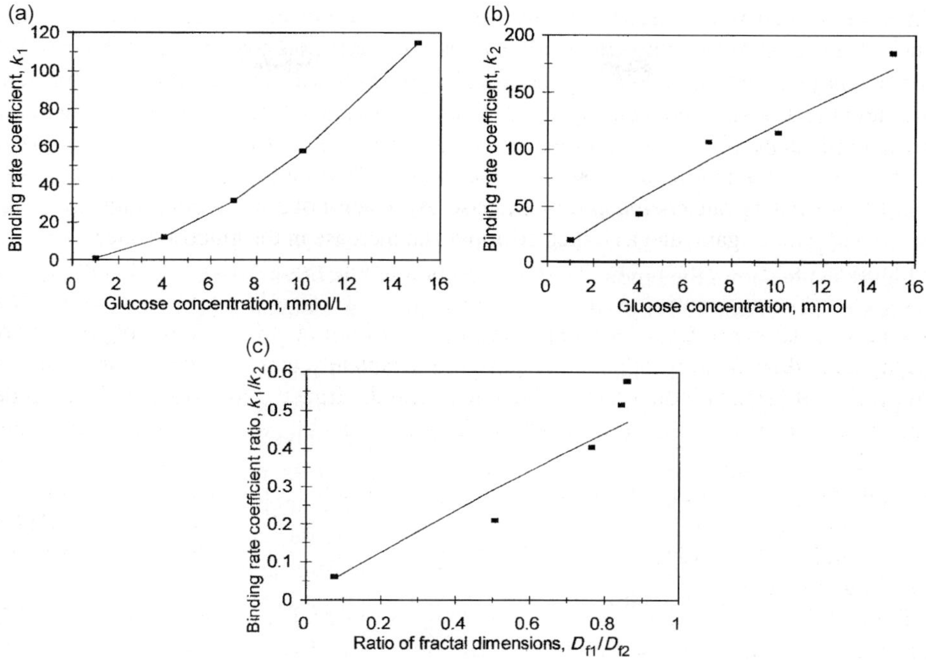

Figure 6.12 (a) Increase in the binding rate coefficient, k_1 with an increase in the glucose concentration (in mmol/l) in solution, (b) increase in the binding rate coefficient, k_2 with an increase in the glucose concentration (in mmol/l) in solution, (c) increase in the ratio of the binding rate coefficients, k_1/k_2 with an increase in the ratio of the fractal dimensions, D_{f1}/D_{f2}.

the ratio of the binding rate coefficients, k_1/k_2 is given by:

$$k_1/k_2 = (0.5371 \pm 0.1469)(D_{f1}/D_{f2})^{0.8812 \pm 0.1175} \tag{6.6c}$$

The fit is quite good. There is some scatter in the data, and this is reflected in the error in the constant (0.5371 ± 0.1469). The ratio of the binding rate coefficients, k_1/k_2 exhibits an order of dependence less than first order (equal to 0.8812) on the ratio of the fractal dimensions, (D_{f1}/D_{f2}).

Lin *et al.* (2004) have recently developed a glucose biosensor based on carbon nanotube (CNT) nanoelectrode ensembles (NEEs). The fabrication of CNT nanotube ensembles has been described in the literature (Tu *et al.*, 2002, 2003). Lin *et al.* (2004) indicate that recently the electrochemical properties of CNT have come into prominence. Considerable research has been done in applying CNTs as electrochemical biosensors (Azmian *et al.*, 2002; Nguyen *et al.*, 2002; Li *et al.*, 2003; Yu *et al.*, 2003; Sotiropoulou and Chaniotakis, 2003; Shim *et al.*, 2002). These authors immobilized glucose oxidase on the CNTs using carbodiimide chemistry. Amide linkages were formed between amine residues and the carboxylic acid groups on the CNT tips. Lin *et al.*, (2004) emphasize that their CNT biosensor is able to detect glucose effectively in the presence of common interferents such as acetaminophen, uric acid, and ascorbic acid. These authors emphasize that their biosensor is able to operate without permselective membrane barriers and artificial electron mediators. This simplifies the design and fabrication of the biosensor.

Lin *et al.* (2004) compared the responses for appropriate physiological levels of glucose, acetaminophen, ascorbic acid, and uric acid at a potential of +0.4 and −0.2 V. These authors noted that at +0.4 V there was significant interference from ascorbic acid, uric acid, and acetaminophen during the detection of glucose. This interference was substantially reduced when operating at −0.2 V. Figure 6.13 shows the fabrication of the Lin *et al.* (2004) glucose biosensor based on a CNT NEE. The two steps involved include: (a) the electrochemical treatment of the CNT NEE for functionalization, and (b) the coupling of the glucose oxidase to the functionalized CNT NEE.

Figure 6.14a shows the binding and dissociation of glucose in solution to the CNT NEE biosensor (Lin *et al.*, 2004). A single fractal analysis is adequate to describe the binding and the dissociation kinetics. The values of the binding rate coefficient, k and the fractal dimension, D_f for a single fractal analysis are given in Table 6.8.

Figure 6.14b shows the binding and dissociation of ascorbic acid in solution to the CNT NEE glucose biosensor (Lin *et al.*, 2004). A dual-fractal analysis is required to adequately describe the binding kinetics. A single-fractal analysis is adequate to describe the dissociation kinetics. The values of (a) the binding rate coefficient, k, and the fractal dimension, D_f for a single-fractal analysis, (b) the binding rate coefficients, k_1 and k_2, and the fractal dimensions, D_{f1} and D_{f2} for a dual-fractal analysis, and (c) the dissociation rate coefficient, k_d and the fractal dimension, D_{fd} are given in Tables 6.8a and b. Note that

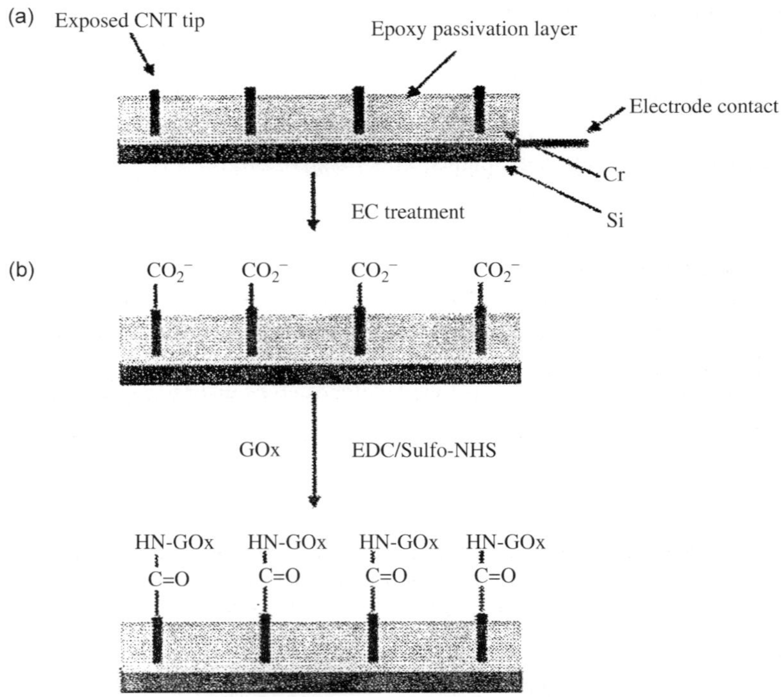

Figure 6.13 Fabrication of the Lin *et al.* (2004) CNT NEE biosensor for glucose: (a) functionalization step involving electrochemical treatment of the CNT NEE, (b) coupling of the glucose oxidase to the carbon nanotube nanoelectrode ensembles.

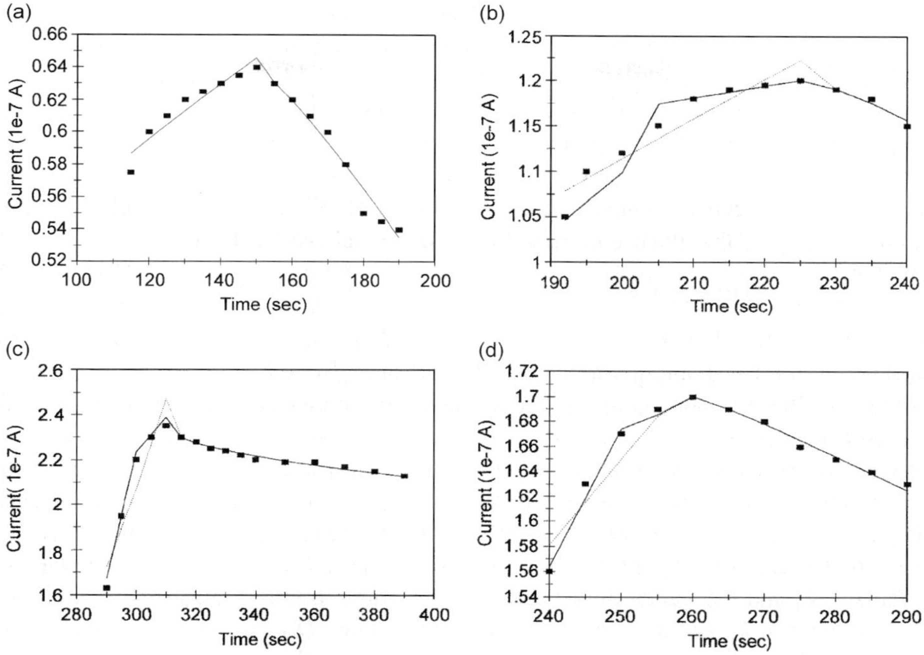

Figure 6.14 Binding of different analytes in solution to the carbon nanotube (CNT) nanoelectrode ensemble (NNE): (a) glucose, (b) ascorbic acid, (c) uric acid, (d) acetaminophen.

a dual-fractal analysis is required to adequately model the binding kinetics for the interferent ascorbic acid, whereas a single-fractal analysis is adequate to describe the binding kinetics for glucose. This indicates a possible change in the binding mechanism.

Figure 6.14c shows the binding and dissociation of acetaminophen acid in solution to the CNT NEE glucose biosensor (Lin *et al.*, 2004). Once again, a dual-fractal analysis is required to adequately describe the binding kinetics of this interferent. A single-fractal analysis is adequate to describe the dissociation kinetics. The values of (a) the binding rate coefficient, k, and the fractal dimension, D_f for a single-fractal analysis, (b) the binding rate coefficients, k_1 and k_2, and the fractal dimensions, D_{f1} and D_{f2} for a dual-fractal analysis, and (c) the dissociation rate coefficient, k_d and the fractal dimension, D_{fd} are given in Table 6.7. It is of interest to note that to describe the binding kinetics of both

Table 6.8a

Binding and dissociation rate coefficients for glucose, ascorbic acid, acetaminophen and uric acid in solution to a nanoelectrode (NEE) glucose biosensor (Lin *et al.*, 2004)

Compound	k	k_1	k_2	k_d
Glucose	0.1057 ± 0.0012	na	na	0.00139 ± 0.00017
Ascorbic acid	0.0169 ± 0.00029	0.000203 ± 0.00002	0.3311 ± 0.0005	0.00095 ± 0.00028
Uric acid	$8.8 \times 10^{-14} \pm 5 \times 10^{-15}$	2.7×10^{-22}	2.3×10^{-5}	0.0229 ± 0.0017
Acetaminophen	0.0052 ± 0.00007	0.000165 ± 0.000001	0.1359 ± 0.0004	0.00166 ± 0.00017

Table 6.8b

Fractal dimensions in the binding and in the dissociation phase for glucose, ascorbic acid, acetaminophen, and uric acid in solution to a nanoelectrode (NEE) glucose biosensor (Lin *et al.*, 2004)

Compound	D_f	D_{f1}	D_{f2}	D_{fd}
Glucose	2.2776 ± 0.0928	na	na	0.6544 ± 0.1252
Ascorbic acid	1.4185 ± 0.2278	0.6185 ± 0.3224	2.5244 ± 0.4756	$0.1782 + 0.6622$
Uric acid	~ 0	~ 0	~ 0	1.9644 ± 0.0514
Acetaminophen	0.9122 ± 0.4238	$0.0 + 0.5360$	2.0912 ± 0.17164	0.7578 ± 0.131

interferents (for the detection of glucose), a dual-fractal analysis is required to describe the binding kinetics, whereas a single-fractal analysis is adequate to describe the binding kinetics for glucose.

Figure 6.14d shows the binding and dissociation of uric acid in solution to the CNT NEE glucose biosensor (Lin *et al.*, 2004). Once again, a dual-fractal analysis is required to adequately describe the binding kinetics of this interferent. A single-fractal analysis is adequate to describe the dissociation kinetics. The values of (a) the binding rate coefficient, k, and the fractal dimension, D_f for a single-fractal analysis, (b) the binding rate coefficients, k_1 and k_2, and the fractal dimensions, D_{f1} and D_{f2} for a dual-fractal analysis, and (c) the dissociation rate coefficient, k_d and the fractal dimension, D_{fd} are given in Tables 6.8a and b. It is of interest to note, once again, that to describe the binding kinetics of all three interferents (for the detection of glucose), a dual-fractal analysis is required to describe the binding kinetics, whereas a single-fractal analysis is adequate to describe the binding kinetics for glucose.

Tables 6.8a and b and Figure 6.15a show that the binding rate coefficient, k_2 increases as the fractal dimension, D_{f2} increases. For the data presented in Figure 6.15a, the binding rate coefficient, k_2 is given by:

$$k_2 = (0.0138 \pm 0.0029)D_{f2}^{3.284 \pm 0.1059} \tag{6.6d}$$

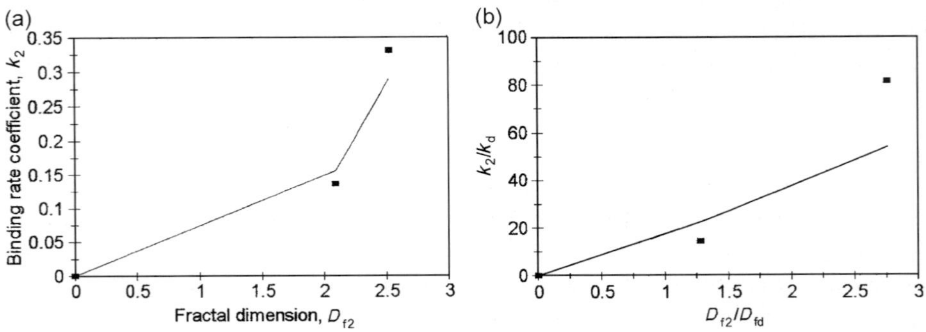

Figure 6.15 (a) Increase in the binding rate coefficient, k_2 with an increase in the fractal dimension, D_{f2}, (b) increase in the ratio of the binding rate coefficient to the dissociation rate coefficient ratio, k_2/k_d with an increase in the fractal dimension ratio, D_{f2}/D_{fd}.

The fit is quite good. Only three data points are available. Tables 6.8a and b indicate that one of the data points is actually $D_{f2} \sim 0$. In order that this point may also be used, an arbitrary very small value of 1.0×10^{-6} was also used. The selection of this very low value point did not make a difference in the fit of the line. The binding rate coefficient, k_2 is sensitive to the degree of heterogeneity that exists on the CNT surface, as noted by the greater than third order dependence on the fractal dimension, D_f. It should be pointed out that the data referred to here is for the interferents, and not glucose itself.

Tables 6.8a and b and Figure 6.15b show that the ratio of the binding and the dissociation rate coefficient, k_2/k_d increases as the ratio of fractal dimensions, D_{f2}/D_{fd} increases. For the data presented in Figure 6.15b, the ratio of the binding to the dissociation rate coefficient, k_2/k_d is given by:

$$k_2/k_d = (16.374 \pm 13.723)(D_{f2}/D_{fd})^{1.156 \pm 0.0603} \quad (6.6e)$$

The fit is quite good. Only three data points are available. Tables 6.8a and b indicate that one of the data points is actually $D_{f2} \sim 0$. In order that this point may also be used, an arbitrary very small value of 1×10^{-6} was also used. The selection of this very low value point did not make a difference in the fit of the line. The ratio of the binding to the dissociation rate coefficient, k_2/k_d exhibits an order of dependence slightly higher than first (equal to 1.156) order on the fractal dimension ratio, D_{f2}/D_{fd}. Once again, it should be pointed out that the data referred to here is for interferents, and not for glucose itself.

Lin *et al.* (2004) emphasize that their biosensor is suitable to detect glucose in sweat, urine, and serum. Furthermore, their biosensor may be adapted to also detect other compounds of medical interest such as cholesterol, alcohol, lactate, and acetylcholine.

Liu and Lu (2003) have recently indicated the need for the development of sensors to detect Pb^{++}. This is particularly important due to its deleterious effects on the health of young children (Needleman, 2002; Godwin, 2001). Liu and Lu (2003) emphasize that the development of simple lead sensors (Telting-Diaz and Bakker, 2002; Chen and Huang, 2002; Blake *et al.*, 2001) could lead to the measurement of Pb^{++} on a real-time basis in household, environmental, and in clinic applications. Liu and Lu (2003) developed a DNAzyme-based biosensor for Pb^{++}. The biosensor exhibited high sensitivity and selectivity at 4 °C.

Liu and Lu (2003) indicate that DNAzymes (DNA enzymes, catalytic DNA or deoxyribozymes) catalyze many of the same reactions as RNA or protein enzymes (Lu, 2002; Sen and Geyer, 1998). The advantage of DNAzymes is that they unlike proteins can be denatured and renatured many times without losing binding ability or activity.

Li and Lu (2000) and Cuenoud and Szostak (1995) summarize the three steps that are involved in the binding kinetics:

(a) Hybridization of the substrate to the enzyme (by a Watson–Crick base pairing).
(b) Cleavage of the substrate by the enzyme. Divalent metal ions are used as cofactors.
(c) In fluorescence-based detection, the fluorescence increase occurs in the third step with the release of the fluorophore-labeled cleavage product.

Figure 6.16 shows the fluorescence increase of the two-quencher DNAzyme-based Pb^{++} sensor at room temperature with 500 nM divalent metal ions (Liu and Lu, 2003).

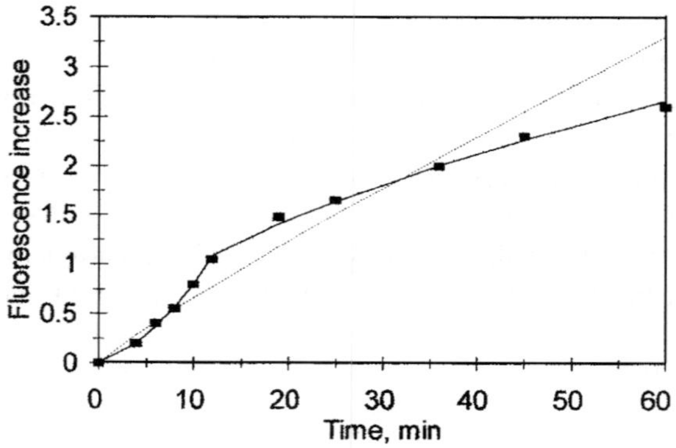

Figure 6.16 Binding (cleavage) kinetics for 500 nM Pb^{++} in solution to a DNAzyme biosensor (Liu and Lu, 2003). (All other six metal ions gave signals similar to background.)

A dual-fractal analysis was required to adequately describe the binding kinetics. The values of (a) the binding rate coefficient, k and the fractal dimension, D_f for a single-fractal analysis, and (b) the binding rate coefficients, k_1 and k_2 and the fractal dimensions, D_{f1} and D_{f2} for a dual-fractal analysis are given in Table 6.9.

The authors indicate that with an increase in temperature to room temperature the DNAzyme biosensor performance decreases considerably. Liu and Lu (2003) emphasize that their new design of a DNAzyme biosensor incorporates inter- and intra-molecular quenchers that help alleviate this problem with temperature, and should also be applicable to other nucleic acid-based biosensors to help minimize background fluorescence.

Fernandez-Sanchez *et al.* (2004a) have recently developed a disposable noncompetitive immunosensor for free and total prostate-specific antigen (PSA). Landis *et al.* (1999) indicate that PSA is a good tumor marker for the early detection of prostate cancer at the organ-confined stage. Fernandez-Sanchez *et al.* (2004a) indicate that PSA is a 33 kDa single chain glycoprotein that circulates in the blood stream. It is mainly complexed to α_1-antichymotrypsin. Panniek *et al.* (1998) indicate the importance of testing for the ratio of free (f-PSA) to total (t-PSA) in the diagnosis of prostate cancer. These authors indicate that the ratio of f/t PSA significantly decreases with the appearance of the disease.

Fernandez-Sanchez *et al.* (2004a) developed a one-step, single-use, lateral flow immunosensor using an electrochemical sensor coated with a pH-sensitive polymer. These authors coated strips of a nitrocellulose membrane with appropriate antibodies for f-PSA and t-PSA, respectively. These antibody-coated nitrocellulose membranes were

Table 6.9

Binding rate coefficients and fractal dimension values for the fluorescent DNAzyme biosensor for Pb^{++} detection at room temperature (Liu and Lu, 2003)

k	k_1	k_2	D_f	D_{f1}	D_{f2}
0.0840 ± 0.021	0.0262 ± 0.0013	0.2756 ± 0.0097	1.2058 ± 0.164	0.0322 + 0.109	1.892 ± 0.0052

used as the solid supports for the designed noncompetitive immunoassay. They emphasize that their developed immunosensor is simple to use, is portable, and robust.

Figure 6.17a shows the binding of 0 ng/ml of free PSA in solution to the anti-free PSA immunostrip (Fernandez-Sanchez *et al.*, 2004a). A single-fractal analysis is adequate to describe the binding kinetics. The values of (a) the binding rate coefficient, k, and the fractal dimension, D_f for a single-fractal analysis are given in Table 6.10.

Figure 6.17b shows the binding of 1 ng/ml f-PSA in solution to the anti-free PSA immunostrip (Fernandez-Sanchez *et al.*, 2004a). A dual-fractal analysis is required

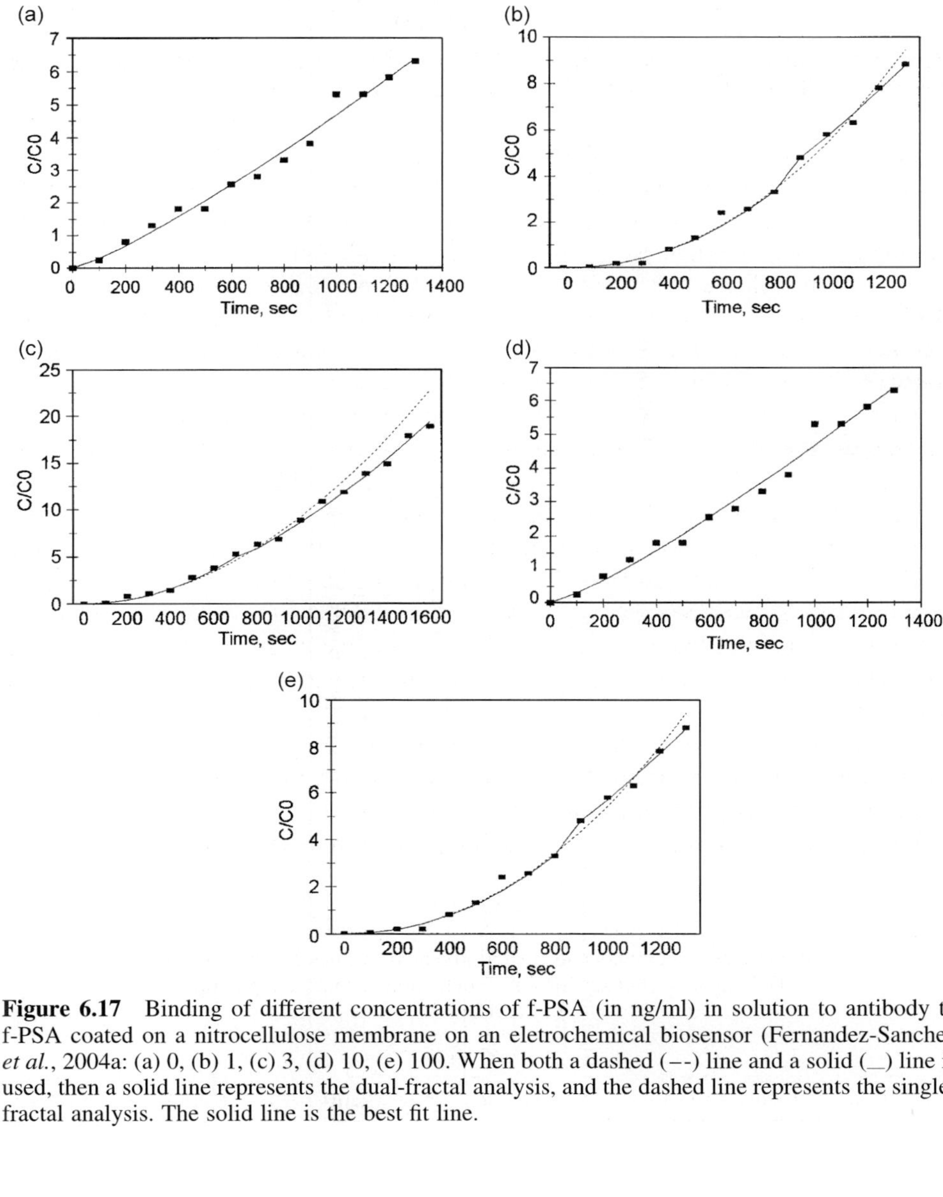

Figure 6.17 Binding of different concentrations of f-PSA (in ng/ml) in solution to antibody to f-PSA coated on a nitrocellulose membrane on an eletrochemical biosensor (Fernandez-Sanchez *et al.*, 2004a: (a) 0, (b) 1, (c) 3, (d) 10, (e) 100. When both a dashed (–-) line and a solid (__) line is used, then a solid line represents the dual-fractal analysis, and the dashed line represents the single-fractal analysis. The solid line is the best fit line.

Table 6.10

Binding rate coefficients and fractal dimension values for free prostate specific antigen (f-PSA) in solution to antibody to f-PSA coated on a nitrocellulose membrane on an electrochemical transducer in a lateral flow immunoassay format. Effect of different f-PSA concentrations (Fernandez-Sanchez *et al.*, 2004a)

Free PSA concentration (ng/ml)	k	k_1	k_2	D_f	D_{f1}	D_{f2}
0	0.00123 ± 0.000155	na	na	0.6132 ± 0.0892	na	na
1	$2.4 \pm 0.7 \times 10^{-6}$	$2.5 \pm 1.0 \times 10^{-6}$	$6.9 \pm 0.2 \times 10^{-5}$	$0 + 0.191$	$0 + 0.3622$	$0 + 0.239$
3	$1.5 \pm 0.5 \times 10^{-5}$	$7.1 \pm 0.3 \times 10^{-6}$	$5.9 \pm 0.3 \times 10^{-5}$	$0 + 0.1936$	$0 + 0.3304$	$0 + 0.1594$
10	$1.2 \pm 0.3 \times 10^{-6}$	$5.5 \pm 1.5 \times 10^{-7}$	$5.7 \pm 0.1 \times 10^{-5}$	$0 + 0.1486$	$0 + 0.1946$	$0 + 0.1147$
100	0.000145 ± 0.000034	na	na	$0 + 0.1256$	na	na

to adequately describe the binding kinetics. The values of (a) the binding rate coefficient, k, and the fractal dimension, D_f for a single-fractal analysis, and (b) the binding rate coefficients, k_1 and k_2, and the fractal dimensions, D_{f1} and D_{f2} for a dual-fractal analysis are given in Table 6.10.

Figure 6.17c shows the binding of 3 ng/ml f-PSA in solution to the anti-free PSA immunostrip (Fernandez-Sanchez *et al.*, 2004a). Once again, a dual-fractal analysis is required to adequately describe the binding kinetics. The values of (a) the binding rate coefficient, k, and the fractal dimension, D_f for a single-fractal analysis, and (b) the binding rate coefficients, k_1 and k_2, and the fractal dimensions, D_{f1} and D_{f2} for a dual-fractal analysis are given in Table 6.10. The binding mechanism does not seem to change as one goes from 1 to 3 ng/ml free PSA in solution since a dual-fractal analysis is required to adequately describe the binding kinetics in both cases.

Figure 6.17d shows the binding of 10 ng/ml f-PSA in solution to the anti-free PSA immunostrip (Fernandez-Sanchez *et al.*, 2004a). A dual-fractal analysis is required to adequately describe the binding kinetics. The values of (a) the binding rate coefficient, k, and the fractal dimension, D_f for a single-fractal analysis, and (b) the binding rate coefficients, k_1 and k_2, and the fractal dimensions, D_{f1} and D_{f2} for a dual-fractal analysis are given in Table 6.10. The binding mechanism does not seem to change as one goes from 1 to 10 ng/ml free PSA in solution since a dual-fractal analysis is required to adequately describe the binding kinetics in both cases.

Figure 6.17e shows the binding of 100 ng/ml of free PSA in solution to the anti-free PSA immunostrip (Fernandez-Sanchez *et al.*, 2004a). In this case, a single-fractal analysis is adequate to describe the binding kinetics. The values of (a) the binding rate coefficient, k, and the fractal dimension, D_f for a single-fractal analysis are given in Table 6.10. Note that at this highest concentration (100 ng/ml) of f-PSA analyzed, there is apparently a change in the binding mechanism since a single-fractal analysis is adequate to describe the binding kinetics, whereas at the lower concentrations (1–10 mg/ml f-PSA) a dual-fractal analysis is required to describe the binding kinetics. Apparently, the highest concentration of f-PSA used leads to a saturation of the 'receptors' or active sites of the anti-PSA on the immunostrip surface.

Figure 6.18 and Table 6.10 show that as the f-PSA concentration increases from 1 to 10 mg/ml, and where a dual-fractal analysis applies for the binding kinetics, the binding rate coefficient, k_2 decreases, and is given by:

$$k_2 = (6.7 \pm 0.4) \times 10^{-5}[\text{f-PSA}]^{-0.0821 \pm 0.0327} \tag{6.7}$$

The fit is quite good. Only three data points were available. The availability of more data points would lead to a more reliable fit. The binding rate coefficient, k_2 is only mildly sensitive to the f-PSA concentration in solution as noted by the very low order of dependence (equal to -0.0821).

Figure 6.19a shows the binding of 0 ng/ml total PSA in solution to the anti-total PSA immunostrip (Fernandez-Sanchez *et al.*, 2004a). A dual-fractal analysis is required to adequately describe the binding kinetics. The values of (a) the binding rate coefficient, k and the fractal dimension, D_f, for a single fractal analysis, and (b) the binding rate coefficients, k_1 and k_2 and the fractal dimensions, D_{f1} and D_f are given in Tables 6.11a and b.

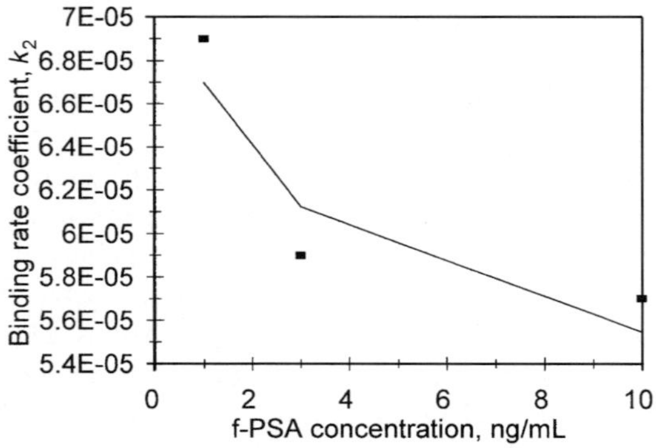

Figure 6.18 Decrease in the binding rate coefficient, k_2 with an increase in the f-PSA concentration in solution.

Figure 6.19b shows the binding of 1 ng/ml of total PSA in solution to the anti-total PSA immunostrip (Fernandez-Sanchez *et al.*, 2004a). A single-fractal analysis is adequate to describe the binding kinetics. The values of the binding rate coefficient, k and the fractal dimension, D_f are given in Tables 6.11a and b.

Figure 6.19c shows the binding of 3 ng/ml of total PSA in solution to the anti-total PSA immunostrip (Fernandez-Sanchez *et al.*, 2004a). Once again, a single-fractal analysis

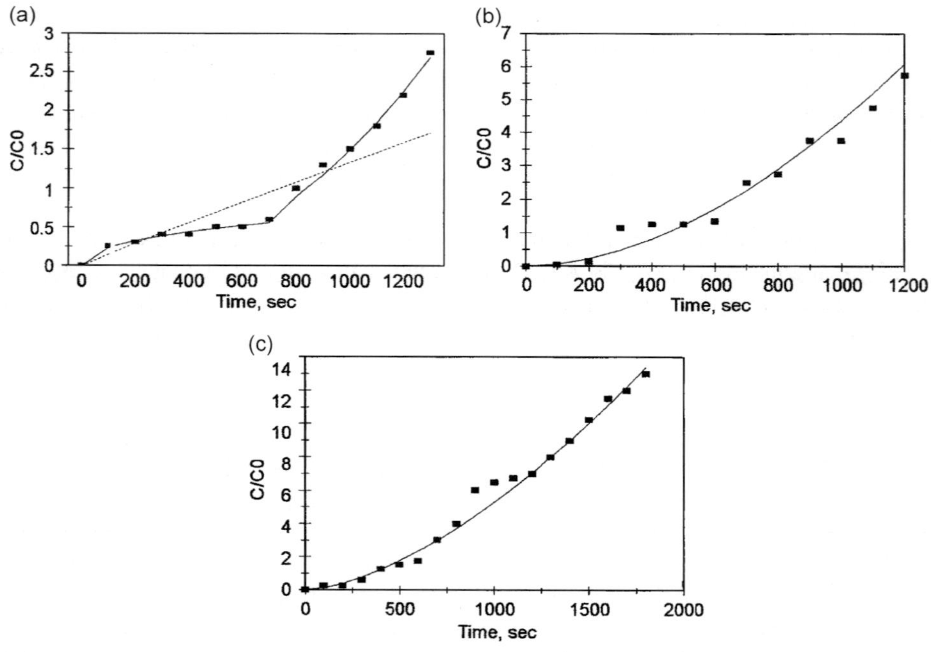

Figure 6.19 Binding of different concentrations of t-PSA (in ng/ml) in solution to antibody to t-PSA coated on a nitrocellulose membrane on an electrochemical biosensor (Fernandez-Sanchez *et al.*, 2004a): (a) 0, (b) 1, (c) 3.

Table 6.11a

Rate coefficients for the binding of total-PSA in solution to anti-total PSA coated on a nitrocellulose membrane on an electrochemical transducer in a lateral flow immunoassay format (Fernandez-Sanchez *et al.*, 2004a)

Total PSA concentration (ng/ml)	k	k_1	k_2
0	0.00189 ± 0.00079	0.0314 ± 0.0023	$(2.3 \pm 0.2) \times 10^{-7}$
1	$(1.4 \pm 0.6) \times 10^{-5}$	na	na
3	$(9.5 \pm 2.6) \times 10^{-5}$	na	na

is adequate to describe the binding kinetics. The values of the binding rate coefficient, k and the fractal dimension, D_f are given in Tables 6.11a and b.

Fernandez-Sanchez *et al.* (2004b) have also recently developed a noncompetitive 'sandwich' type immunoassay for the detection of free prostate antigen (f-PSA) and total prostate antigen (t-PSA). This is complementary to the study just presented. These authors indicate that PSA is a serine protease that is secreted in high concentrations by the prostrate gland into the seminal fluid. Their immunosensor used an impedance measurement of a degradation process occurring on a pH-sensitive polymer coated electrode. As in their previous study (Fernandez-Sanchez *et al.*, 2004a) their immuno-strips used nitrocellulose membranes coated with either anti-f-PSA or anti-t-PSA antibodies. This study is similar to their previous study except for the presence of lateral flow.

Figure 6.20a shows the binding of 0 ng/ml f-PSA in solution to anti-f-PSA coated on the nitrocellulose membrane immunostrip (Fernandez-Sanchez *et al.*, 2004b). A single-fractal analysis is adequate to describe the binding kinetics. The values of the binding rate coefficient, k, and the fractal dimension, D_f, are given in Table 6.12.

Figure 6.20b shows the binding of 1 ng/ml of f-PSA in solution to the anti-f-PSA coated on the nitrocellulose membrane immunostrip. A dual-fractal analysis is required to adequately describe the binding kinetics. The values of (a) the binding rate coefficient, k, and the fractal dimension, D_f for a single-fractal analysis, and (b) the binding rate coefficients, k_1 and k_2 and the fractal dimensions, D_{f1} and D_{f2} for a dual-fractal analysis are given in Table 6.12. It is of interest to note that as the fractal dimension increases by a

Table 6.11b

Fractal dimensions for the binding of total-PSA in solution to anti-total PSA coated on a nitrocellulose membrane on an electrochemical transducer in a lateral flow immunoassay format (Fernandez-Sanchez *et al.*, 2004a)

Total PSA concentration (ng/ml)	D_f	D_{f1}	D_{f2}
0	1.1018 ± 0.2634	2.1284 ± 0.0847	0 ± 0.2978
1	0 + 0.2862	na	na
3	0 + 0.1464	na	na

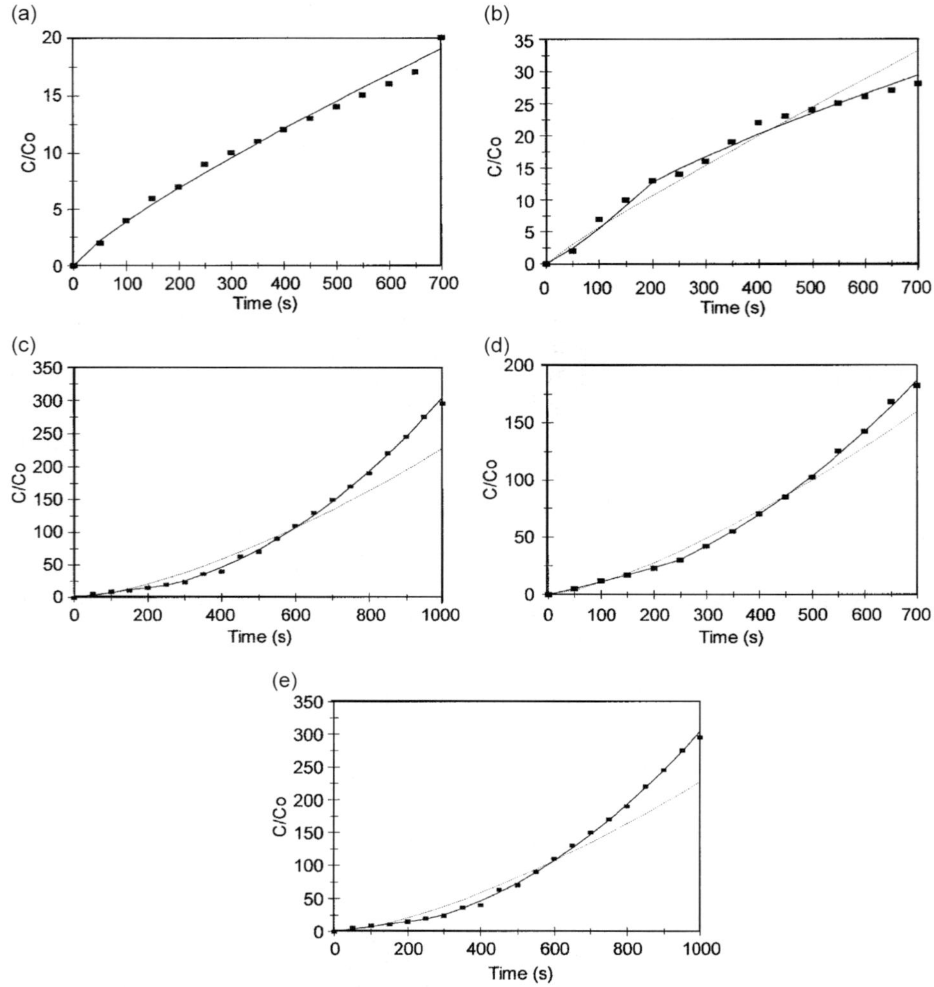

Figure 6.20 Binding of different concentrations of f-PSA (in ng/ml) in solution to antibody f-PSA coated on a nitrocellulose membrane on a lateral flow immunosensor (Fernandez-Sanchez *et al.*, 2004b): (a) 0, (b) 1, (c) 2, (d) 10, (e) 30. When both a dashed (--) line and a solid (—) line is used then a solid line represents the dual-fractal analysis, and the dashed line represents the single-fractal analysis. The solid line is the best fit line.

factor of 2.96 from a value of D_{f1} equal to 0.5649 to D_{f2} equal to 1.6749, the binding rate coefficient increases by a factor of 3.11 from a value of k_1 equal to 0.02023 to k_2 equal to 0.384. An increase in the degree of heterogeneity on the surface (increase in the fractal dimension value) leads to an increase in the binding rate coefficient.

Figure 6.20c shows the binding of 2 ng/ml of f-PSA in solution to the anti-f-PSA coated on the nitrocellulose membrane immunostrip. A dual-fractal analysis is required to adequately describe the binding kinetics. The values of (a) the binding rate coefficient, k, and the fractal dimension, D_f for a single-fractal analysis, and (b) the binding rate coefficients, k_1 and k_2 and the fractal dimensions, D_{f1} and D_{f2} for a dual-fractal analysis

Table 6.12

Binding rate coefficients and fractal dimensions for the binding of free PSA (f-PSA) in solution to anti-f-PSA immobilized on an immunostrip using lateral flow and a noncompetitive sandwich assay (Fernandez-Sanchez *et al.*, 2004b)

f-PSA concentration in solution (ng/ml)	k	k_1	k_2	D_f	D_{f1}	D_{f2}
0	0.09317 ± 0.00574	na	na	1.3756 ± 0.0429	na	na
1	0.08953 ± 0.0726	0.02023 ± 0.0047	0.384 ± 0.0192	1.1952 ± 0.1275	0.5649 ± 0.3278	1.6749 ± 0.096
2	0.02063 ± 0.0028	0.0221 ± 0.0024	0.00478 ± 0.0004	0.6891 ± 0.0919	0.6803 ± 0.2586	0.2123 ± 0.1318
10	0.0176 ± 0.0027	0.0725 ± 0.0040	0.00182 ± 0.00003	0.2139 ± 0.1056	0.8164 ± 0.0852	0 + 0.0423
30	0.00851 ± 0.0031	0.1454 ± 0.018	0.000219 ± 0.00004	0.0488 + 0.176	1.2403 ± 0.1589	0 + 0.0514

are given in Table 6.12. It is of interest to note that as the fractal dimension decreases by a factor of 3.20 from a value of D_{f1} equal to 0.6803 to D_{f2} equal to 0.2123, the binding rate coefficient increases by a factor of 39.83 from a value of k_1 equal to 0.0221 to k_2 equal to 0.00478. In this case, a decrease in the degree of heterogeneity on the surface (increase in the fractal dimension value) leads to a decrease in the binding rate coefficient. Note that changes in the degree of heterogeneity and in the binding rate coefficient are in the same direction.

Figure 6.20d shows the binding of 10 ng/ml of f-PSA in solution to the anti-f-PSA coated on the nitrocellulose membrane immunostrip. A dual-fractal analysis is required to adequately describe the binding kinetics. The values of (a) the binding rate coefficient, k, and the fractal dimension, D_f for a single-fractal analysis, and (b) the binding rate coefficients, k_1 and k_2 and the fractal dimensions, D_{f1} and D_{f2} for a dual-fractal analysis are given in Table 6.12. In this case the fractal dimension in the second phase of binding, D_{f2} is $\sim$0. This indicates that the biosensor surface exists as a 'Cantor' like dust. However, here too, an increase in the degree of heterogeneity on the surface leads to an increase in the binding rate coefficient, as noted by the values of the fractal dimensions and the binding rate coefficients in the first and in the second phase of binding.

Figure 6.20e shows the binding of 30 ng/ml of f-PSA in solution to the anti-f-PSA coated on the nitrocellulose membrane immunostrip. A dual-fractal analysis is required to adequately describe the binding kinetics. The values of (a) the binding rate coefficient, k, and the fractal dimension, D_f for a single-fractal analysis, and (b) the binding rate coefficients, k_1 and k_2 and the fractal dimensions, D_{f1} and D_{f2} for a dual-fractal analysis are given in Table 6.12. In this case too, the fractal dimension in the second phase of binding, D_{f2} is $\sim$0. Once again, this indicates that the biosensor surface exists as a 'Cantor' like dust. Apparently, at the higher f-PSA concentrations in solution (10 and 30 ng/ml), the immunostrip surface acts like a 'Cantor' like dust. At present, no explanation is offered for this behavior. Here too, however, and as noted above, an increase in the degree of heterogeneity on the surface leads to an increase in the binding rate coefficient, as noted by the values of the fractal dimensions and the binding rate coefficients in the first and in the second phase of binding.

Table 6.12 and Figure 6.21a show the increase in the binding rate coefficient, k_1 with an increase in the f-PSA concentration in solution. In the 1–30 ng/ml f-PSA concentration

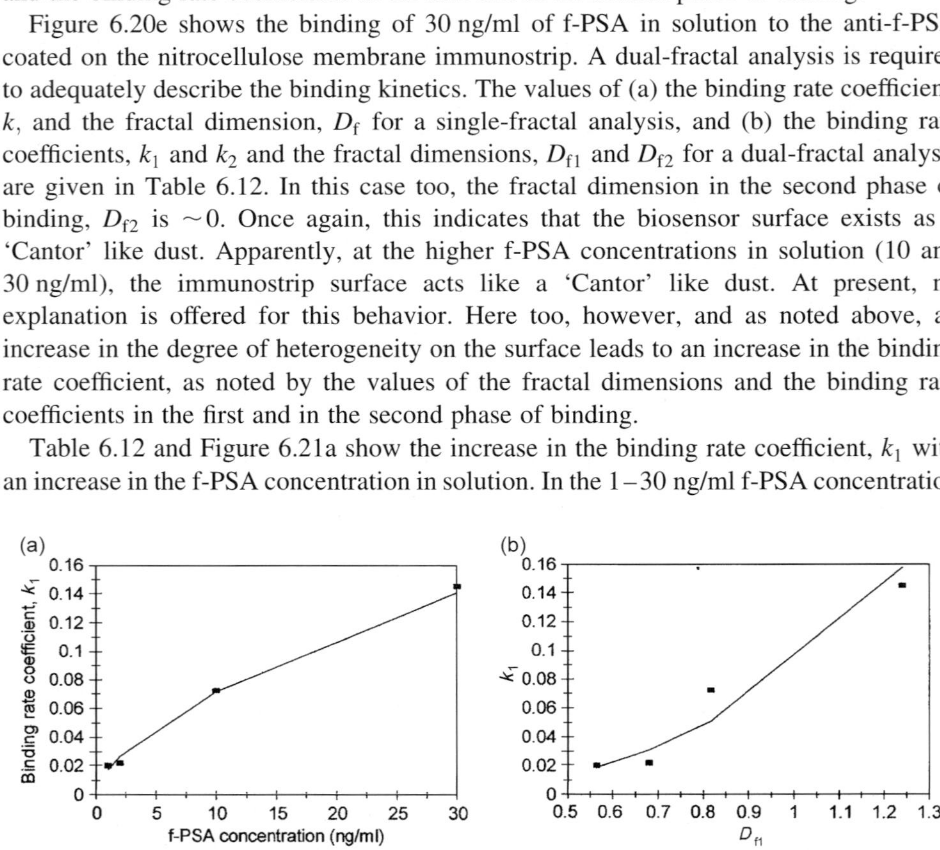

Figure 6.21 Increase in the binding rate coefficient, k_1 with an: (a) increase in the f-PSA concentration (in ng/ml) in solution, (b) increase in the fractal dimension, D_{f1}.

in solution analyzed, the binding rate coefficient, k_1 is given by:

$$k_1 = (0.01744 \pm 0.0033)[\text{f-PSA}]^{0.6144 \pm 0.0641} \tag{6.8a}$$

The fit is good. Only four data points are available. The availability of more data points would lead to a more reliable fit. The binding rate coefficient, k_1 exhibits an order of dependence between one-half and first order (equal to 0.6144) on the f-PSA concentration in solution. The noninteger order of dependence of k_1 on the f-PSA concentration in solution reinforces the fractal nature of the system.

Table 6.12 and Figure 6.21b show the increase in the binding rate coefficient, k_1 with an increase in the fractal dimension, D_{f1}. In the 1–30 ng/ml f-PSA concentration in solution analyzed, the binding rate coefficient, k_1 is given by:

$$k_1 = (0.0883 \pm 0.0377)D_{f1}^{2.709 \pm 0.6099} \tag{6.8b}$$

The fit is good. Only four data points are available. The availability of more data points would lead to a more reliable fit. The binding rate coefficient, k_1 is sensitive to the degree of heterogeneity that exists on the surface, D_{f1} since it exhibits an order of dependence between second and third order (equal to 2.709).

Figure 6.22a shows the binding of 0 ng/ml t-PSA in solution to anti-t-PSA coated on the nitrocellulose membrane immunostrip (Fernandez-Sanchez *et al.*, 2004b). A dual-fractal

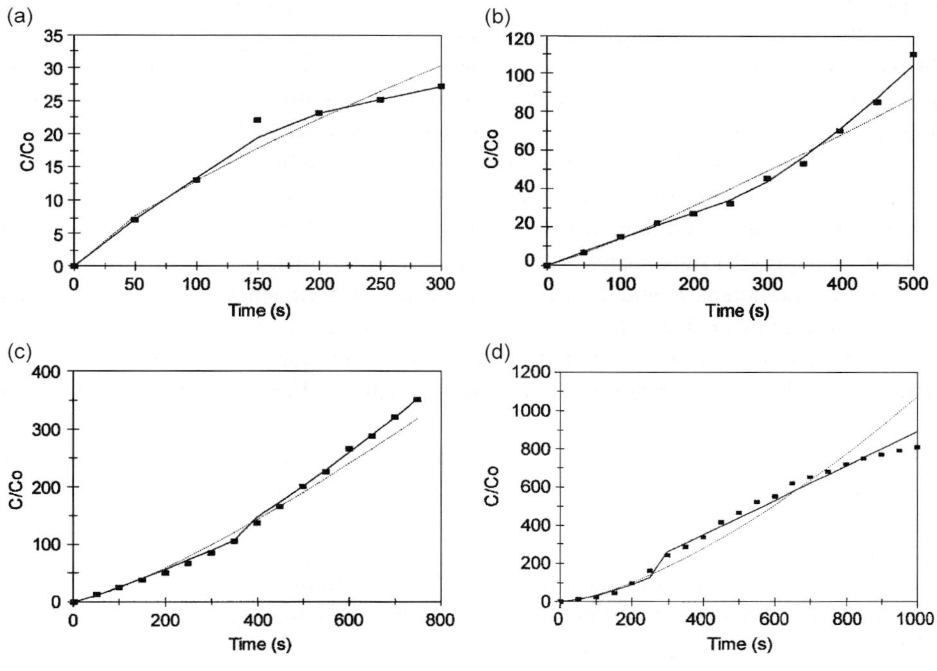

Figure 6.22 Binding of different concentrations of t-PSA (in ng/ml) in solution to antibody t-PSA coated on a nitrocellulose membrane on a lateral flow immunosensor (Fernandez-Sanchez *et al.*, 2004b): (a) 0, (b) 2, (c) 10, (d) 30. When both a dashed (--) line and a solid (—) line is used, then a solid line represents the dual-fractal analysis, and the dashed line represents the single-fractal analysis. The solid line is the best fit line.

analysis is required to adequately describe the binding kinetics. The values of (a) the binding rate coefficient, k, and the fractal dimension, D_f for a single-fractal analysis, and (b) the binding rate coefficients, k_1 and k_2, and the fractal dimensions, D_{f1} and D_{f2} for a dual-fractal analysis are given in Table 6.13.

Figure 6.22b shows the binding of 2 ng/ml t-PSA in solution to anti-t-PSA coated on the nitrocellulose membrane immunostrip (Fernandez-Sanchez *et al.*, 2004b). A dual-fractal analysis is required to adequately describe the binding kinetics. The values of (a) the binding rate coefficient, k, and the fractal dimension, D_f for a single-fractal analysis, and (b) the binding rate coefficients, k_1 and k_2, and the fractal dimensions, D_{f1} and D_{f2} for a dual-fractal analysis are given in Table 6.13. It is of interest to note that as the fractal dimension decreases from D_{f1} equal to 1.1061 to $D_{f2} \sim 0$, the binding rate coefficient decreases by a factor of 78.31 from k_1 equal to 0.1809 to k_2 equal to 0.00231. Changes in the degree of heterogeneity on the biosensor surface and in the binding rate coefficient are in the same direction.

Figure 6.22c shows the binding of 10 ng/ml t-PSA in solution to anti-t-PSA coated on the nitrocellulose membrane immunostrip (Fernandez-Sanchez *et al.*, 2004b). A dual-fractal analysis is required to adequately describe the binding kinetics. The values of (a) the binding rate coefficient, k, and the fractal dimension, D_f for a single-fractal analysis, and (b) the binding rate coefficients, k_1 and k_2, and the fractal dimensions, D_{f1} and D_{f2} for a dual-fractal analysis are given in Table 6.13. It is of interest to note that as the fractal dimension decreases by a factor of 3.287 from D_{f1} equal to 0.7109 to D_{f2} equal to 0.2163, the binding rate coefficient decreases by a factor of 3.672 from k_1 equal to 0.1311 to k_2 equal to 0.0357. Changes in the degree of heterogeneity on the biosensor surface and in the binding rate coefficient are in the same direction. In this case, there is almost a linear decrease, but this is a coincidence. An increase in the t-PSA concentration from 0 to 10 ng/ml in solution leads to a decrease in the binding rate coefficient, k_1.

Figure 6.22d shows the binding of 30 ng/ml t-PSA in solution to anti-t-PSA coated on the nitrocellulose membrane immunostrip (Fernandez-Sanchez *et al.*, 2004b). A dual-fractal analysis is required to adequately describe the binding kinetics. The values of (a) the binding rate coefficient, k, and the fractal dimension, D_f for a single-fractal analysis, and (b) the binding rate coefficients, k_1 and k_2, and the fractal dimensions, D_{f1} and D_{f2} for a dual-fractal analysis are given in Table 6.13. It is of interest to note that as the fractal dimension increases by a factor of 307.23 from D_{f1} equal to 0.0031to D_{f2} equal to 0.9524, the binding rate coefficient increases by a factor of 23.56 from k_1 equal to 0.0321 to k_2 equal to 0.7563. Changes in the degree of heterogeneity on the biosensor surface and in the binding rate coefficient are, once again, in the same direction. Also, an increase in the t-PSA concentration from 0 to 30 ng/ml in solution leads to a decrease in the binding rate coefficient, k_1.

Figure 6.23a and Table 6.13 show the decrease in the binding rate coefficient, k_1 with an increase in the t-PSA concentration in solution. In the 2–30 ng/ml t-PSA concentration in solution, the binding rate coefficient, k_1 is given by:

$$k_1 = (0.3325 \pm 0.2571)[\text{t-PSA}]^{-0.6061 \pm 0.2973} \tag{6.9a}$$

The fit is reasonable. Only three data points are available. The availability of more data points would lead to a more reliable fit. The binding rate coefficient, k_1 decreases as

Table 6.13

Binding rate coefficients and fractal dimensions for the binding of total PSA (t-PSA) in solution to anti-t-PSA immobilized on an immunostrip using lateral flow and a noncompetitive sandwich assay (Fernandez-Sanchez *et al.*, 2004b)

f-PSA concentration in solution (ng/ml)	k	k_1	k_2	D_f	D_{f1}	D_{f2}
0	0.3674 ± 0.0533	0.1997 ± 0.0239	2.8379 ± 0.0112	1.4631 ± 0.179	1.1748 ± 0.2168	2.2107 ± 0.0276
2	0.07428 ± 0.0112	0.1809 ± 0.0116	0.00231 ± 0.00014	0.7252 ± 0.1270	1.1061 ± 0.0984	~0 + 0.1656
10	0.06974 ± 0.0094	0.1311 ± 0.0141	0.0357 ± 0.00053	0.4547 ± 0.0860	0.7109 ± 0.0108	0.2163 ± 0.0942
30	0.0400 ± 0.00922	0.0321 ± 0.0104	0.7563 ± 0.0484	0.0477 + 0.1169	0.0031 + 0.4406	0.9524 ± 0.0884

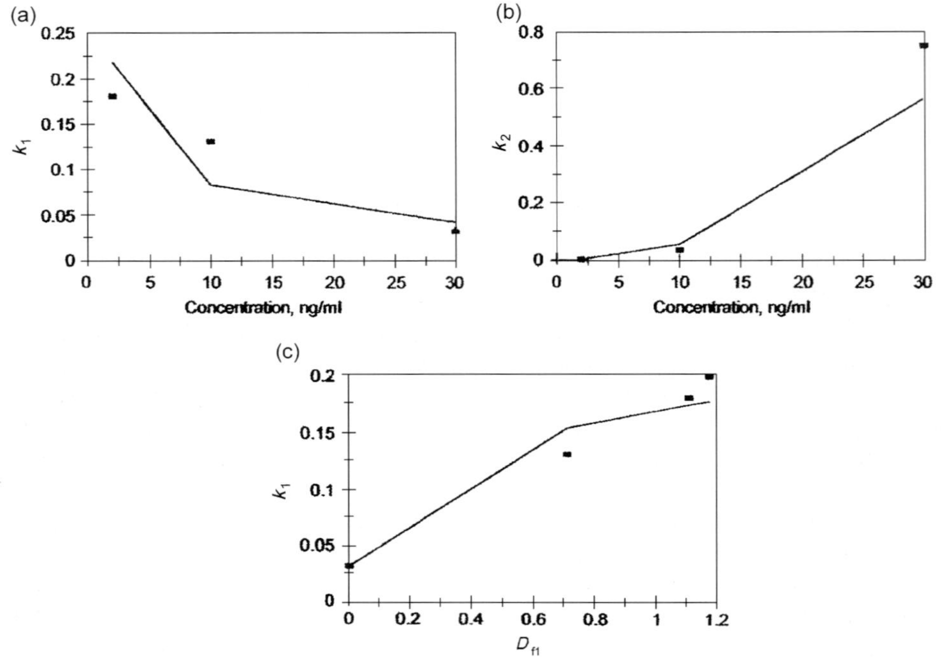

Figure 6.23 (a) Decrease in the binding rate coefficient, k_1 with an increase in the t-PSA concentration (in ng/ml) in solution, (b) increase in the binding rate coefficient, k_2 with an increase in the t-PSA concentration (in ng/ml) in solution, (c) increase in the binding rate coefficient, k_1 with an increase in the fractal dimension, D_{f1} on the immunostrip surface.

the t-PSA concentration in solution increases. The binding rate coefficient, k_1 exhibits a negative order between one-half and first order (equal to -0.6061) on the t-PSA concentration in solution. The noninteger order of dependence exhibited by k_1 on the t-PSA concentration in solution lends support to the fractal nature of the system.

Figure 6.23b and Table 6.13 show the decrease in the binding rate coefficient, k_2 with an increase in the t-PSA concentration in solution. In the 2–30 ng/ml t-PSA concentration in solution, the binding rate coefficient, k_2 is given by:

$$k_2 = (0.000441 \pm 0.000349)[\text{t-PSA}]^{2.1071 \pm 0.3030} \tag{6.9b}$$

The fit is reasonable. Only three data points are available. The availability of more data points would lead to a more reliable fit. The binding rate coefficient, k_2 exhibits between second and two and one-half order (equal to 2.1071) of dependence on the t-PSA concentration in solution. Once again, the noninteger order of dependence exhibited by k_1 on the t-PSA concentration in solution lends support to the fractal nature of the system.

Figure 6.23c and Table 6.13 show the increase in the binding rate coefficient, k_1 with an increase in degree of heterogeneity or the fractal dimension, D_{f1} on the immunostrip (biosensor) surface. In the 2–30 ng/ml t-PSA concentration in solution, the binding rate coefficient, k_1 is given by:

$$k_1 = (0.1699 \pm 0.026) D_{f1}^{0.2891 \pm 0.02827} \tag{6.9c}$$

The fit is quite good. Only four data points are available. The availability of more data points would lead to a more reliable fit. The binding rate coefficient, k_1 exhibits only a mild dependence as noted by the less than one-half order (equal to 0.2891) on the degree of heterogeneity or fractal dimension, D_{f1} on the immunostrip surface.

6.4 CONCLUSIONS

A fractal analysis for the binding of riboflavin binding protein (RBP) in solution to riboflavin (Rf) derivative immobilized on a sensor chip (Caelen *et al.*, 2004) provides values of the binding rate coefficient, k, and the degree of heterogeneity made quantitative by the fractal dimension, D_f, on the sensor chip surface. The fractal analysis is an alternate and convenient means to provide a kinetic analysis of the diffusional-limited reactions occurring on heterogeneous or structured surfaces. The fractal analysis is also applied to the binding of (a) glucose in solution to glucose oxidase immobilized on a microcantilever surface (Pei *et al.*, 2004), (b) endogenous acetylcholine and choline to an acetylcholine and choline sensor, respectively (Mitchell, 2004), and Pb^{++} to a DNAzyme biosensor (Liu and Lu, 2003). Other binding and dissociation reactions are also analyzed.

For example, the data presented by Caelen *et al.* (2004) on the determination of riboflavin in milk samples by a SPR biosensor chip is re-analyzed to provide values of the binding rate coefficient, k, and link the binding rate coefficient, k, with the degree of heterogeneity or fractal dimension, D_f that exists on the biosensor surface. This provides a more complete picture of the reaction kinetics occurring on the sensor chip surface. Both single- as well as dual-fractal analysis are used to adequately model the binding kinetics. The dual-fractal analysis is used only when the single-fractal analysis did not provide an adequate fit (sum of least squares less than 0.97). This was done by regression analysis provided by Quatrro Pro 8.0 (1997).

In accord with the prefactor analysis for fractal aggregates (Sorenson and Roberts, 1997), quantitative (predictive) expressions are developed for (a) the binding rate coefficient, k, as a function of D_f (degree of heterogeneity) that exists on the surface. The fractal dimension for the binding is not a typical independent variable, such as analyte concentration in solution or the receptor (Rf derivative or other) on the biosensor surface that may be directly manipulated. It is estimated from Eqs. 6.1a and b (as the case may be). The predictive relationships presented, for example, for the rate coefficient, k, as a function of the analyte [RBP] concentration or the degree of heterogeneity that exists on the surface (fractal dimension value) provide a means by which these binding rate coefficients may be manipulated by changing either the analyte concentration in solution or the degree of heterogeneity that exists on the surface. Note that a change in the degree of heterogeneity on the surface generally leads to a change in the binding rate coefficient in the same direction.

It is suggested that the fractal surface (roughness) leads to turbulence, which enhances mixing, decreases diffusional limitations, and leads to an increase in the binding rate coefficient (Martin *et al.*, 1993). For this to occur the characteristic length of the turbulent boundary layer may have to extend a few monolayers above the sensor chip surface to affect bulk diffusion to and from the surface. However, given the extremely laminar flow regimes in most biosensors this may not actually take place. The sensor chip surface is characterized by grooves and ridges, and this surface morphology may lead to eddy diffusion. This eddy diffusion can then help to enhance the mixing and extend the characteristic length of the boundary layer to affect the bulk diffusion to and from the surface.

The binding mechanisms for the choline (Ch) sensor and the acetylcholine (ACh) sensor are different since a single-fractal analysis is adequate to model the binding kinetics observed with a Ch sensor whereas a dual-fractal analysis is required for the ACh sensor (Mitchell, 2004). On comparing the binding and dissociation rate coefficients for unstimulated and K^+-evoked increase for the endogenous ACh case, one notes that the K^+ affects the dissociation rate coefficient more than it does the binding rate coefficient.

For the binding of different concentrations in solution to glucose oxidase immobilized on the microcantilever surface (Pei *et al.*, 2004), one notes that (a) the binding rate coefficient, k_2 exhibits a slightly higher than second-order (equal to 2.023) dependence on the fractal dimension, D_{f2} or the degree of heterogeneity on the microcantilever sensor surface.

For the binding of 0.2–10.0 μg/ml RBP in solution to Rf derivative immobilized on a sensor chip surface the binding rate coefficient, k, for a single-fractal analysis is very sensitive to the degree of heterogeneity or fractal dimension, D_f, that exists on the surface as noted by the very high order of dependence (equal to 6.583). This should encourage experimentalists to pay more attention to the nature of the surface, and how it may be manipulated to advantage in desired directions.

The monitoring of glucose for the control of diabetes is and will be an important application of biosensor technology. This is especially true since the occurrence of diabetes amongst the general population is on the increase. Leegsma-Vogt *et al.* (2004) have analyzed insulin and glucose levels at different locations in the body. For example, an analysis of their data indicates that the binding rate coefficient, k_1 and k_2 for glucose are higher in the interstitial adipose tissue than in plasma. As expected, the corresponding fractal dimensions on the probe surface are also higher. Expressions are presented for (a) the dissociation rate coefficient, k_d as function of the fractal dimension for dissociation, D_{fd}, (b) the ratio of the binding and the dissociation rate coefficients, k_1/k_d as a function of the fractal dimension ratio, D_{f1}/D_{fd} and (c) the ratio of the binding and dissociation rate coefficients, k_2/k_d as a function of the fractal dimension ratio, D_{f2}/D_{fd}.

A kinetic analysis is also presented for (a) the detection of glucose using GGBP as the receptor immobilized on a SPR biosensor surface (Hsieh *et al.*, 2004), (b) for the detection of glucose using a wireless, remote query biosensor with a ribbon-like mass sensitive magnetoelastic sensor as a transducer (Cai *et al.*, 2004), and (c) for the detection of free prostate specific antigen (f-PSA) and total prostate specific antigen (t-PSA) in solution (Fernandez-Sanchez *et al.*, 2004a,b). Fernandez-Sanchez *et al.* (2004a,b) emphasize that the current trend is to take the ratio of the f-PSA and the t-PSA to help diagnose the onset of this disease.

The analysis of the different examples of biomedical interest presented in this chapter (and in other chapters of the book) should encourage experimentalists to pay more attention to the nature of the surface, and how it may be manipulated to advantage in desired directions. This is of particular value for the biomedical cases (or community) wherein the earlier one may detect and diagnose the probable onset of diseases the earlier one can begin the medical protocols necessary to help prevent, alleviate, or correct the onset of especially debilitating and intractable diseases.

The diseases analyzed in this chapter are in no way to indicate that these are the only important diseases, but they should just serve as a sample, and as to how the fractal analysis can and would help in providing a better understanding of the onset of these diseases. Any insight that is made available by such an analysis that helps in the management of these intractable diseases should prove invaluable.

REFERENCES

American Diabetic Assoication, Study shows sharp rise in cost of diabetes nationwide, Report released February 7, 2003, www.diabetes.org/for-media/2003-press releases/02-27-03.jsp

Azmian, BR, JJ Davis, KS Coleman, CB Bagshaw and MLH Green, *Journal of the American Chemical Society*, 2002, **124**, 12664–12665.

Blake, DA, RM Jones, RC Blake, AR Pavlov, IA Darwish and H Yu, *Biosenors & Bioelectronics*, 2001, **16**, 79.

Butala, HD, A Ramakrishnan and A Sadana, *BioSystems*, 2003, **70**, 235.

Butala, HD, A Ramakrishnan and A Sadana, *Sensors & Actuators B*, 2003, **88**, 266.

Caelen, I, A Kalman and L Wahlstrom, *Analytical Chemistry*, 2004, **76**, 137.

Cai, Q, K Zeng, C Ruan, TA Desai and CA Grimes, *Analytical Chemistry*, 2004, **76**, 4038–4043.

Caselunghe, MB and J Lindeberg, *Journal of Food Chemistry*, 2000, **70**, 523.

Chen, CT and WP Huang, *Journal of the American Chemical Society*, 2002, **124**, 6246.

Corel Quattro Pro 8.0, Corel Corporation Limited, Ottawa, Canada, 1997.

Cuenoud, B and JW Szostak, *Nature*, 1995, **375**, 611.

Fernandez-Sanchez, C, CJ McNeil, K Rawson and O Nilsson, *Analytical Chemistry*, 2004, **76**, 5649–5656.

Fernandez-Sanchez, C, AM Gallardo-Soto, K Rawson, O Nilsson and CJ McNeill, *Biochemistry Communications*, 2004, **6**, 138–143.

Gerstein, M, AM Lesk and C Chothia, *Biochemistry*, 1994, **33**, 6739–6749.

Godwin, HA, *Current Opinion in Chemical Biology*, 2001, **5**, 223.

Grace, TA and E Stenberg, *Cereal Foods World*, 2002, **47**, 7.

Haines, J, F Aulents and P Patel, *Leatherhead, LFRA*, 2001, 1–13.

Havlin, S, *Molecular diffusion and reaction*. The Fractal Approach to Heterogeneous Chemistry: Surfaces, Colloids, Polymers, pp. 251–269, Wiley, New York, 1989.

Henry, C, *Analytical Chemistry*, 1998, **70**, 549A.

Hsieh, HV, ZA Pfieffer, TA Amiss, DB Sherman and JB Pitner, *Biosensors & Bioelectronics*, 2004, **19**, 653–660.

Indyk, HE, EA Evans, MB Caselunghe, BS Persson, PM Finglas, DC Woolard and EI Filonzi, *Journal of AOAC International*, 2000, **83**, 1141.

Landis, SH, T Murray and S Bolden, *Cancer Journal for Clinicians*, 1999, **49**, 8–31.

Lee, CK and SL Lee, *Surface Science*, 1995, **325**, 294.

Leegsma-Vogt, G, MM Rhemrev-Boom, RG Tiessen, K Venema and J Korf, The potential of biosensor technology in clinical monitoring and experimental research. *Bio-Medical Materials and Engineering*, 2004, **14**, 455–464.

Lemiere, J, DV Gool and R Dom, *Acta Neurology (Belgium)*, 1999, **99**, 96.

Lenigk, R, E Lam, A Lai, H Wang, Y Han, P Carlier and R Renneberg, Enzyme biosensor for studying therapeutics of Alzheimer's disease. *Biosensors & Bioelectronics*, 2000, **15**(9–10), 541–547.

Li, J and Y Lu, *Journal of the American Chemical Society*, 2000, **122**, 10466–10467.

Li, J, HT Ng, A Cassell, W Fan, H Chen, Q Ye, J Koehne, J Han and M Meyyappan, *Nano Letters*, 2003, **3**, 597–602.

Lin, Y, F Lu, Y Tu and Z Ren, *Nano Letters*, 2004, **4**(2), 191–195.

Liu, J and Y Lu, *Analytical Chemistry*, 2003, **75**, 6666.

Lofas, S and B Johnsson, *Journal of the Chemical Society, Chemical Communications*, 1990, 1526.

Lu, Y, *Chemical European Journal*, 2002, **8**, 4588.
Martin, JS, GC Frye, AJ Ricco and AD Senturia, *Analytical Chemistry*, 1993, **65**(20), 2910.
Merril, AM and DB McCormick, *Methods Enzymology*, 1988, **168**, 216.
Mitchell, K, *Analytical Chemistry*, 2004, **76**, 1098.
Needleman, HL, *Human Lead Exposure*. CRC Press, Boca Raton, Fl, 2002.
Nguyen, CV, L Delzeit, AM Cassell, J Li, J Han and M Meyyapan, *Nano Letters*, 2002, **2**, 1079–1081.
Panniek, J, HG Rittenhouse, DW Chan, JJ Epstein, PC Walsh and AW Partin, *Journal of Urology*, 1998, **159**, 1238–1242.
Pei, J, F Tian and T Thundat, *Analytical Chemistry*, 2004, **76**(2), 292.
Pellerin, L and PJ Magistretti, Exicatory amino acids stimulate aerobic glycolysis in astrocytes via an activation of the Na^+/K^+ ATPase. *Developments in Neuroscience*, 1996, **18**, 336–342.
Ramanathan, K, BR Jonnson and BD Danielsson, *Analytica Chimica Acta*, 2001, **427**, 1.
Rhemrev-Boom, R, Biosensor device and ultrafiltration sampling for coninuous invivo monitoring of glucose. *Biocybernetics and Biomedical Engineering*, 1999, **19**, 97–104.
Sadana, A, *Journal of Colloid & Interface Science*, 2003, **263**, 420.
Salins, LLE, RA Ware, CM Ensor and S Daunert, *Analytical Biochemistry*, 2001, **294**, 19–26.
Sen, D and CR Geyer, *Current Opinion in Chemical Biology*, 1998, **2**, 680.
Sensors and Sensor Networks (Sensors), Program Solicitation, NSF 04-522, National Science Foundation, Arlington, Virginia.
Shim, M, NWS Kam, RJ Chen, YM Li and HJ Dai, *Nano Letters*, 2002, **2**, 285–288.
Sorenson, CM and GC Roberts, *Journal of Colloid & Interface Science*, 1997, **186**, 447.
Sotiropoulou, S and NA Chaniotakis, *Analytical and Bioanalytical Chemistry*, 2003, **375**, 103–105.
Telting-Diaz, M and E Bakker, Mass-produced ionophore-based fluorescent microspheres for trace level determination of lead ions. *Analytical Chemistry*, 2002, **74** (20), 5251–5256.
Thundat, T, RJ Warmack, GY Chen and DP Allison, *Applied Physics Letters*, 1994, **64**, 2894.
Thundat, T, EA Wachter, SL Sharp and RJ Warmack, *Applied Physics Letters*, 1995, **66**, 1685.
Tu, Y, ZP Huang, DZ Wang, JG Wen and ZF Ren, *Applied Physics Letters*, 2002, **80**, 4018.
Tu, Y, Y Lin and ZF Ren, *Nano Letters*, 2003, **3**, 107–109.
Yonzon, CR, CL Haynes, X Zhang, T Joseph Jr., and RP van Duyne, A glucose biosensor based on surface-enhanced Raman scattering: improved partition layer, temporal stability, reversibility, and resistance to serum protein interference. *Analytical Chemistry*, 2004, **76**, 78–85.
Yu, X, D Chattopadhay, I Galeska, F Papadimitriakopoulos and JF Rusling, *Electrochemical Communications*, 2003, **5**, 408–411.
Zou, JY, MM Flocco and SL Mowbray, *Journal of Molecular Biology*, 1993, **233**, 739–752.
Zukin, RS, PR Hartig and DE Koshland, Jr, *Proceedings of the National Academy of Sciences USA*, 1997, **74**, 1932–1936.

– 7 –

Fractal Analysis of Human Heart Fatty Acid Binding Protein Binding Kinetics by a Precipitate Enhanced Immunoassay (PEIA)

7.1 INTRODUCTION

In this chapter we deviate from the concept of biosensors, and provide an example for the detection of fatty acid binding protein (FABP) using precipitate-enhanced immunoassay (PEIA) (Speijer *et al.*, 2004). This also permits one to show some variation in the analytical procedures, besides biosensors (the binding kinetics on which) may be analyzed using fractal analysis. FABP is an early marker of ischemic heart disease. Speijer *et al.* (2004) indicate that a standard enzyme immunoassay (EIA) is a sandwich-type procedure. In this procedure, the analyte is sandwiched between a surface-bound catcher antibody, and a tagging antibody conjugated to an enzyme. This enzyme is generally horse-radish peroxidase (HRP). The amount of an analyte captured by the HRP is made quantitative by measuring the color formation on the addition of a chromogenic HRP substrate. Speijer *et al.* (2004) have mentioned the disadvantages of the above procedure which include long assay times and variability in the assay results.

Robers *et al.* (1999) introduced the PEIA procedure. The sandwich procedure of the EIA is followed with the exception that instead of using a chromogenic substance, a precipitate-forming substrate is used. Subsequently, the rate of precipitate formation is measured by ellipsometry. Speijer *et al.* (2004) emphasize that the rate of precipitate formation on a reflecting substrate permits assays of high sensitivity and short residence times.

One-step assays are particularly useful in biological samples. Glatz *et al.* (2002) indicate that FABP is one of the earliest markers of acute myocardial injury. Speijer *et al.* (2004) emphasize that a short assay time for the detection of FABP in a one-step format in plasma samples would be of considerable assistance.

Figure 7.1 shows the schematic diagram of the ellipsometer (Speijer *et al.*, 2004). These authors indicate that the core of their system is the position of the cuvette with the reflecting slide. The first step is the formation of the catcher antibody–antigen conjugate

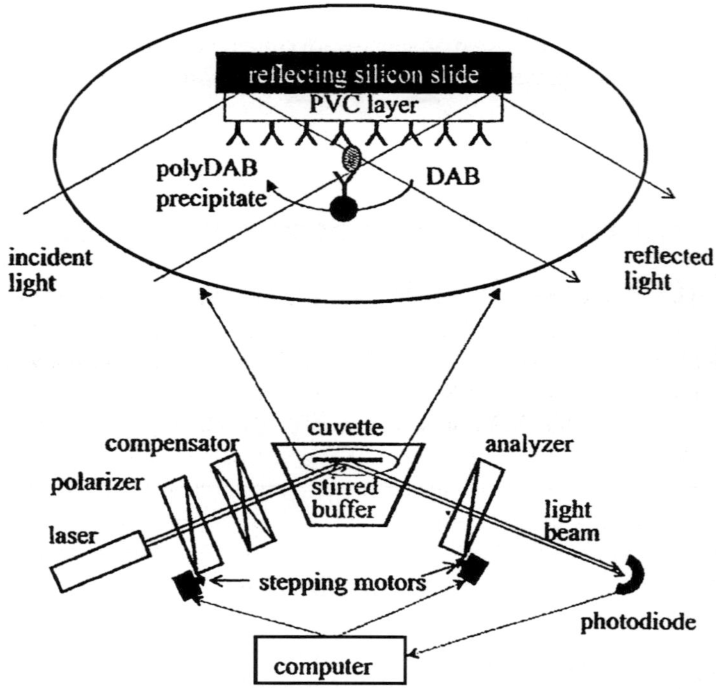

Figure 7.1 Schematic diagram of the ellipsometer (Speijer *et al.*, 2004).

sandwich on the slide surface. Then, the precipitation of the DAB (3,3′-idaminobenzidine tetrahydrochloride)-derived polymer is measured ellipsometrically.

Speijer *et al.* (2004) indicate that the resulting precipitation curves are S-shaped. There is an initial slow increase, followed by a DAB-accelerated phase, and a 'slow-down' phase when the substrate is consumed. Finally, when the substrate is completely consumed the precipitation stops. They used third-degree polynomial curves to fit the data and make quantitative comparisons of the different experimental conditions used. In this chapter we will re-analyze their data using the fractal analysis method presented in the previous chapters, the S-shaped exhibited by the binding curves notwithstanding.

Just like the biosensor examples presented in the previous chapters, here too for the precipitation-enhanced immunoassay, the analyte (FABP) in solution mixed with HRP-labeled anti-FABP tagging antibody has to diffuse through the solution to react with the heterogeneously immobilized FABP antibody on a silicon slide. DAB and H_2O_2 (hydrogen peroxide) were added at time, $t = 0$ to facilitate the precipitation and accurate measurement. The final concentrations were 0.003% hydrogen peroxide and 0.1 mg/ml DAB. As indicated earlier, the FABP antibodies are heterogeneously immobilized on the silicon slide. Thus, a fractal analysis will be used to characterize this heterogeneity and to analyze the kinetics of DAB-enhanced PEIA.

In this chapter, we present an alternate analysis for the binding of different concentrations of FABP in solution to an FABP antibody coated slide. Reproducibility studies of the binding of FABP in solution to FABP antibody coated slides are also presented. Finally, the binding of FABP in solution to FABP antibody coated slides using a one-step procedure and a two-step procedure are compared. In no way are we indicating that the fractal analysis presented is any better then the original analysis presented by Speijer *et al.* (2004). The fractal analysis is presented to provide an alternate analysis to help provide a fresh perspective.

7.2 THEORY

Havlin (1989) has reviewed and analyzed the diffusion of reactants towards fractal surfaces. The details of the theory and the equations involved for the binding and the dissociation phases for analyte–receptor (in our case, FABP in solution to FABP antibody) are available (Ramakrishnan and Sadana, 2001; Sadana and Vo-Dinh, 2001). Here we will attempt to apply these equations to the FABP in solution to FABP antibody interaction.

The basic idea is to fold everything into the lumped parameter, the fractal dimension, D_f. This lumped parameter takes care of the chemical, orientational, and geometrical details. It is also assumed that the kinetic rate depends only on the geometric (static) parameter, D_f. It would also be useful to indicate the low and high limits in angstroms of the fractal character to be presented.

7.2.1 Single-fractal analysis

Binding rate coefficient

Havlin (1989) indicates that the diffusion of a particle (analyte) from a homogeneous solution to a solid surface (e.g. receptor-coated surface) on which it reacts to form a product (analyte–receptor) complex is given by:

$$(\text{Analyte}\cdot\text{Receptor}) \sim \begin{cases} t^{(3-D_{f,\text{bind}})/2} = t^p & (t < t_c) \\ t^{1/2} & (t > t_c) \end{cases} \quad (7.1a)$$

where the analyte–receptor represents the association complex formed on the surface. Here $p = -b$, and D_f is the fractal dimension of the surface. Havlin (1989) states that the crossover value may be determined by $r_c^2 \sim t_c$. Above the characteristic length, r_c, the self-similarity of the surface is lost and the surface may be considered homogeneous. Eq. 7.1a indicates that the concentration of the product [analyte–receptor] on a solid fractal surface scales at short and intermediate times as analyte–receptor $\sim t^p$ with the coefficient $p = (3 - D_f)/2$ at short time scales and $p = 1/2$ at intermediate time scales. Note that D_f, $D_{f,\text{assoc}}$, and $D_{f,\text{bind}}$ are used interchangeably. This equation is associated with the short-term diffusional properties of a random walk on a fractal surface. Note that, in perfectly stirred kinetics on a regular (nonfractal) structure (or surface), k_1 is a constant, that is, it is independent of time. In other words, the limit of regular structures (or surfaces) and the absence of diffusion-limited kinetics leads to k_1 being independent of time. In all other situations, one would expect a scaling behavior given by $k_1 \sim k' t^{-b}$ with $-b = p < 0$. Also, the appearance of the coefficient, p different from $p = 0$ is the consequence of two different phenomena, that is, the heterogeneity (fractality) of the surface and the imperfect mixing (diffusion-limited) condition. Finally, for a homogeneous surface where D_f is equal to 2, and when only diffusional limitations are present, $p = 1/2$ as it should be. Another way of looking at the $p = 1/2$ case (where $D_{f,bind}$ is equal to two) is that the analyte in solution views the fractal object, in our case the receptor-coated biosensor surface, from a 'large distance.' In essence, in the association process, the diffusion of the analyte from the solution to the receptor surface creates a depletion layer of width $(Đt)^{1/2}$ where $Đ$ is the diffusion constant. This gives rise to the fractal power law, (Analyte–Receptor) $\sim t^{(3-D_{f,bind})/2}$.

The values of the parameters k (association or binding rate coefficient), p, and D_f in eq. 7.1a may be obtained for analyte–receptor binding kinetics data. This may be done by a regression analysis using, for example, Quattro Pro (1997) along with eq. 7.1a where (analyte–receptor) $= kt^p$. The fractal dimension may be obtained from the parameter p. Higher values of the fractal dimension would indicate higher degrees of 'disorder' or heterogeneity or inhomogeneity on the surface.

Another way of looking at the diffusive process is that it inherently involves fluctuations at the molecular level that may be described by a random walk (Weiss, 1994). This author indicates that the kinetics of transport on disordered (or heterogeneous) media itself needs to be described by a random-walk model. When both of these are present, that is the diffusion phenomena as well as a fractal surface, then one needs to analyze the interplay of both of these fluctuations. In essence, the disorder on the surface (or a higher fractal dimension, D_f) tends to slow down the motion of a particle (analyte in our case) moving in such a medium. Basically, according to Weiss (Weiss, 1990) the particle (random walker analyte) is trapped in regions in space, as it oscillates for a long time before resuming its motion.

Havlin (1989) indicates that the crossover value may be determined by $r_c^2 \sim t_c$. Above the characteristic length, r_c, the self-similarity of the surface is lost. For time, t greater than t_c the surface may be considered homogeneous, and 'regular' diffusion is now present. One may consider the analysis to be presented as an intermediate 'heuristic' approach in that in the future one may also be able to develop an autonomous (and not time-dependent) model of diffusion-limited kinetics in disordered media. No evidence of fractality is presented.

7.2.2 Dual-fractal analysis

Binding rate coefficient

The single-fractal analysis we have just presented is extended to include two fractal dimensions. At present, the time ($t = t_1$) at which the first fractal dimension 'changes' to the second fractal dimension is arbitrary and empirical. For the major part it is dictated by the data analyzed and the experience gained by handling a single-fractal analysis. The r^2 (regression coefficient) value obtained is also used to determine if a single-fractal analysis is sufficient, or one needs to use a dual-fractal analysis to provide an adequate fit. Only if the r^2 value is less than 0.97 for a single-fractal analysis, do we use a dual-fractal model.

In this case, the analyte–receptor complex is given by:

$$(\text{Analyte}\cdot\text{Receptor}) \sim \begin{cases} t^{(3-D_{\mathrm{f1,bind}})/2} = t^{p_1} & (t < t_1) \\ t^{(3-D_{\mathrm{f2,bind}})/2} = t^{p_2} & (t_1 < t < t_2 = t_c) \\ t^{1/2} & (t > t_c) \end{cases} \tag{7.1b}$$

In catalytic reactions the reactant binds to the active site on the surface and the product is released. In this sense, the catalytic surface exhibits an unchanging fractal surface to the reactant in the absence of fouling and other complications. In the case of analyte–receptor association the biosensor surface exhibits a changing fractal surface to the analyte in solution. This occurs since as each association (binding) reaction takes place, smaller and smaller amounts of 'association' sites or receptors are available on the biosensor surface to which the analyte may bind. Furthermore, as the reaction proceeds, there is an increasing degree of heterogeneity on the biosensor surface for some reaction systems. This is manifested by two degrees of heterogeneity, or two fractal dimensions on the biosensor surface. In the theoretical limit one might envisage a temporal fractal dimension wherein there is a continuous change in the degree of heterogeneity on the surface; though of course, such situations would be very rare, if at all.

Note that, at present, the dual-fractal analysis does not have a basis at the molecular level. This represents two different levels of heterogeneity on the biosensor surface. But, in some examples presented, a single-fractal analysis is clearly inadequate to model the data. Only in these cases does one resort to a dual-fractal analysis.

7.3 RESULTS

Speijer *et al.* (2004) have presented and analyzed data for the one-step PEIA for the binding of different concentrations of FABP in solution to FABP antibody immobilized on a silicon slide. The FABP coated slide was incubated for 10 min with a 10-fold diluted plasma containing HRP-labeled anti-FABP tagging antibody.

Figure 7.2a shows the binding of 1.25 nM FABP in solution to FABP antibody coated on a silicon slide. A single-fractal analysis is adequate to describe the binding kinetics. The values of the binding rate coefficient, k and the corresponding fractal dimension, D_f are given in Table 7.1. The values of the binding rate coefficients and the fractal dimensions presented in Table 7.1 were obtained from a regression analysis using Corel Quattro Pro 8.0 (1997) to model the data using eqs. 7.1a and 7.1b. The binding rate coefficients presented in Table 7.1 are within 95% confidence limits. For example, for the binding of 1.25 ng/ml FABP in solution to FABP antibody coated on a silicon slide the binding rate coefficient, k for a single-fractal analysis is 0.000778 ± 0.000048. The 95% confidence limit indicates that the k value lies between 0.00073 and 0.000826.

Figure 7.2b shows the binding of 2.5 ng/ml FABP in solution to FABP antibody coated on a silicon slide. A dual-fractal analysis is required to adequately describe the binding kinetics. The values of (a) the binding rate coefficient, k and the fractal dimension, D_f for a single-fractal analysis, and (b) the binding rate coefficients, k_1 and k_2 and the fractal dimensions, D_{f1} and D_{f2} for a dual-fractal analysis are given in Table 7.1. There is an

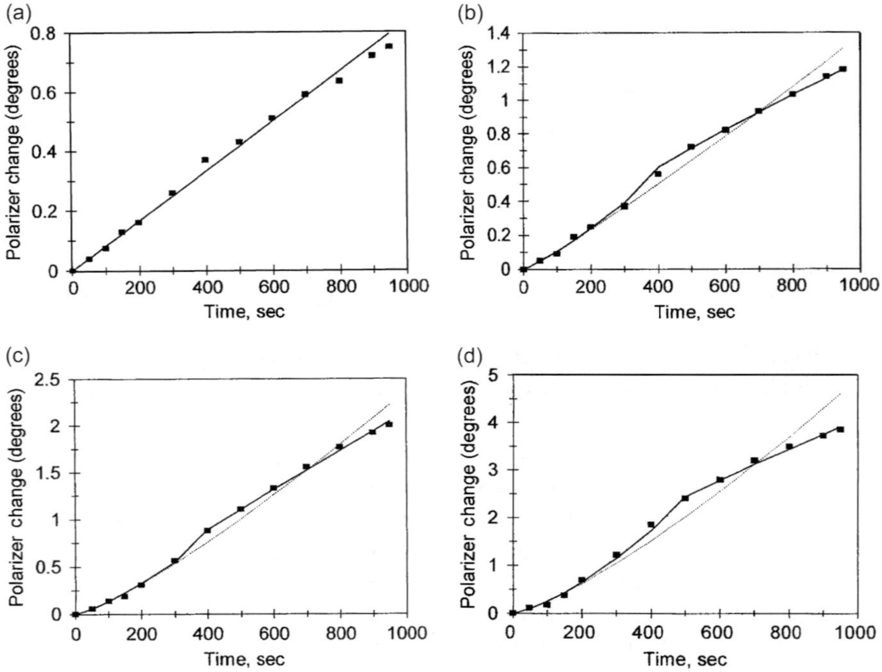

Figure 7.2 Binding of different concentrations of FABP in solution to FABP antibody coated on a silicon slide (One-step PEIA). FABP concentrations (in ng/ml): (a) 1.25, (b) 2.5, (c) 5, (d) 10.

apparent change in the binding mechanism as one goes from the 1.25 ng/ml FABP in solution to 2.5 ng/ml FABP in solution. This is because at the lower concentration (1.25 ng/ml) of FABP in solution a single-fractal analysis is adequate to describe the binding kinetics, whereas at the higher concentration (2.5 ng/ml) of FABP in solution a dual-fractal analysis is required to adequately describe the binding kinetics.

It is also of interest to note that for the 2.5 ng/ml FABP in solution, as the fractal dimension increases by a factor of 2.27 from a value of D_{f1} equal to 0.6352 to D_{f2} equal to 1.4396, the binding rate coefficient increases by a factor of 12.2 from a value of k_1 equal to 0.00046 to k_2 equal to 0.00561. Note that the changes in the fractal dimension (degree of heterogeneity on the surface) and in the binding rate coefficient are in the same direction.

Figure 7.2c shows the binding of 5.0 ng/ml FABP in solution to FABP antibody coated on a silicon slide. A dual-fractal analysis is required to adequately describe the binding kinetics. The values of (a) the binding rate coefficient, k and the fractal dimension, D_f for a single-fractal analysis, and (b) the binding rate coefficients, k_1 and k_2 and the fractal dimensions, D_{f1} and D_{f2} for a dual-fractal analysis are given in Table 7.1. It is also of interest to note that for the 5.0 ng/ml FABP in solution, as the fractal dimension increases by a factor of 2.50 from a value of D_{f1} equal to 0.4358 to D_{f2} equal to 1.0882, the binding rate coefficient increases by a factor of 7.89 from a value of k_1 equal to 0.000369 to k_2 equal to 0.00291. Note that the changes in the fractal dimension (degree of heterogeneity on the surface) and in the binding rate coefficient are once again in the same direction.

Table 7.1

Binding rate coefficient and fractal dimension values for the binding of fatty acid binding protein (FABP) to FABP antibody immobilized on a silicon slide (Speijer *et al.*, 2004). One-step precipitate-enhanced immunoassay (PEIA)

FABP concentration in solution (ng/ml)/ FABP antibody coated silicon slide	k	k_1	k_2	D_f	D_{f1}	D_{f2}
1.25	0.000778 ± 0.000048	na	na	0.9790 ± 0.0376	na	na
2.5	0.000678 ± 0.000071	0.00046 ± 0.000047	0.00561 ± 0.000032	0.7938 ± 0.0626	0.6352 ± 0.0972	1.4396 ± 0.0204
5.0	0.000465 ± 0.000046	0.000369 ± 0.000044	0.00291 ± 0.000056	0.5286 ± 0.0602	0.4358 ± 0.1358	1.0882 ± 0.0491
10.0	0.000828 ± 0.000138	0.000386 ± 0.000085	0.02551 ± 0.00049	0.432 ± 0.011	0.1942 + 0.2004	1.5324 ± 0.0686

Figure 7.2d shows the binding of 10.0 ng/ml FABP in solution to FABP antibody coated on a silicon slide. A dual-fractal analysis is required to adequately describe the binding kinetics. The values of (a) the binding rate coefficient, k and the fractal dimension, D_f for a single-fractal analysis, and (b) the binding rate coefficients, k_1 and k_2 and the fractal dimensions, D_{f1} and D_{f2} for a dual-fractal analysis are given in Table 7.1. It is also of interest to note that for the 10.0 ng/ml FABP in solution, as the fractal dimension increases by a factor of 7.89 from a value of D_{f1} equal to 0.1942 to D_{f2} equal to 1.5324, the binding rate coefficient increases by a factor of 66.1 from a value of k_1 equal to 0.000386 to k_2 equal to 0.0251. Note that the changes in the fractal dimension (degree of heterogeneity on the surface) and in the binding rate coefficient as noted above are in the same direction.

No clear trend is observable for the binding rate coefficients, either k_1 or k_2 as the FABP concentration in solution increases. Only three data points are available. Table 7.2 shows the increase in (a) the ratio of the fractal dimensions in the two phases of binding, D_{f2}/D_{f1} and in (b) the ratio of the binding rate coefficients, k_2/k_1 as the FABP concentration in solution increases. The trend is clearer for the fractal dimension ratio, and not so clear for the binding rate coefficient ratio.

Consider presumably that the binding mechanism comprises of two steps: (a) an initial contact or docking mechanism (of the analyte and the receptor), followed by (b) a rearrangement (making and breaking of bonds, and attainment of the lowest free energy). Then, an increase in the FABP concentration in solution favors step (b) relatively more than step (a) at least in this concentration range of 2.5–10 ng/ml FABP analyzed.

From the data in Table 7.2 and Figure 7.3a, one notes the increase in the ratio, D_{f2}/D_{f1} with an increase in the FABP concentration (in ng/ml) in solution. For the data presented in Table 7.2 and in Figure 7.3a, the ratio of the fractal dimensions is given by:

$$D_{f2}/D_{f1} = (0.836 \pm 0.448)[\mathrm{FABP}]^{0.8986 \pm 0.4384} \qquad (7.2a)$$

The fit is quite good. Only three data points are available. More data points are required to provide a better fit. The ratio of fractal dimensions exhibits an order of dependence (equal to 0.8986) less than first order on the FABP concentration in solution.

From the data in Table 7.2 and Figure 7.3b, one can note the increase in the binding rate coefficient ratio, k_2/k_1 with an increase in the fractal dimension ratio, D_{f2}/D_{f1}. For the data presented in Table 7.2 and in Figure 7.3b, the binding rate coefficient ratio is given by:

$$k_2/k_1 = (2.556 \pm 1.318)(D_{f2}/D_{f1})^{1.5633 \pm 0.4243} \qquad (7.2b)$$

Table 7.2

Increase in the ratio of the fractal dimensions, D_{f2}/D_{f1} and in the binding rate coefficients, k_2/k_1 with an increase in the FABP concentration in solution

FABP concentration (ng/ml)	D_{f2}/D_{f1}	k_2/k_1
2.5	2.27	12.2
5.0	2.50	7.89
10	7.89	66.1

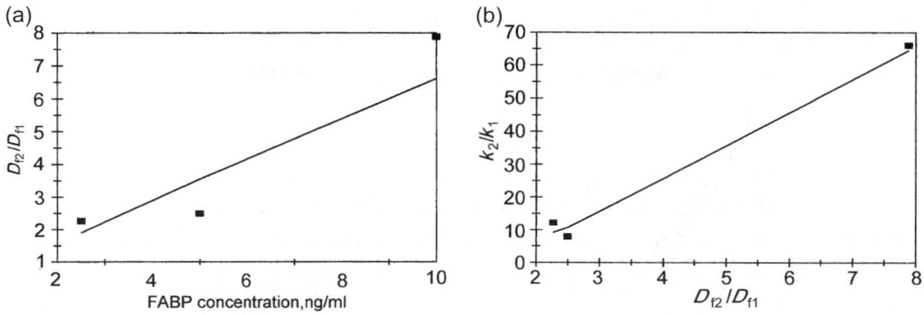

Figure 7.3 (a) Increase in the fractal dimension ratio, D_{f2}/D_{f1} with an increase in the FABP concentration (in ng/ml) in solution. (b) Increase in the ratio of the binding rate coefficients, k_2/k_1 with an increase in the FABP concentration (in ng/ml) in solution.

The fit is good. Only three data points are available. More data points are required to provide a better fit. The ratio of binding rate coefficients is quite sensitive to the ratio of fractal dimensions, as noted by the order of dependence (equal to 1.5633) between first and second order.

Speijer *et al.* (2004) wanted to analyze the reproducibility of the one-step precipitate immunoassay. Figure 7.4a,b show the binding of 0 ng/ml FABP in solution to FABP antibody immobilized on a silicon slide. A dual-fractal analysis is used to adequately model the binding kinetics in both cases. The values of (a) the binding rate coefficient, k and the fractal dimension, D_f for a single-fractal analysis, and (b) the binding rate coefficients, k_1 and k_2 and the fractal dimensions, D_{f1} and D_{f2} for a dual-fractal analysis are given in Table 7.3. Note that increases in the fractal dimension from D_{f1} to D_{f2} lead to increases in the binding rate coefficient, k_1 to k_2. For example, for the binding of 0 ng/ml FABP in solution an increase in the fractal dimension by a factor of 1.48 from a value of D_{f1} equal to 1.4848 to D_{f2} equal to 2.1946 leads to an increase in the binding rate coefficient by a factor of 8.94 from a value of k_1 equal to 0.003357 to k_2 equal to 0.03.

Similarly, for the data presented in the second row of Table 7.3, and for 0 ng/ml FABP in solution, an increase in the fractal dimension by a factor of 1.742 from a value of D_{f1} equal to 1.4732 to D_{f2} equal to 2.5668 leads to an increase in the binding rate coefficient by a factor of 8.24 from a value of k_1 equal to 0.002463 to k_2 equal to 0.0203. The numbers provided do indicate an idea of the reproducibility in the kinetic parameters involved. However, clearly there is a significant difference in the values of the kinetic parameters obtained. For example, the values of k_1 reported are 0.02463 and 0.00357. Similarly, the values of k_2 reported are 0.0203 and 0.030. Surely, more experiments need to be performed in order that one may obtain more reproducible kinetic parameters and results. These results are consistent with the suggestions made by Speijer *et al.* (2004).

Figure 7.4c,d shows the binding of 2 ng/ml FABP in solution to FABP antibody immobilized on a silicon slide. A dual-fractal analysis is used to adequately model the binding kinetics in both cases. The values of (a) the binding rate coefficient, k and the fractal dimension, D_f for a single-fractal analysis, and (b) the binding rate coefficients, k_1 and k_2 and the fractal dimensions, D_{f1} and D_{f2} for a dual-fractal analysis are given in Table 7.3. Note that, once again, increases in the fractal dimension from D_{f1} to D_{f2} lead

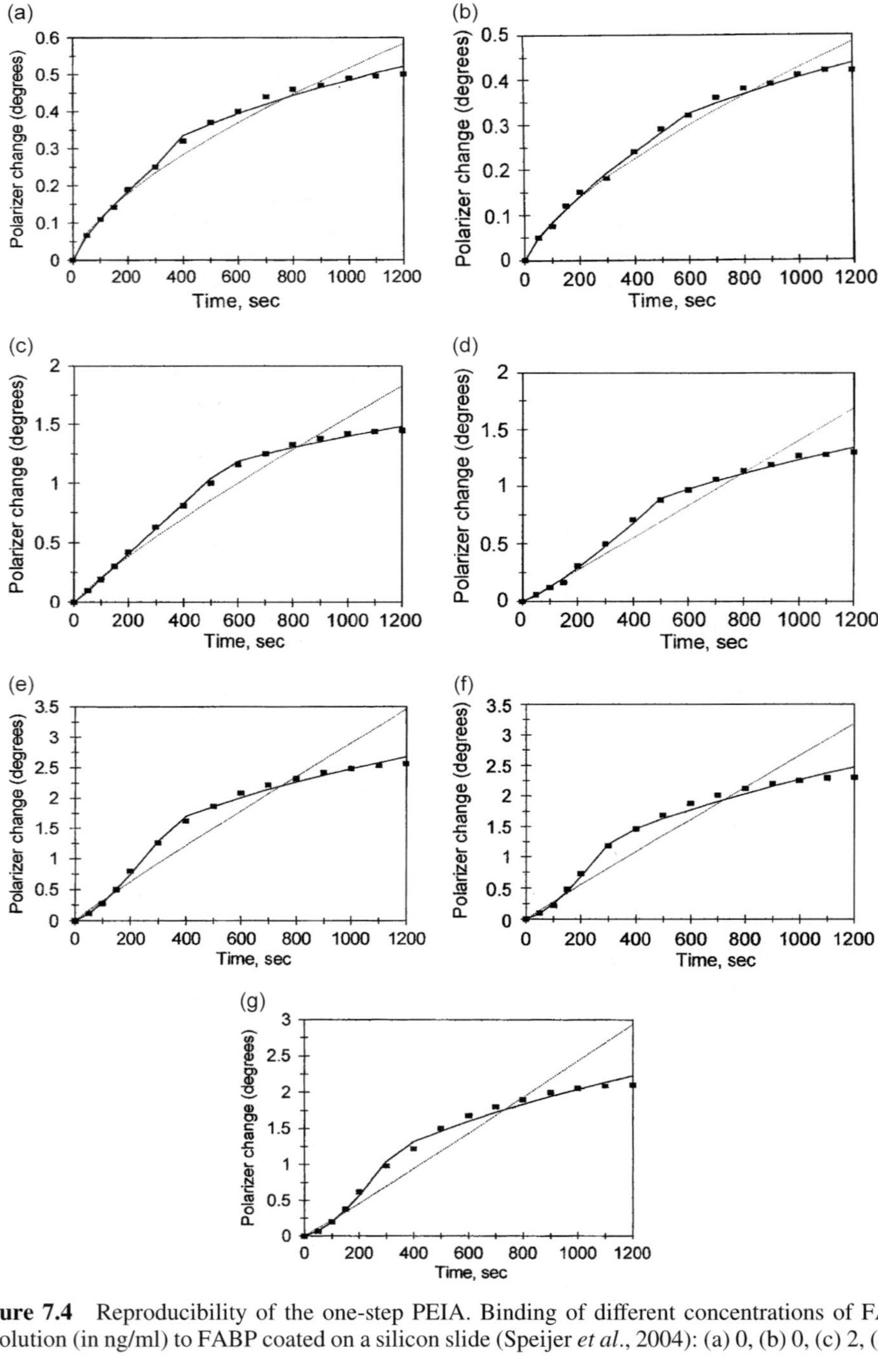

Figure 7.4 Reproducibility of the one-step PEIA. Binding of different concentrations of FABP in solution (in ng/ml) to FABP coated on a silicon slide (Speijer *et al.*, 2004): (a) 0, (b) 0, (c) 2, (d) 2, (e) 4, (f) 4, (g) 4.

Table 7.3

Binding rate coefficient and fractal dimension values for the binding of FABP in solution to FABP antibody immobilized on a silicon slide (Speijer *et al.*, 2004). Reproducibility of the one-step precipitate-enhanced immunoassay (PEIA)

FABP concentration in solution (ng/ml)/ antibody coated silicon slide	CV%	k	k_1	k_2	D_f	D_{f1}	D_{f2}
0	12	0.005571 ± 0.000528	0.003357 ± 0.000095	0.0300 ± 0.0011	1.6878 ± 0.0512	1.4848 ± 0.0328	2.1946 ± 0.0656
0	12	0.003399 ± 0.00029	0.002463 ± 0.000158	0.0203 ± 0.0006	1.6006 ± 0.0512	1.4732 ± 0.0550	2.5668 ± 0.07746
2	6.5	0.001248 ± 0.000246	0.000469 ± 0.000058	0.0507 ± 0.0012	0.9656 ± 0.1018	0.5726 ± 0.1184	2.0756 ± 0.0566
2	6.5	0.003898 ± 0.000570	0.00176 ± 0.00006	0.1485 ± 0.003	1.2650 ± 0.0772	0.9468 ± 0.0414	2.3502 ± 0.0676
4	8.5	0.001935 ± 0.000566	0.00020 ± 0.000013	0.07598 ± 0.00382	0.9333 ± 0.1451	0 + 0.0938	2.0464 ± 0.0936
4	8.5	0.00414 ± 0.00105	0.000624 ± 0.000036	0.1455 ± 0.0050	1.1020 ± 0.1288	0.3246 ± 0.0796	2.1792 ± 0.0648
4	8.5	0.003212 ± 0.000931	0.000362 ± 0.00004	0.0841 ± 0.0045	1.0536 ± 0.1440	0.1530 ± 0.1520	2.0462 ± 0.0762

to increases in the binding rate coefficient, k_1 to k_2. For example, for the binding of 2 ng/ml FABP in solution an increase in the fractal dimension by a factor of 3.573 from a value of D_{f1} equal to 0.5726 to D_{f2} equal to 2.0756 leads to an increase in the binding rate coefficient by a factor of 108.1 from a value of k_1 equal to 0.000469 to k_2 equal to 0.0507.

Similarly, for the data presented in the fourth row of Table 7.3, and for 2 ng/ml FABP in solution, an increase in the fractal dimension by a factor of 2.482 from a value of D_{f1} equal to 0.9468 to D_{f2} equal to 2.3502 leads to an increase in the binding rate coefficient by a factor of 84.375 from a value of k_1 equal to 0.00176 to k_2 equal to 0.1485. The numbers provided in rows three and four of Table 7.3, once again, do indicate an idea of the reproducibility in the kinetic parameters involved. As indicated previously, there are significant differences in the values of (a) the binding rate coefficient values, and in (b) the fractal dimension values. Once again, it may be reiterated that more effort needs to be placed to run experiments that yield reproducible results and kinetic parameters that are closer to each other under similar or identical operating conditions.

Figure 7.4e–g shows the binding of 4 ng/ml FABP in solution to FABP antibody immobilized on a silicon slide. A dual-fractal analysis is used to adequately model the binding kinetics in all three cases. The values of (a) the binding rate coefficient, k and the fractal dimension, D_f for a single-fractal analysis, and (b) the binding rate coefficients, k_1 and k_2 and the fractal dimensions, D_{f1} and D_{f2} for a dual-fractal analysis are given in Table 7.3. Note that increases in the fractal dimension from D_{f1} to D_{f2} lead to increases in the binding rate coefficient, k_1 to k_2. For example, for the binding of 4 ng/ml FABP in solution an increase in the fractal dimension by a factor of 21.82 from a value of D_{f1} equal to 0.0938 (error; actual estimated value is zero) to D_{f2} equal to 2.0464 leads to an increase in the binding rate coefficient by a factor of 379.9 from a value of k_1 equal to 0.00020 to k_2 equal to 0.07598.

Similarly, for the data presented in the sixth row of Table 7.3, and for 4 ng/ml FABP in solution, an increase in the fractal dimension by a factor of 6.71 from a value of D_{f1} equal to 0.3246 to D_{f2} equal to 2.1792 leads to an increase in the binding rate coefficient by a factor of 233.3 from a value of k_1 equal to 0.000624 to k_2 equal to 0.1455. The numbers shown, as indicated above do provide an idea of the reproducibility in the kinetic parameters involved.

Also, for the data presented in the seventh row of Table 7.3, and for 4 ng/ml FABP in solution, an increase in the fractal dimension by a factor of 13.37 from a value of D_{f1} equal to 0.1530 to D_{f2} equal to 2.0462 leads to an increase in the binding rate coefficient by a factor of 232.3 from a value of k_1 equal to 0.000362 to k_2 equal to 0.0841. The numbers provided, once again, do indicate an idea of the reproducibility in the kinetic parameters involved.

For the data presented in Table 7.3 and for 4 ng/ml FABP in solution Figure 7.5a shows the linear increase in the binding rate coefficient, k_1 with an increase in the fractal dimension, D_{f1}. Apparently, even at the same concentration of 4 ng/ml, these three runs exhibit different degrees of heterogeneity (fractal dimension, D_f values) on the biosensor surface. For the data presented in Table 7.3 and in Figure 7.5a the binding rate coefficient, k_1 is given by:

$$k_1 = (0.001784 \pm 0.00022)D_{f1}^{0.900 \pm 0.133} \tag{7.3a}$$

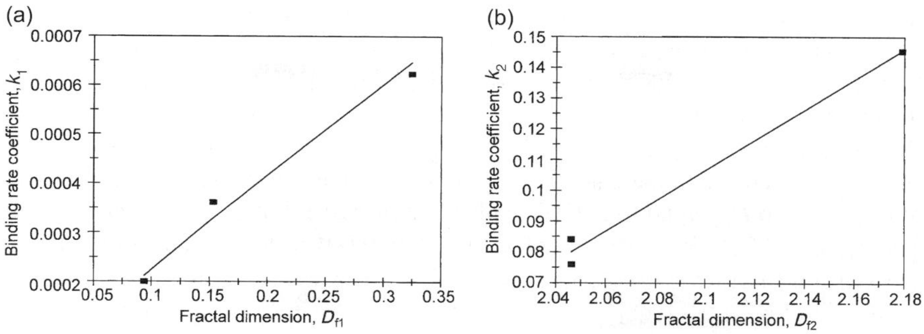

Figure 7.5 (a) Increase in the binding rate coefficient, k_1 with an increase in the fractal dimension, D_{f1} (data for 4 ng/ml FABP only considered). (b) Increase in the binding rate coefficient, k_2 with an increase in the fractal dimension, D_{f2} (data for 4 ng/ml FABP only considered).

The fit is very good. Only three data points are available. The availability of more data points would lead to a more reliable and better fit. The close to first-order (equal to 0.900) dependence exhibited by the binding rate coefficient, k_1 on the fractal dimension, D_{f1} explains the linear increase in the binding rate coefficient with an increase in the degree of heterogeneity on the silicon slide surface.

For the data presented in Table 7.3 and for 4 ng/ml FABP in solution Figure 7.5b shows the increase in the binding rate coefficient, k_2 with an increase in the fractal dimension, D_{f2}. As indicated above, even at the same concentration of 4 ng/ml, these three runs exhibit different degrees of heterogeneity in the second phase of binding (fractal dimension, D_{f2} values) on the silicon slide surface. For the data presented in Table 7.3 and in Figure 7.5b the binding rate coefficient, k_2 is given by:

$$k_2 = (8.8 \times 10^{-5} \pm 0.7 \times 10^{-5}) D_{f2}^{9.514 \pm 1.413} \tag{7.3b}$$

Once again, the fit is very good, even though two points are very close to each other (D_{f2} equal to 2.0462 and 2.0464). This just reinforces the reproducibility of the experiments and the fractal analysis procedure. Once again, only three data points are available. The availability of more data points would lead to a more reliable and better fit. The binding rate coefficient, k_2 is extremely sensitive to the degree of heterogeneity on the biosensor surface (fractal dimension, D_{f2}) as noted by the very high order of dependence (equal to 9.514).

Only three data points were presented in Figures 7.5a,b for 4 ng/ml FABP in solution since only three data points were available. In order to obtain a better and more reliable fit, more data points are required (as indicated earlier). One way of accomplishing this is to combine the data points presented in Table 7.3 (two for 2 ng/ml FABP in solution and three for 4 ng/ml FABP in solution).

For the data presented in Table 7.3 and for 2 and 4 ng/ml FABP in solution Figure 7.6a shows an increase in the binding rate coefficient, k_1 with an increase in the fractal dimension, D_{f1}. For the data presented in Table 7.3 and in Figure 7.6a the binding rate coefficient, k_1 is given by:

$$k_1 = (0.00129 \pm 0.00066) D_{f1}^{0.762 \pm 0.220} \tag{7.4a}$$

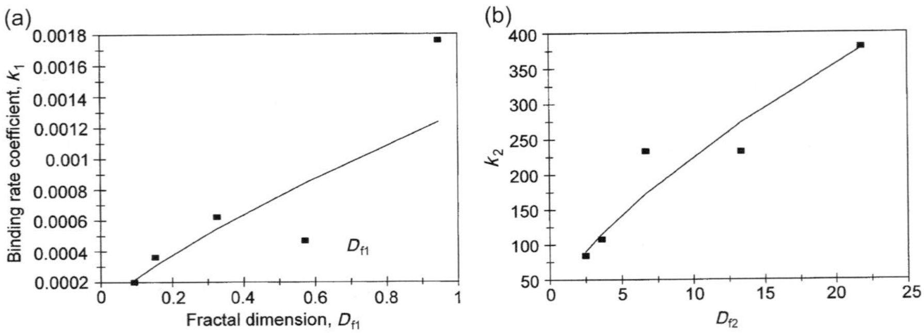

Figure 7.6 (a) Increase in the binding rate coefficient, k_1 with an increase in the fractal dimension, D_{f1} (data for 2 and 4 ng/ml FABP only considered). (b) Increase in the binding rate coefficient, k_2 with an increase in the fractal dimension, D_{f2} (data for 2 and 4 ng/ml FABP only considered).

There is some scatter in the data. Presumably, this is because data from two different data sets (2 and 4 ng/ml FABP in solution) have been used. The binding rate coefficient, k_1 exhibits an order of dependence that is less than first order (equal to 0.762). It is of interest to note that the order of dependence exhibited by the binding rate coefficient, k_1 on the fractal dimension, D_{f1} (equal to 0.762) is less than (equal to 0.900) when both data sets (2 and 4 ng/ml FABP in solution; five data points total) are used than when only a single data set (4 ng/ml FABP in solution; three data points total) is used.

For the data presented in Table 7.3 and for 2 and 4 ng/ml FABP in solution Figure 7.6b shows the increase in the binding rate coefficient, k_2 with an increase in the fractal dimension, D_{f2}. For the data presented in Table 7.3 and in Figure 7.6b the binding rate coefficient, k_2 is given by:

$$k_2 = (49.453 \pm 11.075)D_{f2}^{0.6599 \pm 0.1126} \tag{7.4b}$$

The fit is very good considering that data points from two different data sets are chosen. Surprisingly, when data points from the two different data sets are plotted together, the binding rate coefficient, k_2 exhibits only a mild order of dependence (equal to 0.6599) on the degree of heterogeneity that exists on the surface. No explanation is offered for this sharp drop in the order of dependence (equal to 9.514) when only the 4 ng/ml data set is used when compared to the order of dependence (equal to 0.699) exhibited in eq. 7.4b above when data from 2 and 4 ng/ml data sets are used.

Table 7.4 clearly shows the trend in D_{f2}/D_{f1} and in k_2/k_1 with an increase in the FABP concentration in solution. Here since we are analyzing reproducibility studies, averages of the D_{f2}/D_{f1} and k_2/k_1 ratios may be noted.

Speijer *et al.* (2004) wanted to compare the DAB precipitation rates in the one-step with the two-step PEIA. In the one-step PEIA procedure FABP-coated silicon slides were incubated for 20 min in buffer C (50 mM Tris-HCl, pH 7.5 with 0.1 M NaCl and 5 g/l skimmed milk powder) containing 5 ng/ml FABP and HRP-labeled anti-FABP tagging antibody. At time, $t = 0$ DAB and H_2O_2 were added without flushing. Figure 7.7a shows the binding of the 5 ng/ml FABP in solution to the antibody coated silicon slide. This is curve one in Table 7.5 and is the one-step PEIA. A dual-fractal analysis is required to adequately describe the binding kinetics. The values of (a) the binding rate coefficient, k and the fractal

Table 7.4

Increase in the ratio of the fractal dimensions, D_{f2}/D_{f1} and in the binding rate coefficients, k_2/k_1 with an increase in the FABP concentration in solution. Reproducibility studies

FABP concentration (ng/ml)	D_{f2}/D_{f1}	k_2/k_1	D_{f2}/D_{f1} average	k_2/k_1 average
0	1.48	8.94	1.61	8.59
0	1.742	8.24	1.61	8.59
2	3.573	108.1	3.028	96.24
2	2.482	84.38	3.028	96.24
4	21.82	379.9	13.96	281.8
4	6.71	233.3	13.96	281.8
4	13.37	232.3	13.96	281.8

dimension, D_f for a single-fractal analysis, and (b) the binding rate coefficients, k_1 and k_2 and the fractal dimensions, D_{f1} and D_{f2} for a dual-fractal analysis are given in Table 7.5.

Figure 7.7b shows the binding of the 5 ng/ml FABP in solution to the antibody coated silicon slide. The procedure is the same as in Figure 7.7a with the exception that the DAB and H_2O_2 were added with flushing. This is the two-step PEIA, and is curve number two in Table 7.5. Once again a dual-fractal analysis is adequate to describe the binding kinetics. The values of (a) the binding rate coefficient, k and the fractal dimension, D_f for

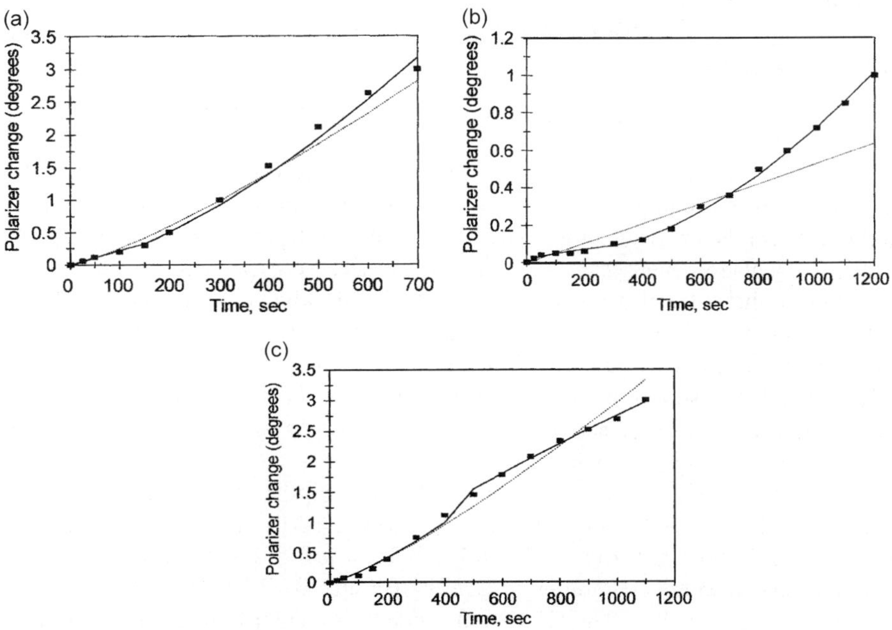

Figure 7.7 Binding of the fatty acid binding protein (FABP) in solution to a FABP-coated silicon slide. Comparison of the one-step procedure with the two-step procedure (Speijer *et al.*, 2004): (a) curve 1 (one-step procedure); (b) curve 2 (two-step procedure); (c) curve 3 (two-step procedure).

Table 7.5

Binding rate coefficient and fractal dimension values for the binding of FABP in solution to FABP-antibody coated silicon slides. Comparison of the one-step procedure (curve 1) with the two-step procedure (curve 2 and 3) (Speijer *et al.*, 2004)

Curve number	Analyte in solution/ receptor on surface: experimental condition	k	k_1	k_2	D_f	D_{f1}	D_{f2}
One	5 ng/ml FABP and HRP-labeled anti-FABP tagging antibody/FABP antibody coated silicon slide. DAB and H_2O_2 added without flushing	0.00087 ± 0.00018	0.00266 ± 0.000034	0.000214 ± 0.000021	0.5340 ± 0.1066	1.0766 ± 0.1416	0.0676 + 0.1184
Two	5 ng/ml FABP and HRP-labeled anti-FABP tagging antibody/FABP antibody coated silicon slide. DAB and H_2O_2 added without flushing	0.000469 ± 0.000321	0.00333 ± 0.00062	$1.5 \times 10^{-6} \pm 0.1 \times 10^{-6}$	0.9640 ± 0.180	1.8400 ± 0.1414	~0
Three	5 ng/ml FABP and HRP-labeled anti-FABP tagging antibody/FABP antibody coated silicon slide. Flushing followed by addition of a non-specific antibody (125 ng/ml HRP-labeled antibody against tissue-type plasiminogen activator). Thereafter, DAB and H_2O_2 added.	0.000641 ± 0.00012	0.000562 ± 0.00013	0.00938 ± 0.00017	0.5566 ± 0.0782	0.500 ± 0.1342	1.3550 ± 0.071

a single-fractal analysis, and (b) the binding rate coefficients, k_1 and k_2 and the fractal dimensions, D_{f1} and D_{f2} for a dual-fractal analysis are given in Table 7.5.

It is of interest to compare the binding rate coefficient and fractal dimension values of the one-step PEIA (curve number 1) with the two-step PEIA (curve number 2). As one goes from the one-step PEIA to the two-step PEIA, there is (a) an increase in the fractal dimension, D_{f1} by a factor of 1.71 from a value of 1.0766 to 1.84, and (b) to an increase in the binding rate coefficient, k_1 by a factor of 1.24 from a value of 0.00266 to 0.0033. However, there is (a) a decrease in the fractal dimension value in the second phase of binding, D_{f2} from a value of 0.0676 to approximately zero, and to (b) a decrease in the binding rate coefficient, k_2 by a factor of 142.67 from a value of 0.00021 to 1.5×10^{-6}. The binding rate coefficient, k_1 predominates and thus curve 1 (one-step PEIA) exhibits a higher rate of precipitation then curve 2 (two-step PEIA).

Figure 7.7c shows the binding of the 5 ng/ml FABP in solution to the antibody coated silicon slide. The procedure is the same as in Figure 7.7b with the exception that after the washing step an irrelevant antibody–HRP conjugate was added to the two-step assay. This is curve three in Table 7.5. The addition of the irrelevant antibody–HRP conjugate converted the DAB, which enhanced the precipitation. Once again a dual-fractal analysis is adequate to describe the binding kinetics. The values of (a) the binding rate coefficient, k and the fractal dimension, D_f for a single-fractal analysis, and (b) the binding rate coefficients, k_1 and k_2, and the fractal dimensions, D_{f1} and D_{f2} for a dual-fractal analysis are presented in Table 7.5.

It is of interest to compare the binding rate coefficient and fractal dimension values of the curves 2 (no addition of irrelevant antibody) and 3 (addition of irrelevant antibody). On comparing curves two and three one notes that there is (a) a decrease in the fractal dimension, D_{f1} by a factor of 3.68 from a value of 1.84 to 0.5; (b) a decrease in the binding rate coefficient, k_1 by a factor of 5.92 from a value of 0.0033 to 0.000562; (c) an increase in the fractal dimension value in the second phase of binding, D_{f2} from a value of approximately 0 to 1.3550; and (d) an increase in the binding rate coefficient, k_2 by a factor of 6253 from a value of 1.5×10^{-6} to 0.00938. Apparently, in this case the binding rate coefficient, k_2 predominates since the rate of precipitation is higher for curve three than it is for curve two.

For the data presented in Table 7.5 and in Figure 7.8 for the one-step and the two-step PEIA the binding rate coefficient, k_1 is given by:

$$k_1 = (0.001715 \pm 0.000875) D_{f1}^{1.411 \pm 0.444} \tag{7.5a}$$

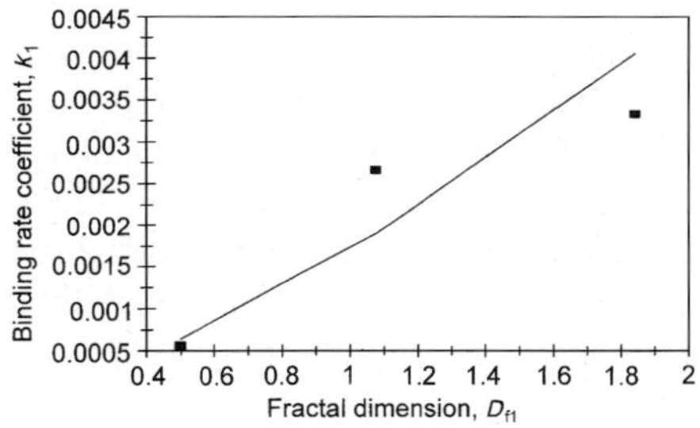

Figure 7.8 Increase in the binding rate coefficient, k_1 with an increase in the fractal dimension, D_{f1} or the degree of heterogeneity on the silicon slide.

The fit is very good. Only three data points are available. The availability of more data points would lead to a more reliable and better fit. The close to one and one-half-order (equal to 1.411) dependence exhibited by the binding rate coefficient, k_1 on the fractal dimension, D_{f1} explains the non-linearity in the binding rate coefficient with an increase in the degree of heterogeneity on the silicon slide surface.

7.4 CONCLUSION

A fractal analysis for the binding of FABP in solution mixed with HRP-labeled anti-FABP tagging antibody to FABP antibody immobilized on a silicon slide provides a quantitative indication of the state of disorder or the degree of heterogeneity or roughness on the silicon slide surface. A PEIA is used to provide quantitative measurements ellipsometrically (Speijer *et al.*, 2004). This is different from the biosensor examples analyzed, and it provides some variety besides indicating the flexibility of the fractal analysis to help understand the kinetics of non-biosensor type analytical or detection systems. The fractal analysis is offered as an alternate way to improve the understanding of the kinetics in the heterogeneous case with diffusion-limited reactions occurring on structured surfaces.

Data taken from the literature (Speijer *et al.*, 2004) is re-analyzed to include the degree of heterogeneity present on the surface and its influence on the binding kinetics. As indicated in the previous chapters, the fractal dimension provides a quantitative measure of the degree of heterogeneity that exists on the silicon slide surface. Both types of FABP binding are presented wherein either a single- or a dual-fractal analysis is adequate to describe the binding kinetics. The dual-fractal analysis is used only when the single-fractal analysis does not provide an adequate fit (sum of least squares less than 0.97). This was done by regression analysis provided by Corel Quattro Pro 8.0 (1997).

In accordance with the prefactor analysis for fractal aggregates (Sorenson and Robetrs, 1997) quantitative (predictive) expressions are developed for (a) the ratio of the fractal dimensions, D_{f2}/D_{f1} as a function of the FABP concentration in solution, (b) the ratio of the binding rate coefficients, k_2/k_1 as a function of the ratio of the fractal dimensions, D_{f2}/D_{f1}, and (c) the binding rate coefficients, k_1 and k_2 as a function of the fractal dimensions, D_{f1} and D_{f2}, respectively. As indicated in the earlier chapters, the fractal dimension is not a typical independent variable, such as analyte (FABP) concentration in solution or the FABP antibody on the silicon slide surface that may be manipulated. It is estimated from eqs. 7.1a and 7.1b, and one may consider it as a derived variable. The different types of predictive equations presented may it be for the binding rate coefficient(s), or the ratio of the fractal dimensions in the two phases of binding provide a means by which these binding rate coefficients, and the ratio of either the binding rate coefficients, or the fractal dimensions may be manipulated by changing either the analyte (FABP) concentration in solution or the degree of heterogeneity that exists on the silicon slide surface. It may be suggested, as done in earlier chapters, that here too the fractal surface (roughness) leads to turbulence which enhances mixing, decreases diffusional limitations, and leads to an increase in the binding rate coefficient (Martin *et al.*, 1991).

Speijer *et al.* (2004) also wanted to analyze the reproducibility of their procedure. Even for similar conditions (for example, 4 ng/ml FABP in solution; three runs), the degree of heterogeneity on the silicon slide surface was slightly different. However, the binding rate coefficients, k_1 and k_2 correlated well with the degree of heterogeneity, D_{f1} and D_{f2}, respectively on the surface. The binding rate coefficient, k_1 exhibits a less than first order (equal to 0.900) dependence on the degree of heterogeneity of fractal dimension, D_{f1} present on the silicon chip surface. However, the binding rate coefficient, k_2 is extremely sensitive (as noted by the order of dependence equal to 9.514) to the degree of heterogeneity or the fractal dimension, D_{f2} on the silicon chip surface.

Finally, the fractal analysis was developed to analyze the binding (and dissociation) kinetics for analyte–receptor reactions occurring on biosensor surfaces. In this chapter we apply the fractal analysis to the PEIA initially proposed by Robers *et al.* (1999), and applied by Speijer *et al.* (2004) for the detection of FABP in solution to FABP antibody immobilized on a silicon slide surface using a sandwich type assay wherein HRP labeled anti-FABP tagging antibody is used.

The analysis reinforces the view that fractal analysis may be applied not necessarily only for the kinetics of binding (and dissociation) of analyte–receptor reactions occurring on a biosensor surface, but also may be effectively applied to other analyte–receptor systems, such as for the PEIA on an ellipsometric silicon slide surface. It would be worthwhile to extend the fractal analysis to delineate the binding (and dissociation, if applicable) kinetics for analyte–receptor reactions occurring on other analytical devices that may be used to detect analytes of interest. There is presently a current 'explosion' in the development of detection devices, and the fractal analysis is one method that may be amenable to analyze the kinetics of binding and dissociation of different analytes on the surface(s) of these devices.

REFERENCES

Corel Corporation, Corel Quattro Pro 8.0, Ottawa, Canada, 1997.

Glatz, JF, D van der Voort and WT Hermens, *Journal of Clinical Ligand Assay*, 2002, **25**, 1667–1677.

Havlin, S, Molecular Diffusion and Reactions, in *The Fractal Approach to Heterogeneous Chemistry: Surfaces, Colloids, Polymers*, (ed. D Avnir), Wiley, New York, 1989, pp. 251–269.

Martin, SJ, VE Granstaff and GC Frye, *Analytical Chemistry*, 1991, **65**, 1620–1626.

Ramakrishnan, A and A Sadana, *Automedica*, 2001, **20**(3/4), 313–340.

Robers, M, IJ Rensink, CE Hack, LA Aarden, CP Reutelingsperger, JF Glatz and WT Hermens, *Biophysical Journal*, 1999, **76**, 2769–2776.

Sadana, A and T Vo-Dinh, *Biotechnology & Applied Biochemistry*, 2001, **33**, 17–31.

Sorenson, CM and GC Roberts, *Journal of Colloid & Interface Science*, 1997, **186**, 447–453.

Speijer, H, RH Laterveer-Vreeswijk, JFC Glatz, W Nieuwenhuizen and WT Hermens, *Analytical Biochemistry*, 2004, **326**, 257–261.

Weiss, GH, *Fractals in Science*, Berlin: Springer-Verlag, 1994.

– 8 –

Fractal Analysis of Binding and Dissociation Interactions of P38α Mitogen-Activated Protein Kinase Occurring on Biosensor Surfaces

8.1 INTRODUCTION

P38α MAP (mitogen-activated protein) kinase is a well-characterized target for inflammatory disease therapy (Herlaar *et al.*, 1999; Lee *et al.*, 2000). Lee *et al.* (1994) indicate that these MAP kinases are intracellular serine–threonine kinases. They are activated by a broad range of extracellular stimuli. These stimuli include cytokines, osmotic shock, environmental stresses, and bacterial lipopolysaccharides (Han *et al.*, 1994). Casper *et al.* (2004) very recently indicate that p38α sits at the regulatory juncture of a number of important pro-inflammatory mediators. The binding of p38α to its substrates is complicated (LoGrosso *et al.*, 1997; Chen *et al.*, 2000). It would be of interest to obtain kinetic rate coefficients for binding and dissociation of p38α to its substrates in order to begin to obtain a better idea of the regulation of these pro-inflammatory inhibitors.

Figure 8.1 shows the chemical structures of SB-203580 and SKF-86002 (Casper *et al.*, 2004). These authors indicate that SB-203580 blocks pro-inflammatory cytokine production in cells (Lee *et al.*, 1994), and selectively inhibits p38α MAP kinase activity (Cuenda *et al.*, 1995). SB-203580 is an ATP-competitive inhibitor of p38α (Young *et al.*, 1997).

Casper *et al.* (2004) have recently presented binding and dissociation data of p38α to the above two known small molecule p38α inhibitors (SB-203580 and SKF-86002) using a Biacore SPR biosensor. They were able to fully characterize the p38α affinity, kinetics, and thermodynamics over the 5–35 °C temperature range. They have used the phenomenon of ligand stabilization of protein structure on a biosensor surface. These authors further indicate that this procedure may be used to help characterize other members of the protein kinase family.

In this chapter we re-analyze the binding and dissociation kinetics of SB-203580 in solution to p38α immobilized on a Biacore sensor chip surface. As done in the previous

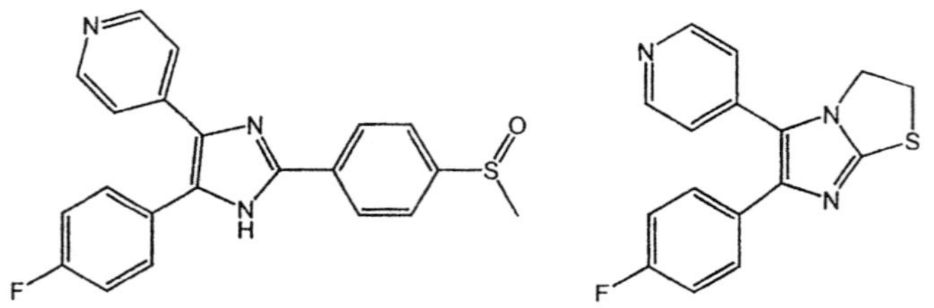

SB-203580 SKF-86002

Figure 8.1 Chemical structures of two small molecule inhibitors of p38α mitogen-activated kinase (Casper *et al.*, 2004): SB203580 and SKF-86002.

chapters we present a fractal analysis of the kinetic data for binding and dissociation. Binding and dissociation rate coefficients, and affinity values will be presented. Wherever possible we will relate these binding and dissociation rate coefficients to the degree of heterogeneity or the fractal dimension, D_f on the sensor chip surface. As indicated in the earlier chapters, in no way are we indicating that the fractal analysis is in any way better than the original analysis. One may consider it as an alternate analysis that provides fresh physical insights into the binding and dissociation kinetics, with special emphasis on the degree of heterogeneity present on the biosensor chip surface. The degree of heterogeneity that is present on the biosensor chip surface along with diffusion effects is often neglected in a kinetic analysis of analyte–receptor reactions occurring on biosensor surfaces.

8.2 THEORY

Havlin (1989) has reviewed and analyzed the diffusion of reactants towards fractal surfaces. The details of the theory and the equations involved for the binding and the dissociation phases for analyte–receptor are available (Ramakrishnan and Sadana, 2001; Sadana and Vo-Dinh, 2001). Here we will attempt to apply these equations to the p38α MAP kinase-small molecule inhibitor, SB-203580 interaction.

The basic idea is to fold everything into the lumped parameter, the fractal dimension, D_f. This lumped parameter takes care of the chemical, orientational, and geometrical details. It is assumed that the kinetic rate depends only on the geometric (static) parameter, D_f. It would also be useful to indicate the low and high limits in angstroms of the fractal character to be presented.

8.2.1 Single-fractal analysis

Binding rate coefficient

Havlin (1989) indicates that the diffusion of a particle (analyte) from a homogeneous solution to a solid surface (e.g. receptor-coated surface) on which it reacts to form

a product (analyte–receptor) complex is given by:

$$(\text{Analyte}\cdot\text{Receptor}) \sim \begin{cases} t^{(3-D_{\mathrm{f,bind}})/2} = t^{p} & (t < t_{\mathrm{c}}) \\ t^{1/2} & (t > t_{\mathrm{c}}) \end{cases} \tag{8.1a}$$

where the analyte–receptor represents the association complex formed on the surface. Here $p = -b$, and D_{f} is the fractal dimension of the surface. Havlin (1989) states that the crossover value may be determined by $r_{\mathrm{c}}^2 \sim t_{\mathrm{c}}$. Above the characteristic length, r_{c}, the self-similarity of the surface is lost and the surface may be considered homogeneous. Eq. (8.1a) indicates that the concentration of the product [analyte–receptor] on a solid fractal surface scales at short and intermediate times as analyte–receptor $\sim t^p$ with the coefficient $p = (3 - D_{\mathrm{f}})/2$ at short time scales and $p = 1/2$ at intermediate time scales. Note that D_{f}, $D_{\mathrm{f,assoc}}$, and $D_{\mathrm{f,bind}}$ are used interchangeably. This equation is associated with the short-term diffusional properties of a random walk on a fractal surface. Note that, in perfectly stirred kinetics on a regular (nonfractal) structure (or surface), k_1 is a constant, that is, it is independent of time. In other words, the limit of regular structures (or surfaces) and the absence of diffusion-limited kinetics leads to k_1 being independent of time. In all other situations, one would expect a scaling behavior given by $k_1 \sim k' t^{-b}$ with $-b = p < 0$. Also, the appearance of the coefficient, p different from $p = 0$ is the consequence of two different phenomena, that is, the heterogeneity (fractality) of the surface and the imperfect mixing (diffusion-limited) condition. Finally, for a homogeneous surface where D_{f} is equal to 2, and when only diffusional limitations are present, $p = 1/2$ as it should be. Another way of looking at the $p = 1/2$ case (where $D_{\mathrm{f,bind}}$ is equal to two) is that the analyte in solution views the fractal object, in our case the receptor-coated biosensor surface, from a 'large distance'. In essence, in the association process, the diffusion of the analyte from the solution to the receptor surface creates a depletion layer of width $(Đt)^{1/2}$ where $Đ$ is the diffusion constant. This gives rise to the fractal power law, $(\text{Analyte}\cdot\text{Receptor}) \sim t^{(3-D_{\mathrm{f,bind}})/2}$.

The values of the parameters k (association rate coefficient), p, and D_{f} in Eq. (8.1a) may be obtained for analyte–receptor binding kinetics data. This may be done by a regression analysis using, for example, Corel Quattro Pro (1997) along with Eq. (8.1a) where $(\text{analyte}\cdot\text{receptor}) = kt^{p}$. The fractal dimension may be obtained from the parameter p. Higher values of the fractal dimension would indicate higher degrees of 'disorder' or heterogeneity or inhomogeneity on the surface.

Another way of looking at the diffusive process is that it inherently involves fluctuations at the molecular level that may be described by a random walk (Weiss, 1994). This author indicates that the kinetics of transport on disordered (or heterogeneous) media itself needs to be described by a random-walk model. When both of these are present, that is the diffusion phenomena as well as a fractal surface, then one needs to analyze the interplay of both of these fluctuations. In essence, the disorder on the surface (or a higher fractal dimension, D_{f}) tends to slow down the motion of a particle (analyte in our case) moving in such a medium. Basically, according to Weiss (Weiss, 1994) the particle (random walker analyte) is trapped in regions in space, as it oscillates for a long time before resuming its motion.

Havlin (1989) indicates that the crossover value may be determined by $r_{\mathrm{c}}^2 \sim t_{\mathrm{c}}$. Above the characteristic length, r_{c}, the self-similarity of the surface is lost. Above t_{c}, the surface may be considered homogeneous, and 'regular' diffusion is now present. One may consider the analysis to be presented as an intermediate 'heuristic' approach in that in the future one may also be able to develop an autonomous (and not time-dependent) model of diffusion-limited kinetics in disordered media. No evidence of fractality is presented.

Dissociation rate coefficient

The diffusion of the dissociated particle (receptor or analyte) from the solid surface (e.g. analyte–receptor complex coated surface) into solution may be given, as a first approximation by:

$$\begin{aligned}(\text{Analyte}\cdot\text{Receptor}) &\sim -t^{(3-D_{\mathrm{f,diss}})/2}, \qquad (t > t_{\mathrm{diss}}) \\ &= -k_{\mathrm{diss}}t^{(3-D_{\mathrm{f,diss}})/2},\end{aligned} \tag{8.1b}$$

Here $D_{\mathrm{f,diss}}$ is the fractal dimension of the surface for the dissociation step. t_{diss} represents the start of the dissociation step. This corresponds to the highest concentration of the analyte–receptor complex on the surface. Henceforth, its concentration only decreases. $D_{\mathrm{f,assoc}}$ may or may not be equal to $D_{\mathrm{f,diss}}$. k_{d} and k_{diss}, and $D_{\mathrm{f,d}}$ and $D_{\mathrm{f,diss}}$ are used interchangeably in the chapter.

8.2.2 Dual-fractal Analysis

Binding rate coefficient

The single-fractal analysis we have just presented is extended to include two fractal dimensions. At present, the time ($t = t_1$) at which the first fractal dimension 'changes' to the second fractal dimension is arbitrary and empirical. For the most part it is dictated by the data analyzed and the experience gained by handling a single-fractal analysis. The r^2 (regression coefficient) value obtained is also used to determine if a single-fractal analysis is sufficient, or one needs to use a dual-fractal analysis to provide an adequate fit. Only if the r^2 value is less than 0.97 for a single-fractal analysis, do we use a dual-fractal model. In this case, the analyte–receptor complex is given by:

$$(\text{Analyte}\cdot\text{Receptor}) \sim \begin{cases} t^{(3-D_{\mathrm{f1,bind}})/2} = t^{p_1} & (t < t_1) \\ t^{(3-D_{\mathrm{f2,bind}})/2} = t^{p_2} & (t_1 < t < t_2 = t_{\mathrm{c}}) \\ t^{1/2} & (t > t_{\mathrm{c}}) \end{cases} \tag{8.1c}$$

In analyte–receptor association the analyte–receptor associates with the active site on the surface and the product is released. In this sense the catalytic surface exhibits an unchanging fractal surface to the reactant in the absence of fouling and other complications. In the case of analyte–receptor association the biosensor surface exhibits a changing fractal surface to the analyte in solution. This occurs since as each association (binding) reaction takes place, smaller and smaller amounts of 'association' sites or receptors are available on the biosensor surface to which the analyte may bind. Furthermore, as the reaction proceeds, there is an increasing degree of heterogeneity on the biosensor surface for some reaction systems. This is manifested by two degrees of heterogeneity, or two fractal dimensions on the biosensor surface. In the theoretical limit one might envisage a temporal fractal dimension wherein there is a continuous change in the degree of heterogeneity on the surface; though of course, such situations would be very rare, if at all.

Note that, at present, the dual-fractal analysis does not have a basis at the molecular level. This represents two different levels of heterogeneity on the biosensor surface. But, in some examples presented, a single-fractal analysis is clearly inadequate to model the data. Only in these cases does one resort to a dual-fractal analysis.

Dissociation rate coefficient

In this case the dissociation rate coefficient is given by:

$$(\text{Analyte}\cdot\text{Receptor}) \sim \begin{cases} -t^{(3-D_{\text{f1,diss}})/2} & (t_{\text{diss}} < t < t_{\text{d1}}) \\ -t^{(3-D_{\text{f2,diss}})/2} & (t_{\text{d1}} < t < t_{\text{d2}}) \end{cases} \tag{8.1d}$$

Here $D_{\text{f,diss}}$ is the fractal dimension of the surface for the dissociation step. t_{diss} represents the start of the dissociation step. This corresponds to the highest concentration of the analyte·receptor on the surface. Henceforth, its concentration only decreases. $D_{\text{f,bind}}$ or $D_{\text{f,assoc}}$ may or may not be equal to $D_{\text{f,diss}}$.

8.3 RESULTS

A fractal analysis will be applied to the data obtained for MAP kinase p38α-small molecule inhibitor, SB-203580 interaction data taken from the literature (Casper *et al.*, 2004). The fractal analysis is only one possible approach to analyzing the diffusion-limited binding kinetics assumed to be present in the systems analyzed.

Understandably, alternate expressions for fitting the data that include saturation, first-order reaction, and no diffusion limitations are available, but these expressions are apparently deficient in describing the heterogeneity that inherently exists on the surface. Another advantage of this technique is that the analyte–receptor binding (as well as the dissociation reaction) is a complex reaction, and the fractal analysis via the fractal dimension and the rate coefficient provides a useful lumped parameter(s) analysis of the diffusion-limited reaction occurring on a heterogeneous surface.

In the classical situation to demonstrate fractality, one should make a log–log plot, and one should definitely have a large amount of data. It may be useful to compare the fit to some other forms, such as an exponential form or to the one involving saturation. At present, we do not present any independent proof or physical evidence of fractals in the examples presented. It is a convenient means (since it provides a lumped parameter) to make the degree of heterogeneity that exists on the surface more quantitative. Thus, there is some arbitrariness in the fractal model to be presented. One might justifiably argue that appropriate modeling may be achieved by using a Langmuirian or other approach. The Langmuirian approach has a major drawback because it does not allow for or accommodate the heterogeneity that exists on the surface.

Casper *et al.* (2004) have analyzed the binding and dissociation kinetics of SB-203580 in solution to p38α immobilized on a Biacore surface plasmon resonance (SPR) biosensor surface. They indicate that a control experiment demonstrated the lack of binding of SB-203580 in solution to a reference surface (dextran layer of the CM5 biosensor chip) without p38α immobilized on it. They presented data for 5, 15, 25, and 35 °C. In this chapter we will re-analyze their binding and dissociation data for SB-203580 in the concentration range 1.23×10^{-8}–1.11×10^{-7} M in solution at 5, 15, and 25 °C to p38α immobilized on a SPR biosensor chip surface.

Figure 8.2a shows the binding and dissociation of 1.23×10^{-8} M SB-203580 in solution at 5 °C to p38α immobilized on a sensor chip surface. A single-fractal analysis is adequate to describe the binding and dissociation kinetics. The values of (a) the binding rate coefficient, k and the fractal dimension in the binding phase, D_f for a single-fractal analysis are given in Table 8.1a and 8.1b, respectively, and (b) the dissociation rate coefficient, k_d and the fractal dimension in the dissociation phase, D_{fd} for a single-fractal analysis are also given in Table 8.1a and 8.1b, respectively. The values of the binding and dissociation rate coefficients presented in Table 8.1a and 8.1b were obtained from a regression analysis using Corel Quattro Pro 8.0 (Corel Quattro Pro, 1997) to model the data using Eqs. (8.1a and b) wherein (Analyte·Receptor) $= kt^{(3-D_f)/2}$ for a single-fractal analysis for the binding phase, and (Analyte·Receptor) $= -kt^{(3-D_{fd})/2}$ for the dissociation phase. The binding and dissociation rate coefficient values presented in Table 8.1a are within 95% confidence limits. For example, for the binding of 1.23×10^{-8} M SB-203580 in solution to p38α immobilized on a sensor chip surface, the binding rate coefficient, k is equal to 0.0267 ± 0.0023.

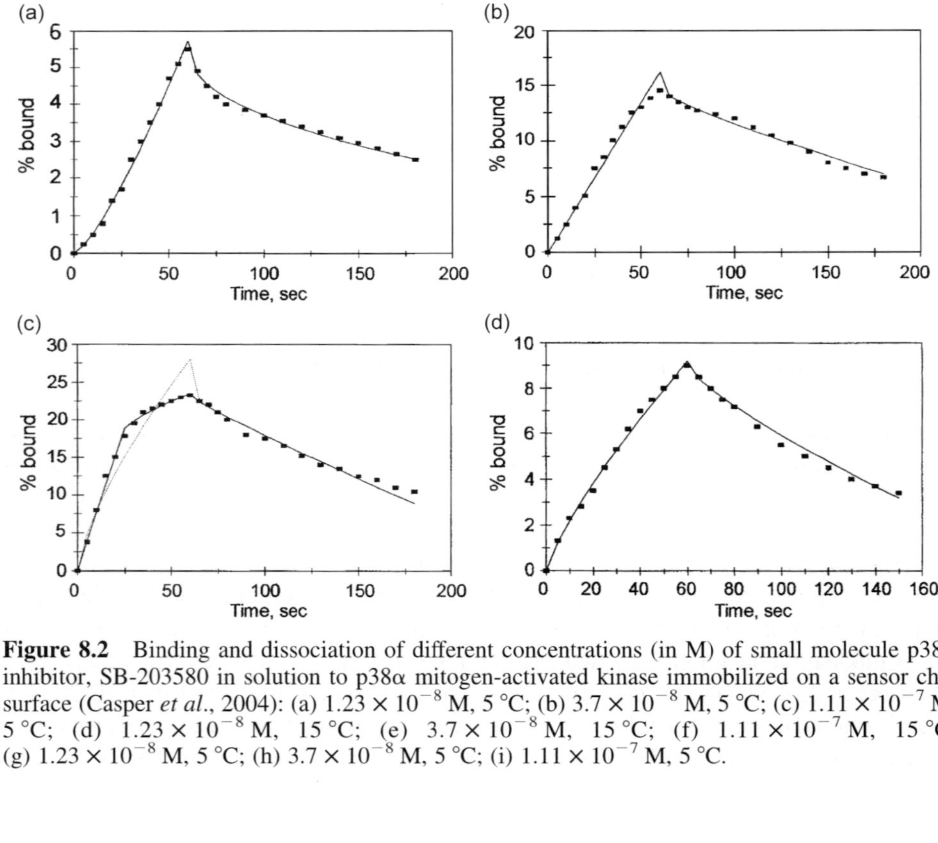

Figure 8.2 Binding and dissociation of different concentrations (in M) of small molecule p38α inhibitor, SB-203580 in solution to p38α mitogen-activated kinase immobilized on a sensor chip surface (Casper *et al*., 2004): (a) 1.23×10^{-8} M, 5 °C; (b) 3.7×10^{-8} M, 5 °C; (c) 1.11×10^{-7} M, 5 °C; (d) 1.23×10^{-8} M, 15 °C; (e) 3.7×10^{-8} M, 15 °C; (f) 1.11×10^{-7} M, 15 °C; (g) 1.23×10^{-8} M, 5 °C; (h) 3.7×10^{-8} M, 5 °C; (i) 1.11×10^{-7} M, 5 °C.

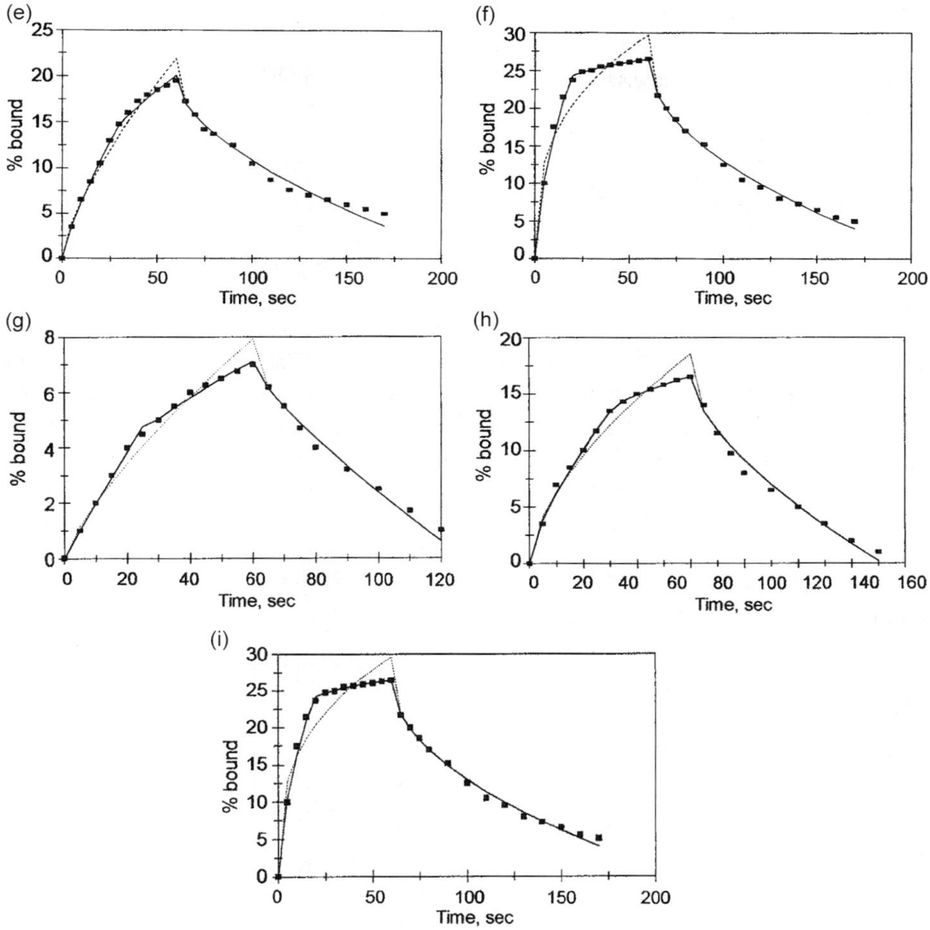

Figure 8.2 Continued.

The 95% confidence limit indicates that the k value lies between 0.0244 and 0.029. This indicates that the values are precise and significant.

Figure 8.2b shows the binding and dissociation of 3.7×10^{-8} M in solution at 5 °C to p38α immobilized on a sensor chip surface. Once again, a single-fractal analysis is adequate to describe the binding and dissociation kinetics. The values of (a) the binding rate coefficient, k and the fractal dimension in the binding phase, D_f for a single-fractal analysis are given in Table 8.1a and 8.1b, respectively, and (b) the dissociation rate coefficient, k_d and the fractal dimension in the dissociation phase, D_{fd} for a single-fractal analysis are also given in Table 8.1a and 8.1b, respectively. It is of interest to note that as the SB-205380 concentration in solution increases by a factor of three from 1.23×10^{-8} to 3.7×10^{-8} M, (a) the binding rate coefficient, k increases by a factor of 9.42 from 0.0267 to 0.2514, (b) and the dissociation rate coefficient, k_d increases by a factor of 4.06 from a value of 0.3239 to 1.3156. Also, there is (a) an increase in the fractal dimension in

Table 8.1a

Binding and dissociation rate coefficients for SB-203580 in solution to p38α immobilized on a Biacore sensor chip surface (Casper *et al.*, 2004)

Temperature, °C	Concentration, M	K	k_1	k_2	k_d
5	1.23×10^{-8}	0.0267 ± 0.0023	na	na	0.3239 ± 0.0412
5	3.7×10^{-8}	0.2514 ± 0.0185	na	na	1.3156 ± 0.0698
5	1.11×10^{-7}	1.5701 ± 0.2590	0.8132 ± 0.0583	8.8174 ± 0.1182	0.1910 ± 0.0237
15	1.23×10^{-8}	0.3477 ± 0.0170	na	na	1.4128 ± 0.0526
15	3.7×10^{-8}	1.2785 ± 0.0995	0.9894 ± 0.0340	3.1431 ± 0.0875	1.7908 ± 0.0428
15	1.11×10^{-7}	7.4543 ± 0.9912	3.7697 ± 0.3031	19.336 ± 0.0476	1.9882 ± 0.0198
25	1.23×10^{-8}	0.3455 ± 0.0378	0.2184 ± 0.0099	0.9093 ± 0.0175	1.4006 ± 0.0708
25	3.7×10^{-8}	1.5923 ± 0.1565	1.2289 ± 0.0901	5.7064 ± 0.0291	1.7292 ± 0.0774
25	1.11×10^{-7}	7.7369 ± 1.3877	3.1298 ± 0.4745	20.562 ± 2.018	2.0178 ± 0.0426

that the binding rate coefficient, k_2 does exhibit an increase with temperature on going from 5 to 25 °C. In this case, the changes in both the temperature and in the degree of heterogeneity on the sensor chip surface lead to changes in the binding rate coefficient, k_2 in the same direction.

Table 8.1a and Figure 8.4 show the increase in the binding rate coefficient, k_2 with an increase in temperature in the narrow temperature range of 5–25 °C. A classical van't Hoff plot (of ln k_2 versus reciprocal temperature) is not attempted due to the fractal nature of the sensor chip surface. In lieu of that a simple plot of $k_2 = a\,(1000/T)^b$ is attempted. Here a and b are coefficients to be determined by regression. In the temperature range 5 °C (278 K) to 25 °C (298 K), and for 1.11×10^{-7} M SB-203580 concentration in solution, the binding rate coefficient, k_2 is given by:

$$k_2 = (6.7 \times 10^7 \pm 2.5 \times 10^7)(1000/T)^{-12.28 \pm 6.390} \tag{8.3}$$

The temperature is in degrees Kelvin (K). There is some scatter in the data. Only three data points are available. The availability of more data points would lead to a more reliable fit. Since only three data points are presented no further concluding statements are made with regard to the plot shown in Figure 8.4.

It would be of interest to see if the fractal dimension in either the binding phase, D_f or in the dissociation phase, D_{fd} changes with a change in the SB-203580 concentration in solution. Note that the fractal dimension is based on a logarithmic scale, thus one may anticipate changes in the fractal dimension that are generally small.

At 5 °C and in the 1.23×10^{-8} to 1.11×10^{-7} M SB-203580 concentration range, Table 8.1b and Figure 8.5a show that the fractal dimension, D_{fd} in the dissociation phase is given by:

$$D_{fd} = (106.66 \pm 16.86)[\text{SB-203580}]^{0.2498 \pm 0.0943} \tag{8.4a}$$

Only three data points are available. The fit is quite good. The availability of more data points would lead to a more reliable fit. The fractal dimension, D_{fd} exhibits only a mild dependence (order equal to 0.2498) on the SB-203580 concentration in solution. Once again, one must remember, and as indicated above, the fractal dimension is based on a

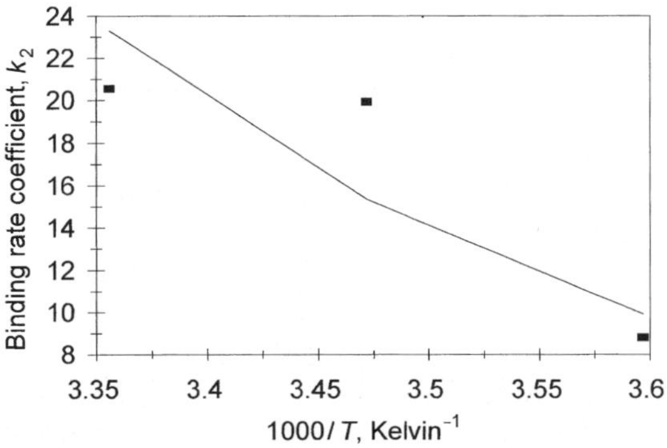

Figure 8.4 Increase in the binding rate coefficient, k_1 with an increase in temperature (in K) at 1.11×10^{-7} M SB-203580 M concentration in solution.

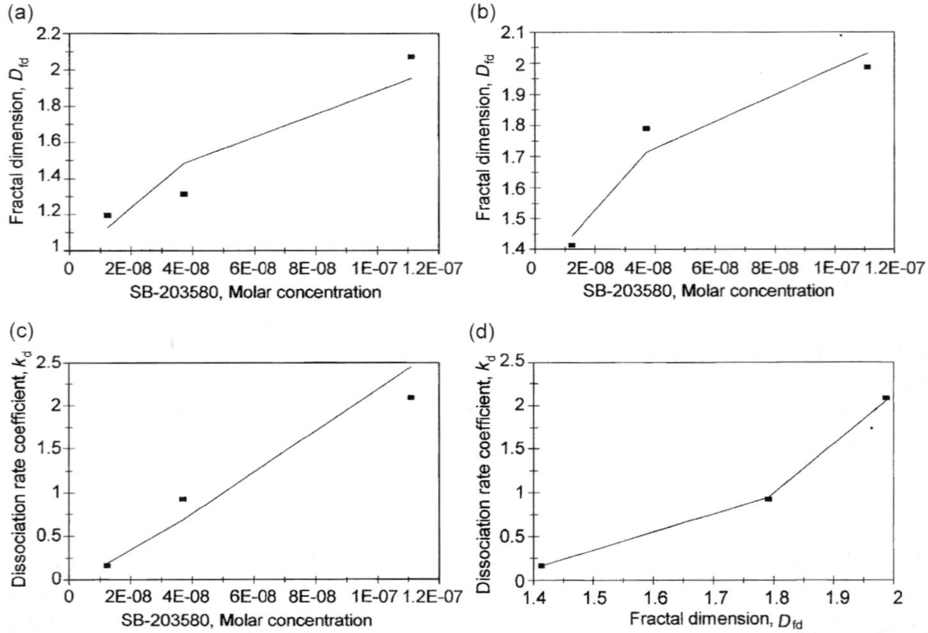

Figure 8.5 (a) Increase in the fractal dimension for dissociation, D_{fd} with an increase in the SB-203580 concentration (in M) in solution at 5 °C; (b) increase in the fractal dimension for dissociation, D_{fd} with an increase in the SB-203580 concentration (in M) in solution at 15 °C; (c) increase in the dissociation rate coefficient, k_d with an increase in the SB-203580 concentration (in M) in solution at 15 °C; (d) increase in the dissociation rate coefficient, k_d with an increase in the fractal dimension in the dissociation phase, D_{fd} at 15 °C.

logarithmic scale, and even small changes in the value of the fractal dimension may indicate significant changes in the degree of heterogeneity on the sensor chip surface.

At 15 °C and in the 1.23×10^{-8} to 1.11×10^{-7} M SB-203580 concentration range, Table 8.1b and Figure 8.5b show that the fractal dimension, D_{fd} in the dissociation phase is given by:

$$D_{fd} = (29.055 \pm 0.671)[\text{SB-203580}]^{0.1659 \pm 0.0146} \tag{8.4b}$$

Only three data points are available. The fit is quite good. The availability of more data points would lead to a more reliable fit. The fractal dimension, D_{fd} exhibits only a mild dependence (order equal to 0.1659) on the SB-203580 concentration in solution. The fractal dimension, D_{fd} in the dissociation phase exhibits a lower order of dependence on SB-203580 concentration in solution at 15 °C (equal to 0.1659) than at 5 °C (equal to 0.2498).

At 15 °C and in the 1.23×10^{-8} to 1.11×10^{-7} M SB-203580 concentration range, Table 8.1a and Figure 8.5c show that the dissociation rate coefficient, k_d is given by:

$$k_d = (2.7 \times 10^8 \pm 1.2 \times 10^8)[\text{SB-203580}]^{1.1564 \pm 0.241} \tag{8.5a}$$

Only three data points are available. The fit is good. The availability of more data points would lead to a more reliable fit. The dissociation rate coefficient, k_d exhibits a slightly higher than first-order (equal to 1.1564) dependence on the SB-203580 concentration in solution.

At 15 °C and in the 1.23×10^{-8} to 1.11×10^{-7} M SB-203580 concentration range, Table 8.1a and Figure 8.5d show that the dissociation rate coefficient, k_d is given by:

$$k_d = (0.0125 \pm 0.0003)D_{fd}^{7.426 \pm 0.1090} \quad (8.5b)$$

Only three data points are available. The fit is good. The availability of more data points would lead to a more reliable fit. The dissociation rate coefficient, k_d is extremely sensitive to the degree of heterogeneity or fractal dimension, D_{fd} present in the dissociation phase as noted by the 7.426 order of dependence exhibited.

At 25 °C and in the 1.23×10^{-8} to 1.11×10^{-7} M SB-203580 concentration range, Table 8.1a and Figure 8.6a show that the binding rate coefficient, k_1 is given by:

$$k_1 = (9.3 \times 10^8 \pm 0.4 \times 10^8)[\text{SB-203580}]^{1.2103 \pm 0.2071} \quad (8.6a)$$

Only three data points are available. The fit is good. The availability of more data points would provide a more reliable fit. The binding rate coefficient, k_1 exhibits a greater than first-order dependence (equal to 1.2103) on the SB-203580 concentration in solution.

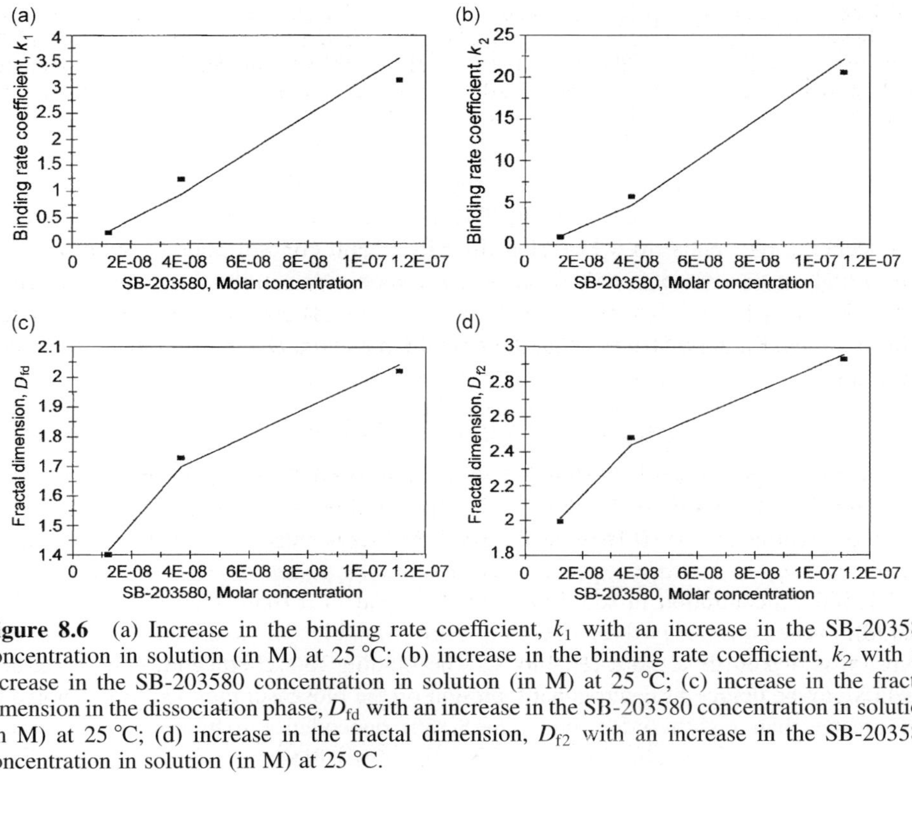

Figure 8.6 (a) Increase in the binding rate coefficient, k_1 with an increase in the SB-203580 concentration in solution (in M) at 25 °C; (b) increase in the binding rate coefficient, k_2 with an increase in the SB-203580 concentration in solution (in M) at 25 °C; (c) increase in the fractal dimension in the dissociation phase, D_{fd} with an increase in the SB-203580 concentration in solution (in M) at 25 °C; (d) increase in the fractal dimension, D_{f2} with an increase in the SB-203580 concentration in solution (in M) at 25 °C.

The non-integer order of dependence exhibited by the binding rate coefficient, k_1 on the SB-203580 concentration in solution reinforces the fractal nature of the system.

At 25 °C and in the 1.23×10^{-8}–1.11×10^{-7} M SB-203580 concentration range, Table 8.1a and Figure 8.6b show that the binding rate coefficient, k_2 is given by:

$$k_2 = (1.6 \times 10^{11} \pm 0.4 \times 10^{11})[\text{SB-203580}]^{1.4176 \pm 0.1445} \quad (8.6b)$$

Only three data points are available. The fit is good. The availability of more data points would provide a more reliable fit. The binding rate coefficient, k_2 exhibits a greater than first-order dependence (in this case equal to 1.4176) on the SB-203580 concentration in solution. The non-integer order of dependence exhibited by the binding rate coefficient, k_2 on the SB-203580 concentration in solution, once again reinforces the fractal nature of the system. The binding rate coefficient, k_2 is slightly more sensitive (order equal to 1.4176) to the SB-203580 concentration in solution than k_1 (order equal 1.2103).

At 25 °C and in the 1.23×10^{-8}–1.11×10^{-7} M SB-203580 concentration range, Table 8.1a and Figure 8.6c show that the fractal dimension in the dissociation phase, D_{fd} is given by:

$$D_{fd} = (29.055 \pm 0.671)[\text{SB-203580}]^{0.1659 \pm 0.01469} \quad (8.7a)$$

Only three data points are available. The fit is good. The availability of more data points would provide a more reliable fit. The fractal dimension, D_{fd} exhibits only a slight dependence (order equal to 0.1659) on the SB-203580 concentration in solution.

At 25 °C and in the 1.23×10^{-8}–1.11×10^{-7} M SB-203580 concentration range, Table 8.1a and Figure 8.6d show that the fractal dimension in the second phase of binding, D_{f2} is given by:

$$D_{f2} = (47.964 \pm 1.033)[\text{SB-203580}]^{0.1741 \pm 0.0137} \quad (8.7b)$$

Only three data points are available. The fit is good. The availability of more data points would provide a more reliable fit. The fractal dimension, D_{f2} exhibits only a slight dependence (order equal to 0.1741) on the SB-203580 concentration in solution. Note that in both of the above cases the fractal dimension either in the dissociation phase or in the second phase of binding exhibits only a slight dependence on the SB-203580 concentration in solution. In other words, the SB-203580 concentration in solution does not significantly affect the degree of heterogeneity on the sensor chip surface (a) either during the second phase of binding or (b) in the dissociation phase, at least for this case.

At 25 °C, Table 8.1a and 8.1b and Figure 8.7 show that the binding rate coefficient, k_2 increases as the degree of heterogeneity or the fractal dimension, D_{f2} increases. For the data presented in Table 8.1a and 8.1b for the three SB-203580 concentrations in solution analyzed at 25 °C, the binding rate coefficient, k_2 is given by:

$$k_2 = (0.003323 \pm 0.000175)D_{f2}^{8.156 \pm 0.189} \quad (8.8)$$

The fit is very good. Only three data points are available. The availability of more data points would provide a more reliable fit. The binding rate coefficient, k_2 is extremely sensitive to the degree of heterogeneity present on the biosensor chip surface as noted by the greater than eighth order (equal to 8.156) dependence exhibited on the fractal dimension, D_{f2}. This underscores the importance that biosensorists need to pay more

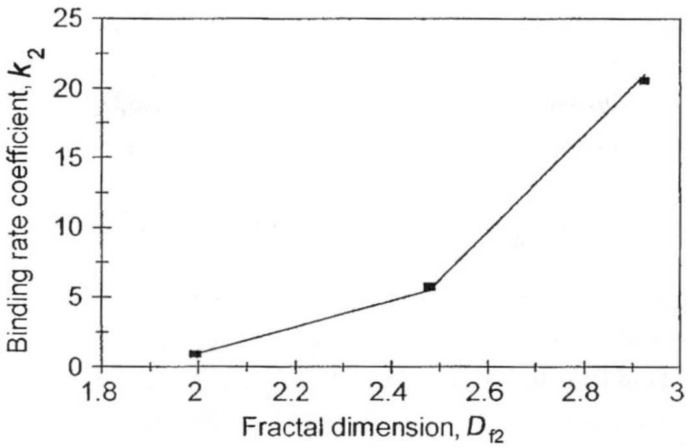

Figure 8.7 Increase in the binding rate coefficient, k_2 with an increase in the fractal dimension, D_{f2}.

attention to the degree of heterogeneity on the biosensor surface, and how it may be manipulated in desired directions to advantage to help improve biosensor performance parameters.

Affinity values are of interest to practicing biosensorists. At 25 °C, Table 8.1a and 8.1b and Figure 8.8a show that the affinity, $K_1(=k_1/k_d)$ increases as the ratio of fractal dimensions in the binding and in the dissociation phase (D_{f1}/D_{fd}) increases. For the data presented in Table 8.1a and 8.1b and in Figure 8.8a and for the three SB-203580 concentrations in solution analyzed at 25 °C, the affinity, K_1 is given by:

$$K_1(=k_1/k_d) = (1.2149 \pm 0.138)(D_{f1}/D_{fd})^{0.897 \pm 0.796} \quad (8.9a)$$

There is a lot of scatter in the data. Only three data points are available. The availability of more data points would provide a more reliable fit. The affinity, K_1 exhibits a less than first-order (equal to 0.897) dependence on the ratio of fractal dimension in the binding and dissociation phases.

At 25 °C, Table 8.1a and 8.1b and Figure 8.8b show that the affinity, $K_2(=k_2/k_d)$ increases as the ratio of fractal dimension in the binding and in the dissociation phase (D_{f2}/D_{fd}) increases. For the data presented in Table 8.1a and 8.1b and in Figure 8.8b and

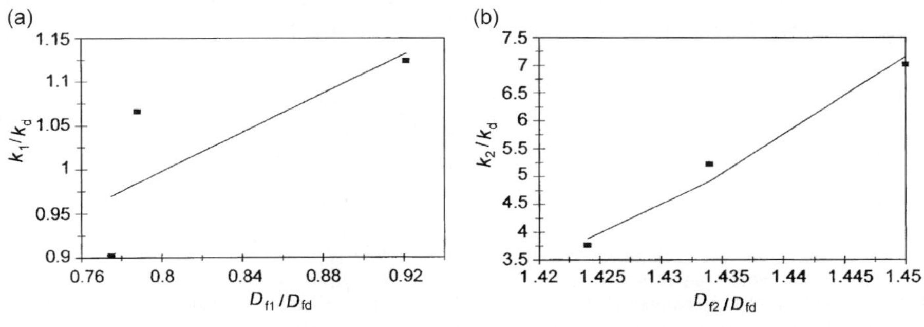

Figure 8.8 (a) Increase in the affinity, K_1 with an increase in the ratio of fractal dimensions, in the binding and in the dissociation phase (D_{f1}/D_{fd}) at 25 °C; (b) increase in the affinity, K_2 with an increase in the ratio of fractal dimensions, in the binding and in the dissociation phase (D_{f2}/D_{fd}) at 25 °C.

for the three SB-203580 concentrations in solution analyzed at 25 °C, the affinity, K_2 is given by:

$$K_2(= k_2/k_d) = (2.6 \times 10^{-5} \pm 0.2 \times 10^{-5})(D_{f2}/D_{fd})^{33.709 \pm 5.486} \qquad (8.9b)$$

The fit for K_2 is very good. It is much better than that observed for K_1. Once again, only three data points are available. The availability of more data points would lead to a more reliable fit. The affinity, K_2 is extremely sensitive (in fact inordinately so) to the degree of heterogeneity present on the biosensor chip surface as noted by the 33.709 order of dependence exhibited on the ratio, D_{f2}/D_{fd}. This, once again, underscores the importance of the nature of the biosensor chip surface, wherein even small or even very small changes can significantly affect the affinity, K_2 and perhaps other biosensor performance parameters.

Table 8.1a and 8.1b and Figure 8.9a show that the ratio of the fractal dimensions, D_{f1}/D_{fd} increases as the SB-203580 concentration in solution increases. For the data presented in Table 8.1a and 8.1b and in Figure 8.9a, the ratio, D_{f1}/D_{fd} is given by:

$$D_{f1}/D_{fd} = (3.159 \pm 0.0281)[\text{SB-203580}]^{0.078 \pm 0.037} \qquad (8.10a)$$

The fit is quite good. Only three data points are available. The availability of more data points would make the fit more reliable. The ratio of the fractal dimensions, D_{f1}/D_{fd} is only slightly dependent on the SB-203580 concentration in solution as noted by the low (equal to 0.078) order of dependence.

Table 8.1a and 8.1b and Figure 8.9b show that the ratio of the fractal dimensions, D_{f2}/D_{fd} increases as the SB-203580 concentration in solution increases. For the data presented in Table 8.1a and 8.1b and in Figure 8.9b, the ratio, D_{f2}/D_{fd} is given by:

$$D_{f2}/D_{fd} = (1.653 \pm 0.0028)[\text{SB-203580}]^{0.0082 \pm 0.0011} \qquad (8.10b)$$

The fit is quite good. Only three data points are available. The availability of more data points would make the fit more reliable. The ratio of the fractal dimensions, D_{f2}/D_{fd} is

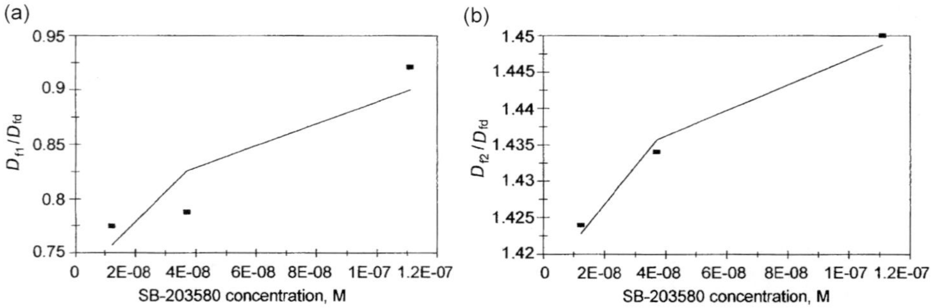

Figure 8.9 (a) Increase in the ratio of fractal dimensions in the binding and in the dissociation phase (D_{f1}/D_{fd}) with an increase in the SB-203580 concentration in solution; (b) increase in the ratio of fractal dimensions in the binding and in the dissociation phase (D_{f2}/D_{fd}) with an increase in the SB-203580 concentration in solution.

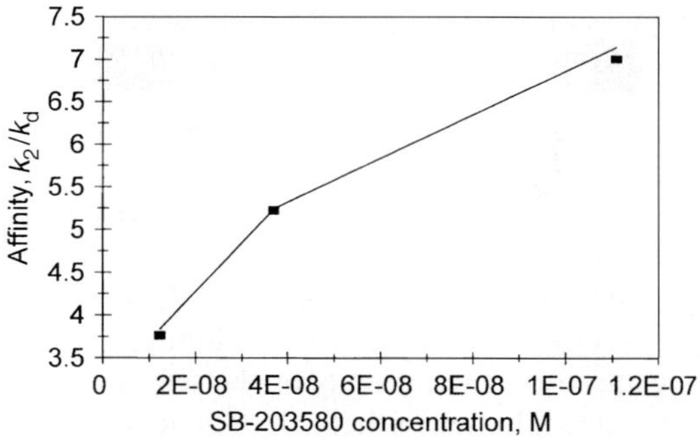

Figure 8.10 Increase in the affinity, K_2 with an increase in the SB-203580 concentration in solution.

only slightly dependent on the SB-203580 concentration in solution as noted by the low (equal to 0.0082) order of dependence. It is quite apparent from Eqs. (8.10a and b) that the SB-203580 concentration in solution does not affect or only mildly affects the degree of heterogeneity on the sensor chip surface.

Table 8.1a and 8.1b and Figure 8.10 show that the affinity, $K_2(= k_2/k_d)$ increases with an increase in the SB-203580 concentration in solution. For the data presented in Table 8.1a and in Figure 8.10, the affinity, K_2 is given by:

$$K_2 = (656.47 \pm 8.912)[\text{SB-203580}]^{0.283\pm0.0087} \tag{8.11}$$

The fit is quite good. It is much better than that observed for K_1. Once again, only three data points are available. The availability of more data points would lead to a more reliable fit. The affinity, K_2 is only mildly sensitive to the SB-203580 concentration in solution as noted by the low (equal to 0.283) order of dependence exhibited.

It would be interest to see how the affinity, $K_1(= k_1/k_d)$ or $K_2(= k_2/k_d)$ changes with temperature. Table 8.1a and Figure 8.11a show the increase in affinity, $K_1(= k_1/k_d)$ with a decrease in temperature. Once again, since heterogeneity on the sensor chip is involved, a classical van't Hoff plot is not attempted. In lieu of that and as done before in this chapter a plot of $K_1(= k_1/k_d) = a(1/T)^b$ is presented. Here a and b are coefficients that need

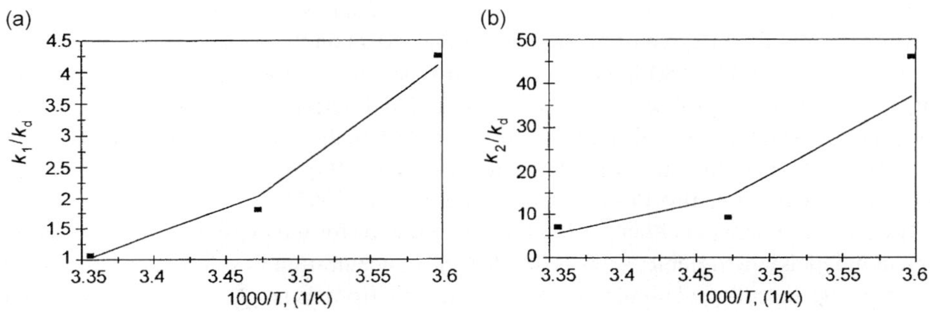

Figure 8.11 (a) Increase in the affinity, K_1 with an increase in reciprocal temperature (K^{-1}); (b) increase in the affinity, K_2 with an increase in reciprocal temperature (K^{-1}).

to be determined by regression. The temperature is in degrees Kelvin (K). For the data presented in Table 8.1a and in Figure 8.11a, the affinity K_1 is given by:

$$K_1 = (3.2 \times 10^{-11} \pm 0.4 \times 10^{-11})(1000/T)^{19.982 \pm 2.526} \quad (8.12a)$$

Only three data points are available. The availability of more data points would lead to a more reliable fit. The affinity, K_1 is extremely sensitive to the changes in temperature as noted by the very high order of dependence (equal to 19.982) exhibited.

Table 8.1a and Figure 8.11b show the increase in affinity, $K_2(= k_2/k_d)$ with a decrease in temperature. Once again, since heterogeneity on the sensor chip is involved, a classical van't Hoff plot is not attempted. In lieu of that and as done before in this chapter a plot of $K_2(= k_2/k_d) = a(1/T)^b$ is presented. The temperature is in degrees Kelvin. For the data presented in Table 8.1a and in Figure 8.11b, the affinity K_2 is given by:

$$K_2 = (2.4 \times 10^{-14} \pm 1.7 \times 10^{-14})(1000/T)^{27.3222 \pm 10.719} \quad (8.12b)$$

Only three data points are available. The availability of more data points would lead to a more reliable fit. The affinity, K_2 is extremely sensitive to the changes in temperature as noted by the very high order of dependence (equal to 27.322) exhibited. The affinity, K_2 is more sensitive to temperature than K_1.

8.4 CONCLUSION

A fractal analysis is presented for the binding and dissociation of a small molecule inhibitor, SB 203580 in solution to p38α kinase immobilized on a sensor chip surface. p38α kinase sits at the regulatory juncture of a number of pro-inflammatory mediators (Casper *et al.*, 2004). The analysis of both the binding as well as the dissociation steps provides a more complete picture of the reaction occurring on the sensor chip surface besides providing for values of affinities, K_i wherever possible. This is the ratio of the rate coefficients in the binding step, k_{bi} to the dissociation step, k_{di}. Here 'I' is a generic step. The analysis should help facilitate an understanding of the complicated binding kinetics that are involved with p38α kinase and its substrates (LoGrosso *et al.*, 1997; Cuenda *et al.*, 1995). The analysis could also be extended to, and as hinted by Casper *et al.* (2004), to other members of the kinase family and the interactions with their respective substrates.

As indicated in the previous chapters the fractal dimension value provides a quantitative measure of the degree of heterogeneity that exists on the biosensor chip surface for the p38α kinase-SB203580 interactions. Both types of examples are presented wherein either a single- or a dual-fractal analysis is required to describe the binding and/or dissociation kinetics. The dual-fractal analysis is used only when the single-fractal analysis did not provide an adequate fit (sum of least squares less than 0.98). This was done by regression provided by Corel Quattro Pro 8.0 (Corel Quattro Pro, 1997).

The fractal dimension either for the binding phase or for the dissociation phase is not a typical independent variable, such as SB-203580 concentration in solution or temperature that may be directly manipulated. It is estimated from Eqs. (8.1a–d), and one may consider it as a derived variable. The predictive equations presented for the binding and the dissociation rate coefficients and the fractal dimensions as a function of the analyte,

SB-203580 concentration in solution provide a means by which these rate coefficients may be manipulated. The predictive equations presented for the binding and dissociations rate coefficients presented as a function of their respective fractal dimensions, D_f of D_{fd}, respectively are of considerable interest since they provide a fresh avenue whereby these coefficients may be manipulated in desired directions. This is particularly so, since these rate coefficients are very sensitive to the degree of heterogeneity that exists on the surface.

The predictive relationships presented for the binding rate coefficients as a function of temperature are of particular interest since they provide a quantitative measure of the influence of temperature on these rate coefficients. A classical van't Hoff plot is not attempted due to the involvement of heterogeneity on the sensor chip surface.

Only a couple of experimental data sets were available wherein one could compare the influence of the degree of heterogeneity on the sensor chip surface and that of temperature on the binding rate coefficients. In one case temperature predominated, and in the other both the degree of heterogeneity and temperature exhibited increases with an increase in the binding rate coefficient. It is quite possible that temperature affects the degree of heterogeneity on the sensor chip surface, which subsequently influences biosensor performance parameters such as binding and dissociation rate coefficients and affinity values. Affinity values are of particular interest to practicing biosensorists. The predictive equations presented for the affinity as a function of (a) the ratio of the fractal dimensions in the binding phase and the dissociation phase (for example, D_f/D_{fd}), and (b) temperature are of particular interest. These relationships provide and suggest a means by which the affinities may be manipulated in desired directions, at least for the p38α kinase-SB 203580 interactions. Similar types of predictive relationships have been presented for other analyte–receptor reactions occurring on biosensor chip surfaces or biosensor surfaces in other chapters.

As indicated in earlier chapters one needs to begin to pay more and more attention to the nature of the sensor chip surface, and analyze its importance on biosensor performance parameters such as binding and dissociation rate coefficients, and affinity values. If this 'single' message has been conveyed in this chapter and in the other chapters of this book, then apparently the time and effort spent in writing this treatise and publishing it would be worth while.

REFERENCES

Casper, D, M Bukhtiyarova and EB Springman, *Analytical Biochemistry*, 2004, **325**, 126–136.

Chen, G, MD Porter, JR Bristol, MJ Fitzgibbon and S Pazhanisamy, Kinetic mechanism of the p38-α MAP kinase: phosphoryl transfer to synthetic peptides, *Biochemistry*, 2000, **39**, 2079–2087.

Corel Quattro Pro 8.0, Corel Corporation Limited, Ottawa, Canada, 1997.

Cuenda, A, J Rouse, YN Doza, R Meier, P Cohen, TF Gallagher, PR Young and JC Lee, *FEBS Letters*, 1995, **364**, 229–233.

Han, J, JD Lee, L Bibbs and RJ Ulevitch, *Science*, 1994, **265**, 808–811.

Havlin, S, *The Fractal Approach to Heterogeneous Chemistry: Surfaces, Colloids, Polymers*, Wiley, New York, 1989, pp. 251–269.

Herlaar, E and Z Brown, p38 MAPK signaling cascades in inflammatory disease, *Molecular Medicine Today*, 1999, **5**, 439–447.

Lee, JC, S Kumar, DE Griswold, DC Underwood, BJ Votta and JL Adams, Inhibition of p38 MAP kinase as a therapeutic strategy, *Immunopharmacology*, 2000, **47**, 185–201.

Lee, JC, JT Laydon, PC McDonnell, TF Gallagher, S Kumar, D Green, D McNulty, MM Blumenthal, IR Siemens, SM Fisher, GP Livi, JR White, JL Adams and PR Young, *Nature*, 1994, **372**, 739–746.

LoGrosso, PV, B Frantz, AM Rolando, SJ O'Keefe, JD Hermes and EA O'Neill, *Biochemistry*, 1997, **36**, 10422–10427.

Ramakrishnan, A and A Sadana, *Automedica*, 2001, **20**(3/4), 313–340.

Sadana, A and T Vo-Dinh, *Biotechnology & Applied Biochemistry*, 2001, **33**, 17–31.

Weiss, GH, *Fractals in Science*. Springer, Berlin, 1994.

Young, PR, MM McLaughlin, S Kumar, ML Doyle, D McNulty, TF Gallagher, S Fisher, PC McDonnell, SC Carr, MJ Huddleston, G Seibel, TG Porter, PG Livi, JL Adams and JC Lee, *Journal of Biological Chemistry*, 1997, **272**, 12116–12121.

– 9 –

Fractal Analysis of Heparin–Protein Interaction Studies Occurring on Biosensor Surfaces

9.1 INTRODUCTION

Heparin interacts with a wide variety of proteins and peptides (Lindahl, 1989). Zhang *et al.* (2002) indicate that it is a proteoglycan that is composed of highly sulfated linear polysaccharides of alternating uronic acid and glucosamine. In the human body heparin and its structurally related heparan sulfate are the most acidic polysaccharides. Heparin binds to growth factors (Mach *et al.*, 1993; Faham *et al.*, 1996). Other interactions of heparin include (a) its influence of angiogenesis and other-proliferation-dependent processes, and (b) its binding to the ectodomain of proteins of pathogens which influence infection (Swaitzky, 1996).

The surface plasmon resonance (SPR) biosensor has been used to measure protein–protein or other biomolecular reactions, and heparin–protein interactions in particular (Mach *et al.*, 1993; Capila *et al.*, 2001; Capila and Linhardt, 2002; Dong *et al.*, 2001; Marks *et al.*, 2001). Protein–DNA, protein–sugar, and protein–lipid interactions have also been analyzed by the SPR technique (Nagata and Handa, 2000). Zhang *et al.* (2002) indicate that heparin was previously immobilized on a sensor chip surface through an avidin–biotin bridge system. They emphasize that heparin binding proteins react specifically with avidin (Marks *et al.*, 2001). Zhang *et al.* (2002) noted further that the streptavidin–biotin heparin chip was also unstable. They along with Van Delden *et al.* (1997) have since then analyzed heparin–protein interactions using immobilized albumin–heparin conjugates. Zhang *et al.* (2002) have recently analyzed heparin–protein interactions on a biochip using a preformed albumin–heparin conjugate.

The system by its very design is heterogeneous (for example, the receptors immobilized on the biosensor surface may exhibit some heterogeneity that is, surface roughness), and often other factors such as mass transport limitations (unless they are carefully eliminated or minimized) play a significant role and further complicate the design (especially its kinetic aspects) of the assay or the correct interpretation of the assay results. One possible way of monitoring for the presence of diffusional limitations and the heterogeneity that exists on the surface is by using fractals. A characteristic feature of fractals is self-similarity at different levels of scale. Fractals exhibit dilatational symmetry. Fractals are disordered systems, and the disorder is described by nonintegral dimensions (Pfeifer and Obert, 1989). Fractals have previously been used to analyze the binding and dissociation kinetics of a wide variety of analyte–receptor systems (Sadana, 2001). Fractals are particularly useful for this type of analysis because they help to characterize the heterogeneity that exists on the surface by a lumped parameter, the fractal dimension.

In this chapter we provide an alternate analysis for the binding and dissociation of heparin binding proteins in solution to covalently immobilized heparin on a SPR biochip using a preformed albumin–heparin conjugate (Zhang *et al.*, 2002). Binding and dissociation rate coefficients, as well as fractal dimension values for the binding and the dissociation phases will be provided wherever applicable. We offer the fractal analysis as an alternate analysis to help improve the understanding; we do not imply that this is better than the original (SPR-based software) analysis.

9.2 THEORY

Havlin (1989) has reviewed and analyzed the diffusion of reactants towards fractal surfaces. The details of the theory and the equations involved for the binding and the dissociation phases for analyte–receptor (in our case, heparin binding protein–heparin) are available (Ramakrishnan and Sadana, 2001; Sadana and Vo-Dinh, 2001). Here we will attempt to apply these equations to the heparin binding protein–heparin interaction.

The basic idea is to fold everything into the lumped parameter, the fractal dimension, D_f. This lumped parameter takes care of the chemical, orientational, and geometrical details. It is also assumed that the kinetic rate depends only on the geometric (static) parameter, D_f. It would also be useful to indicate the low and high limits in angstroms of the fractal character to be presented.

9.2.1 Single-fractal analysis

Binding rate coefficient

Havlin (1989) indicates that the diffusion of a particle (analyte) from a homogeneous solution to a solid surface (e.g. receptor-coated surface) on which it reacts to form a product (analyte–receptor) complex is given by:

$$(\text{Analyte}\cdot\text{Receptor}) \sim \begin{cases} t^{(3-D_{\mathrm{f,bind}})/2} = t^p & (t < t_c) \\ t^{1/2} & (t > t_c) \end{cases} \tag{9.1a}$$

where the analyte–receptor represents the association (binding) complex formed on the surface. Here $p = -b$, and D_f is the fractal dimension of the surface. Havlin (1989) states that the crossover value may be determined by $r_c^2 \sim t_c$. Above the characteristic length, r_c, the self-similarity of the surface is lost and the surface may be considered homogeneous. Eq. (9.1a) indicates that the concentration of the product [analyte–receptor] on a solid fractal surface scales at short and intermediate times as analyte–receptor $\sim t^p$ with the coefficient $p = (3 - D_f)/2$ at short time scales and $p = 1/2$ at intermediate time scales. Note that D_f, $D_{f,assoc}$, and $D_{f,bind}$ are used interchangeably. This equation is associated with the short-term diffusional properties of a random walk on a fractal surface. Note that, in perfectly stirred kinetics on a regular (nonfractal) structure (or surface), k_1 is a constant, that is, it is independent of time. In other words, the limit of regular structures (or surfaces) and the absence of diffusion-limited kinetics leads to k_1 being independent of time. In all other situations, one would expect a scaling behavior given by $k_1 \sim k' t^{-b}$ with $-b = p < 0$. Also, the appearance of the coefficient, p different from $p = 0$ is the consequence of two different phenomena, that is, the heterogeneity (fractality) of the surface and the imperfect mixing (diffusion-limited) condition. Finally, for a homogeneous surface where D_f is equal to 2, and when only diffusional limitations are present, $p = 1/2$ as it should be. Another way of looking at the $p = 1/2$ case (where $D_{f,bind}$ is equal to two) is that the analyte in solution views the fractal object, in our case the receptor-coated biosensor surface, from a 'large distance'. In essence, in the association process, the diffusion of the analyte from the solution to the receptor surface creates a depletion layer of width $(Đ_t)^{1/2}$ where $Đ$ is the diffusion constant. This gives rise to the fractal power law, (Analyte·Receptor) $\sim t^{(3-D_{f,bind})/2}$.

The values of the parameters k (association rate coefficient), p, and D_f in Eq. (9.1a) may be obtained for analyte–receptor binding kinetics data. This may be done by a regression analysis using, for example, Corel Quattro Pro (1997) along with Eq. (9.1a) where (analyte·receptor) $= kt^p$. The fractal dimension may be obtained from the parameter p. Higher values of the fractal dimension would indicate higher degrees of 'disorder' or heterogeneity or inhomogeneity on the surface.

Another way of looking at the diffusive process is that it inherently involves fluctuations at the molecular level that may be described by a random walk (Weiss, 1994). This author indicates that the kinetics of transport on disordered (or heterogeneous) media itself needs to be described by a random walk model. When both of these are present, that is the diffusion phenomena as well as a fractal surface, then one needs to analyze the interplay of both of these fluctuations. In essence, the disorder on the surface (or a higher fractal dimension, D_f) tends to slow down the motion of a particle (analyte in our case) moving in such a medium. Basically, according to Weiss (Weiss, 1994) the particle (random walker analyte) is trapped in regions in space, as it oscillates for a long time before resuming its motion.

Havlin (1989) indicates that the crossover value may be determined by $r_c^2 \sim t_c$. Above the characteristic length, r_c, the self-similarity of the surface is lost. Above t_c, the surface may be considered homogeneous, and 'regular' diffusion is now present. One may consider the analysis to be presented as an intermediate 'heuristic' approach in that in the future one may also be able to develop an autonomous (and not time-dependent) model of diffusion-limited kinetics in disordered media. No evidence of fractality is presented.

Dissociation rate coefficient

The diffusion of the dissociated particle (receptor or analyte) from the solid surface (e.g. analyte–receptor complex coated surface) into solution may be given, as a first approximation by:

$$\begin{aligned}(\text{Analyte}\cdot\text{Receptor}) &\sim -t^{(3-D_{\text{f,diss}})/2} \quad (t > t_{\text{diss}}) \\ &= -k_{\text{diss}} t^{(3-D_{\text{f,diss}})/2}\end{aligned} \tag{9.1b}$$

Here $D_{\text{f,diss}}$ is the fractal dimension of the surface for the dissociation step. t_{diss} represents the start of the dissociation step. This corresponds to the highest concentration of the analyte–receptor complex on the surface. Henceforth, its concentration only decreases. $D_{\text{f,assoc}}$ may or may not be equal to $D_{\text{f,diss}}$. k_{d} and k_{diss}, and $D_{\text{f,d}}$ and $D_{\text{f,diss}}$ are used interchangeably in the chapter.

9.2.2 Dual-fractal analysis

Binding rate coefficient

The single-fractal analysis we have just presented is extended to include two fractal dimensions. At present, the time ($t = t_1$) at which the first fractal dimension 'changes' to the second fractal dimension is arbitrary and empirical. For the most part it is dictated by the data analyzed and the experience gained by handling a single-fractal analysis. The r^2 (regression coefficient) value obtained is also used to determine if a single-fractal analysis is sufficient, or one needs to use a dual-fractal analysis to provide an adequate fit.

Only if the r^2 value is less than 0.97 for a single-fractal analysis, do we use a dual-fractal model. In this case, the analyte–receptor complex is given by:

$$(\text{Analyte}\cdot\text{Receptor}) \sim \begin{cases} t^{(3-D_{\text{f1,bind}})/2} = t^{p_1} & (t < t_1) \\ t^{(3-D_{\text{f2,bind}})/2} = t^{p_2} & (t_1 < t < t_2 = t_c) \\ t^{1/2} & (t > t_c) \end{cases} \tag{9.1c}$$

In analyte–receptor association the analyte–receptor associates with the active site on the surface and the product is released. In this sense the catalytic surface exhibits an unchanging fractal surface to the reactant in the absence of fouling and other complications. In the case of analyte–receptor association the biosensor surface exhibits a changing fractal surface to the analyte in solution. This occurs since as each association (binding) reaction takes place, smaller and smaller amounts of 'association' sites or receptors are available on the biosensor surface to which the analyte may bind. Furthermore, as the reaction proceeds, there is an increasing degree of heterogeneity on the biosensor surface for some reaction systems. This is manifested by two degrees of heterogeneity, or two fractal dimensions on the biosensor surface. In the theoretical limit one might envisage a temporal fractal dimension wherein there is a continuous change in the degree of heterogeneity on the surface; though of course, such situations would be very rare, if at all.

Note that, at present, the dual-fractal analysis does not have a basis at the molecular level. This represents two different levels of heterogeneity on the biosensor surface. But, in some examples presented, a single-fractal analysis is clearly inadequate to model the data. Only in these cases does one resort to a dual-fractal analysis.

Dissociation rate coefficient

In this case the dissociation rate coefficient is given by:

$$(\text{Analyte}\cdot\text{Receptor}) \sim \begin{cases} -t^{(3-D_{\text{f1,diss}})/2} & (t_{\text{diss}} < t < t_{\text{d1}}) \\ -t^{(3-D_{\text{f2,diss}})/2} & (t_{\text{d1}} < t < t_{\text{d2}}) \end{cases} \tag{9.1d}$$

Here $D_{\text{f,diss}}$ is the fractal dimension of the surface for the dissociation step. t_{diss} represents the start of the dissociation step. This corresponds to the highest concentration of the analyte.receptor on the surface. Henceforth, its concentration only decreases. $D_{\text{f,bind}}$ or $D_{\text{f,assoc}}$ may or may not be equal to $D_{\text{f,diss}}$.

9.3 RESULTS

A fractal analysis will be applied to the binding and dissociation of heparin binding proteins in solution to covalently immobilized heparin on a SPR biochip using a preformed albumin–heparin conjugate (Zhang *et al.*, 2002). The fractal analysis is only one possible approach to analyzing the diffusion-limited binding kinetics assumed to be present in the system analyzed. Zhang *et al.* (2002) have used the software to obtain values of the k_{on}, k_{off}, and the affinity, K_{d} values.

Figure 9.1a shows the binding of 13.3 nM acidic fibroblast growth factor (aFGF) in solution to heparin immobilized on a SPR optical biosensor chip surface (Zhang *et al.*, 2002). These authors have used a nonlinear least squares fitting method provided by

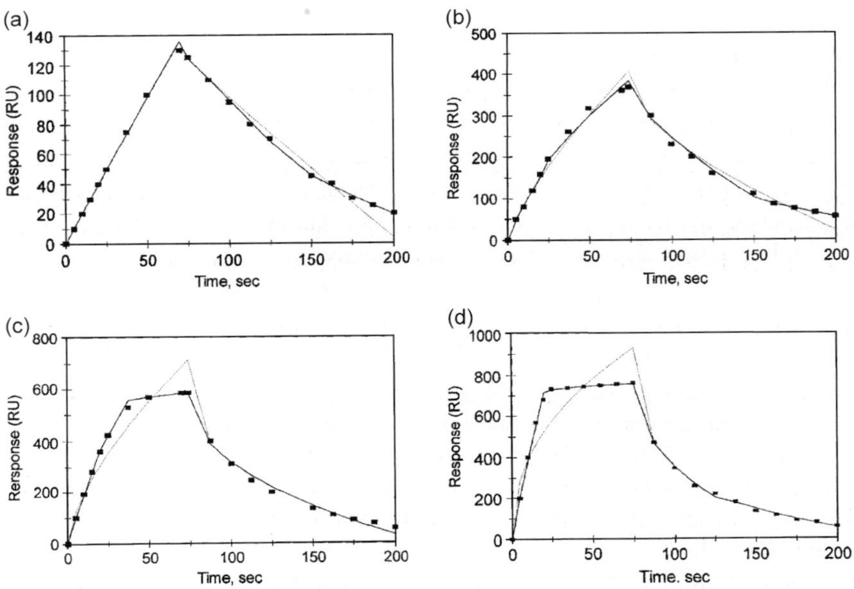

Figure 9.1 Binding of different concentrations of acidic fibroblast growth factor (aFGF) in solution (in nM) to heparin immobilized on a biochip surface (Zhang *et al.*, 2002): (a) 13.3, (b) 33.3, (c) 66.7, (d) 133.

Table 9.1a

Rate coefficient values for the binding and the dissociation phase for acidic fibroblast growth factor (aFGF) and Heparin interaction on a biochip (Zhang *et al.*, 2002)

aFGF in solution (in nM)/Heparin on chip surface	k	k_1	k_2	k_d	k_{d1}	k_{d2}
13.3	2.0842 ± 0.050	na	na	1.262 ± 0.141	1.0074 ± 0.066	7.457 ± 0.114
33.3	15.01 ± 1.28	12.701 ± 0.796	24.71 ± 1.36	14.465 ± 1.484	11.199 ± 1.068	78.81 ± 0.157
66.7	44.78 ± 8.29	22.757 ± 0.369	420.97 ± 0.782	61.523 ± 3.00	na	na
133	1908 + 2109	48.58 ± 2.76	647.52 ± 1.67	123.08 ± 5.955	92.61 ± 3.24	188.96 ± 3.10

the Biaevaluation 3.0 software (Biacore, Uppsala, Sweden). Individual rate coefficient values for the binding and the dissociation phases were obtained separately. As indicated earlier we will re-analyze the data using fractal analysis. A single-fractal analysis is required to adequately describe the binding kinetics. A dual-fractal analysis is, however, required to describe the dissociation kinetics. The value of the binding and the dissociation rate coefficient(s), and the fractal dimension values for the binding and the dissociation phases presented in Table 9.1b were obtained from a regression analysis using Eqs. (9.1a) and (9.1c) for the binding phase, and Eqs. (9.1b) and (9.1d) for the dissociation phase (Corel Quattro Pro, 1997).

The binding and dissociation rate coefficients presented in Table 9.1a,b are within 95% confidence limits. For example, for the binding of 13.3 nM aFGF in solution to heparin immobilized on a sensor chip surface, the binding rate coefficient, k_1 value is 2.0842 ± 0.050. The 95% confidence limit indicates that the k_1 value lies between 2.0342 and 2.1342.

Figure 9.1b shows the binding and dissociation of 33.3 nM aFGF in solution to heparin immobilized on a sensor chip surface. A dual-fractal analysis is required to adequately describe the binding and the dissociation kinetics. The values of (a) the binding rate coefficient, k, and (b) the binding rate coefficients, k_1 and k_2 for a dual-fractal analysis, (c) the dissociation rate coefficient, k_d for a single-fractal analysis, and (d) the dissociation rate coefficients, k_{d1} and k_{d2} are given in Table 9.1a. The values of (a) fractal dimension for the binding phase, D_f, for a single-fractal analysis (b) the fractal dimensions for

Table 9.1b

Fractal dimension values for the binding and the dissociation phase for acidic fibroblast growth factor and heparin interaction on a biochip (Zhang *et al.*, 2002)

aFGF in solution (in nM)/ Heparin on chip surface	D_f	D_{f1}	D_{f2}	D_{fd}	D_{fd1}	D_{fd2}
13.3	1.967 ± 0.021	na	na	1.076 ± 0.0686	0.9446 ± 0.057	1.892 ± 0.079
33.3	1.468 ± 0.0634	1.341 ± 0.1166	2.911 ± 0.0889	1.6884 ± 0.0914	1.5344 ± 0.116	2.430 ± 0.015
66.7	1.714 ± 0.132	1.149 ± 0.031	2.845 ± 0.012	2.091 ± 0.0450	na	na
133	3.0 − 0.321	1.206 ± 0.106	2.927 ± 0.0056	2.2618 ± 0.043	2.083 ± 0.066	2.456 ± 0.040

the binding phase, D_{f1} and D_{f2} for a dual-fractal analysis, (c) the fractal dimension for the dissociation phase, D_{fd} for a single-fractal analysis, and (d) the fractal dimensions, D_{fd1} and D_{fd2} are given in Table 9.1b.

Note, once again that as the fractal dimension value increases from D_{f1} to D_{f2} the binding rate coefficient value increases from k_1 to k_2. An increase in the fractal dimension value by a factor of 1.88 from 1.2062 to 2.9271 leads to an increase in the binding rate coefficient by a factor of 13.33 from 48.582 to 647.51. Also, an increase in the fractal dimension for dissociation by a factor of 1.18 from D_{fd1} equal to 2.0828 to D_{fd2} equal to 2.4556 leads to an increase in the dissociation rate coefficient value by a factor of 2.04 from k_{d1} equal to 92.610 to k_{d2} equal to 188.96.

Apparently, there is a change in the binding mechanism as one increases the aFGF concentration in solution from 13.3 to 33.3 nM. At the lower concentration a single-fractal analysis is adequate to describe the binding kinetics, whereas at the higher concentration a dual-fractal analysis is required to describe the binding kinetics.

Figure 9.1c shows the binding and dissociation of 66.7 nM aFGF in solution to heparin immobilized on a sensor chip surface. A dual-fractal analysis is required to adequately describe the binding kinetics. The dissociation kinetics is adequately described by a single-fractal analysis. The values of (a) the binding rate coefficient, k, for a single-fractal analysis, (b) the binding rate coefficient, k_1 and k_2 for a dual-fractal analysis, and (c) the dissociation rate coefficient, k_d for a single-fractal analysis are given in Table 9.1a. The values of (a) fractal dimension for the binding phase, D_f, for a single-fractal analysis (b) the fractal dimensions for the binding phase, D_{f1} and D_{f2} for a dual-fractal analysis, and (c) the fractal dimension for the dissociation phase, D_{fd} for a single-fractal analysis are given in Table 9.1b.

It is of interest to note that as the analyte (aFGF) concentration in solution doubles from 33.3 to 66.7 nM, and for a dual-fractal analysis, (a) the binding rate coefficient, k_1 increases by a factor of 1.79 from a value of 12.7 to 22.76, and (b) the binding rate coefficient, k_2 increases by a factor of 17.04 from a value of 24.71 to 420.97. Apparently, the binding rate coefficient, k_2 is more sensitive than the binding rate coefficient, k_1 on the change in the aFGF concentration in solution, at least in this concentration range.

Figure 9.1d shows the binding and dissociation of 133 nM aFGF in solution to heparin immobilized on a sensor chip surface. A dual-fractal analysis is required to adequately describe the binding and the dissociation kinetics. The values of (a) the binding rate coefficient, k, and (b) the binding rate coefficient, k_1 and k_2 for a dual-fractal analysis, (c) the dissociation rate coefficient, k_d for a single-fractal analysis, and (d) the dissociation rate coefficients, k_{d1} and k_{d2} are given in Table 9.1a. The values of (a) fractal dimension for the binding phase, D_f, for a single-fractal analysis (b) the fractal dimensions for the binding phase, D_{f1} and D_{f2} for a dual-fractal analysis, (c) the fractal dimension for the dissociation phase, D_{fd} for a single-fractal analysis, and (d) the fractal dimensions, D_{fd1} and D_{fd2} for a dual-fractal analysis are given in Table 9.1b.

Note that as the fractal dimension value increases from D_{f1} to D_{f2} the binding rate coefficient value increases from k_1 to k_2. An increase in the fractal dimension value by a factor of 2.43 from 1.2062 to 2.9271 leads to an increase in the binding rate coefficient by a factor of 13.33 from 48.58 to 647.52. Also, an increase in the fractal dimension for dissociation by a factor of 1.18 from D_{fd1} equal to 2.0828 to D_{fd2} equal to 2.4556 leads to an increase in the dissociation rate coefficient value by a factor of 2.04 from k_{d1} equal to 92.610 to k_{d2} equal to 188.96.

It is worth while comparing the binding and the dissociation rate coefficients with changes in the aFGF concentration in solution from 33.3 to 133 nM. In both of these cases, a dual-fractal analysis is required to describe the binding and the dissociation kinetics. An increase in the aFGF concentration in solution from 33.3 to 133 nM in solution leads to (a) an increase in the binding rate coefficient, k_1 by a factor of 3.825 from a value of 12.7 to 48.58, (b) an increase in the binding rate coefficient, k_2 by a factor of 26.2 from a value of 24.71 to 647.52, (c) an increase in the dissociation rate coefficient, k_{d1} by a factor of 9.2 from a value of 11.199 to 92.61, and (d) an increase in the dissociation rate coefficient, k_{d2} by a factor of 2.534 from a value of 78.81 to 188.96.

In the 33.3 to 133 nM aFGF concentration range, and for a dual-fractal analysis, Figure 9.2a shows the increase in the binding rate coefficient, k_1 with an increase in the aFGF concentration in solution. In this concentration range, the binding rate coefficient, k_1 is given by:

$$k_1 = (0.4132 \pm 0.0314)[\mathrm{aFGF}]^{0.9686 \pm 0.075} \tag{9.2a}$$

The fit is very good. Only three data points are available. The availability of more data points would lead to a better and more reliable fit. The non-integer dependence of the binding rate coefficient, k_1 on the aFGF concentration in solution lends support

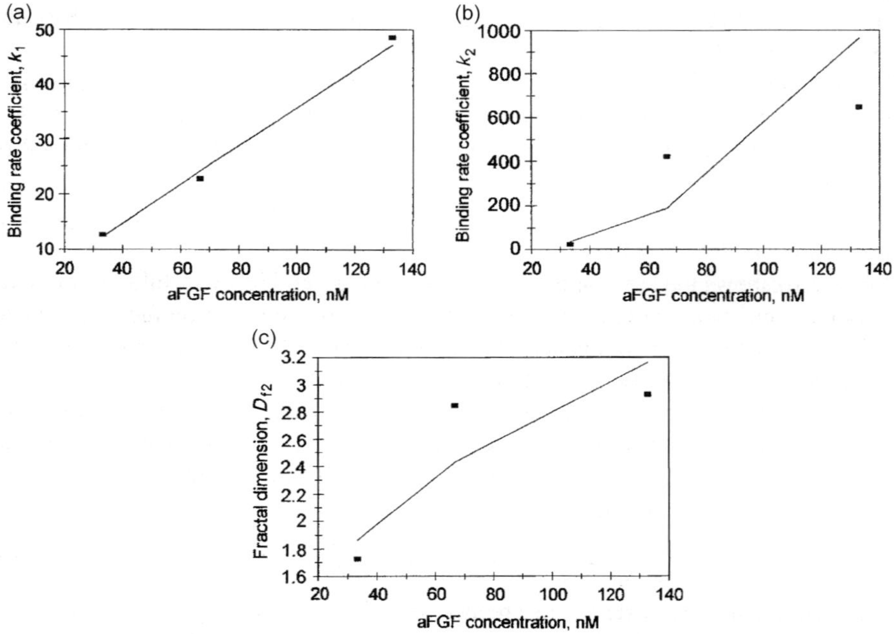

Figure 9.2 (a) Increase in the binding rate coefficient, k_1 with an increase in the aFGF concentration in solution; (b) increase in the binding rate coefficient, k_2 with an increase in the aFGF concentration in solution; (c) increase in the fractal dimension, D_{f2} with an increase in the aFGF concentration in solution.

to the fractal nature of the system. The binding rate coefficient, k_1, exhibits close to a first-order (equal to 0.9686) dependence on the aFGF concentration in solution.

In the 33.3–133 nM aFGF concentration range, and for a dual-fractal analysis, Figure 9.2b shows the increase in the binding rate coefficient, k_2 with an increase in the aFGF concentration in solution. In this concentration range, the binding rate coefficient, k_2 is given by:

$$k_2 = (0.00938 + 0.0155)[\text{aFGF}]^{2.36 \pm 0.998} \tag{9.2b}$$

The fit is reasonable. Only three data points are available. The availability of more data points would lead to a better and more reliable fit. Only the positive error is given for the coefficient since one cannot have a negative value for the binding rate coefficient. The binding rate coefficient, k_2 is quite sensitive to the aFGF concentration in solution as noted by the higher than second order (equal to 2.36) dependence. Once again, the non-integer dependence of the binding rate coefficient, k_2 on the aFGF concentration in solution lends support to the fractal nature of the system.

In the 33.3–133 nM aFGF concentration range, and for a dual-fractal analysis, Figure 9.2c shows the increase in the fractal dimension, D_{f2} with an increase in the aFGF concentration in solution. In this concentration range, the fractal dimension, D_{f2} is given by:

$$D_{f2} = (0.4893 \pm 0.1034)[\text{aFGF}]^{0.3818 \pm 0.1958} \tag{9.2c}$$

The fit is reasonable. Only three data points are available. The availability of more data points would lead to a better and more reliable fit. The fractal dimension, D_{f2} is only mildly sensitive to the aFGF concentration in solution as noted by the less than one-half order dependence (equal to 0.3818).

For the binding of aFGF in solution to heparin immobilized on a sensor chip surface, Figure 9.3a shows the increase in the binding rate coefficient, k_2 with an increase in the fractal dimension, D_{f2}. For the data in Table 9.1a,b and in Figure 9.3a, the binding rate coefficient, k_2 is given by:

$$k_2 = (0.9502 \pm 0.1928) D_{f2}^{5.956 \pm 0.4396} \tag{9.3a}$$

Only three data points are available. The fit is quite good. The availability of more data points would lead to a more reliable and better fit. The binding rate coefficient, k_2 is very sensitive to the degree of heterogeneity that exists on the surface as noted by the very high value (equal to 5.956) of the order of dependence of k_2 on D_{f2}.

For the binding of aFGF in solution to heparin immobilized on a sensor chip surface, Figure 9.3b shows the increase in the dissociation rate coefficient, k_{d1} with an increase in the fractal dimension, D_{fd1}. For the data in Table 9.1a,b and in Figure 9.3b, the binding rate coefficient, k_{d1} is given by:

$$k_{d1} = (1.2671 \pm 0.4359) D_{fd1}^{5.6489 \pm 0.5245} \tag{9.3b}$$

Once again, only three data points are available. The fit is quite good. The availability of more data points would lead to a more reliable and better fit. The dissociation rate coefficient, k_{d1} is very sensitive to the degree of heterogeneity that exists on the surface as noted by the very high value (equal to 5.6489) of the order of dependence of k_{d1} on D_{fd1}. It is also of interest to note that the order of dependence exhibited by both k_2 on D_{f2} and

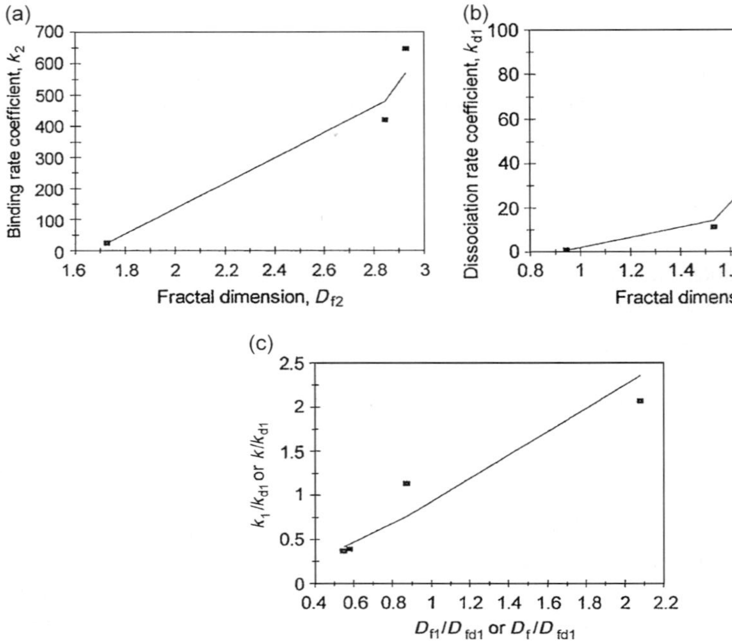

Figure 9.3 (a) Increase in the binding rate coefficient, k_2 with an increase in the fractal dimension, D_{f2}; (b) increase in the dissociation rate coefficient, k_{d1} with an increase in the fractal dimension, D_{fd1}; (c) increase in the affinity, K ($= k_1/k_{d1}$ or k/k_{d1}) with an increase in the fractal dimension ratio in the binding and in the dissociation phase ($= D_{f1}/D_{fd1}$ or D_f/D_{fd1}).

k_{d1} on D_{fd1} lies between 5.5 and 6.0. This, may just be coincidental since one is for the binding (or association) phase, and the other is for the dissociation phase.

Affinity, K values, as indicated earlier, are of interest to practicing biosensorists. For the binding of aFGF in solution to heparin immobilized on a sensor chip surface, Figure 9.3c shows the increase in the affinity, $K(= k_1/k_{d1}$ or $k/k_{fd1})$ with an increase in the fractal dimension ratio (D_{f1}/D_{fd1} or D_f/D_{fd1}). For the data in Table 9.2a,b and in Figure 9.3c,

Table 9.2a

Rate coefficient values for the binding and the dissociation phase for dengue virus envelope protein and heparin interaction (Zhang *et al.*, 2002)

Dengue virus envelope protein (analyte in nM) in solution/Heparin (receptor) on sensor chip surface	k_1	k_d
400	2.6425 ± 0.0707	9.844 ± 0.103
500	4.8282 ± 0.1373	0.3244 ± 0.053
800	11.307 ± 0.505	0.8876 ± 0.0065
1000	11.228 ± 0.539	0.5747 ± 0.0644

Table 9.2b

Fractal dimension values for the binding and the dissociation phase for dengue virus envelope protein and heparin interaction (Zhang *et al.*, 2002)

Dengue virus envelope protein (analyte in nM) in solution/Heparin (receptor) on sensor chip surface	D_f	D_{fd}
400	1.3926 ± 0.0210	2.0484 ± 0.0899
500	1.4870 ± 0.0222	1.4938 ± 0.1356
800	1.5998 ± 0.0348	1.6018 ± 0.0273
1000	1.5798 ± 0.0372	1.3808 ± 0.0956

the affinity, K is given by:

$$K(= k_1/k_{d1} \text{ or } k/k_{d1}) = (0.9912 \pm 0.3441)(D_{f1}/D_{fd1} \text{ or } D_f/D_{fd1})^{1.295 \pm 0.2998} \quad (9.3c)$$

Only four data points are available. The fit is very good considering that two data sets have been plotted together. The availability of more data points would lead to a more reliable and better fit. The affinity, K exhibits an order of dependence between 1 and $1\frac{1}{2}$ on the ratio of fractal dimensions in the binding and in the dissociation phase. If one would prefer to increase the affinity values, then one needs to increase the ratio, D_{f1}/D_{fd1} or D_f/D_{fd1}. This may require a little ingenuity in selectively increasing the fractal dimension for the binding phase, D_f or D_{f1} whilst simultaneously increasing only to a small or to a lesser extent the fractal dimension for the dissociation phase, D_{fd} or D_{fd1}.

The Dengue virus is a mosquito-transmitted virus that causes a febrile disease in humans (Chen *et al.*, 1997). The authors further indicate that the dengue virus envelope protein utilizes a sulfated form of heparin as a receptor. Figure 9.4a shows the binding and dissociation of 400 nM Dengue virus envelope protein in solution to heparin immobilized on a sensor chip (Zhang *et al.*, 2002). Zhang *et al.* (2002) do indicate that there is tight binding between the Dengue virus envelope protein and the heparin. The binding and the dissociation phase shown in Figure 9.4a can be adequately modeled using a single-fractal analysis. The values of (a) the binding rate coefficient, k for a single-fractal analysis, and (b) the dissociation rate coefficient, k_d for a single-fractal analysis are given in Table 9.2a. The values of (a) the fractal dimension for the binding phase, D_f, for a single-fractal analysis, and (b) the fractal dimension, D_{fd} for the dissociation phase and for a single-fractal analysis are given in Table 9.2b.

Figure 9.4b shows the binding and dissociation of 500 nM Dengue virus envelope protein in solution to heparin immobilized on a sensor chip (Zhang *et al.*, 2002). The binding and the dissociation phase shown in Figure 9.4b can once again be adequately modeled using a single-fractal analysis. The values of (a) the binding rate coefficient, k for a single-fractal analysis, and (b) the dissociation rate coefficient, k_d for a single-fractal analysis are given in Table 9.2a. The values of (a) the fractal dimension for the binding phase, D_f, for a single-fractal analysis, and (b) the fractal dimension, D_{fd} for the dissociation phase and for a single-fractal analysis are given in Table 9.2b. It is of interest to note that as the Dengue virus envelope protein concentration in solution increases by 25% from 400 to 500 nM (a) the binding rate coefficient, k increases by a factor of 1.83

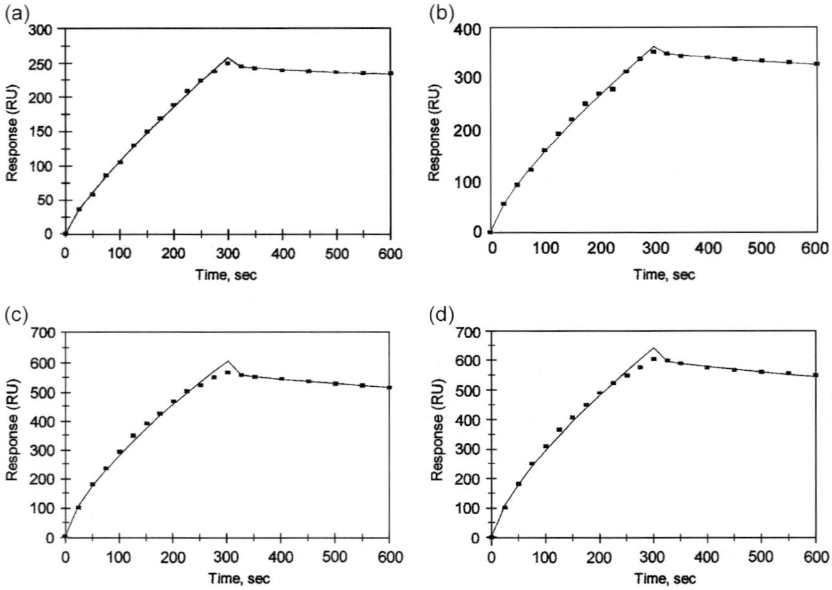

Figure 9.4 Binding of different concentrations of Dengue virus envelope protein concentration in solution (in nM) to heparin immobilized on a sensor chip surface (Zhang *et al.*, 2002): (a) 400, (b) 500, (c) 800, (d) 1000.

from a value of 2.643 to 4.828, and (b) the fractal dimension in the binding phase, D_f increases by 6.78% from a value of 1.3926 to 1.4870.

Figure 9.4c shows the binding and dissociation of 800 nM Dengue virus envelope protein in solution to heparin immobilized on a sensor chip (Zhang *et al.*, 2002). The binding and the dissociation phase shown in Figure 9.4c can be adequately modeled using a single-fractal analysis. The values (a) the binding rate coefficient, k for a single-fractal analysis, and (b) the dissociation rate coefficient, k_d for a single-fractal analysis are given in Table 9.2a. The values of (a) the fractal dimension for the binding phase, D_f, for a single-fractal analysis, and (b) the fractal dimension, D_{fd} for the dissociation phase and for a single-fractal analysis are given in Table 9.2b. It is of interest to note that as the Dengue virus envelope protein concentration in solution increases by a factor of two from 400 to 800 nM (a) the binding rate coefficient, k increases by as factor of 4.28 from a value of 2.643 to 11.31, and (b) the fractal dimension in the binding phase, D_f increases by a factor of 1.15 from a value of 1.3926 to 1.5998. Changes in the binding rate coefficient appear to be in the same direction as changes in the Dengue virus envelope protein concentration and in the fractal dimension or the degree of heterogeneity on the sensor chip surface.

Figure 9.4d shows the binding and dissociation of 1000 nM Dengue virus envelope protein in solution to heparin immobilized on a sensor chip (Zhang *et al.*, 2002). The binding and the dissociation phase shown in Figure 9.4d can be adequately modeled using a single-fractal analysis. The values of (a) the binding rate coefficient, k for a single-fractal analysis, and (b) the dissociation rate coefficient, k_d for a single-fractal analysis are given in Table 9.2a. The values of (a) the fractal dimension for the binding phase, D_f, for a single-fractal analysis, and (b) the fractal dimension for the dissociation phase,

D_{fd} for a single-fractal analysis are given in Table 9.2b. It is of interest to note that as the Dengue virus envelope protein concentration in solution increases by a factor of 2.5 from 400 to 1000 nM (a) the binding rate coefficient, k increases by a factor of 4.25 from a value of 2.643 to 11.23, and (b) the fractal dimension in the binding phase, D_f increases by a factor of 1.13 from a value of 1.3926 to 1.5798.

Note the slight decrease in the binding rate coefficient, k and in the fractal dimension, D_f with a 25% increase in the Dengue virus envelope protein concentration in solution from 800 to 1000 nM. No explanation is offered at present for this. Apparently, changes in the binding rate coefficient, k follow the changes in the fractal dimension, D_f or the degree of heterogeneity on the biosensor chip surface more closely than they do for changes in the Dengue virus envelope protein concentration in solution.

For the binding of 400–1000 nM Dengue virus envelope protein in solution to heparin immobilized on a sensor chip surface, Figure 9.5a shows the increase in the binding rate coefficient, k with an increase in the fractal dimension, D_f. For the data in Table 9.2a,b and in Figure 9.5a, the binding rate coefficient, k is given by:

$$k = (0.0657 \pm 0.0066)D_f^{11.0457 \pm 0.8714} \tag{9.4a}$$

Only four data points are available. The fit is very good. The availability of more data points would lead to a more reliable and better fit. The binding rate coefficient for a single-fractal analysis, k is extremely sensitive to the degree of heterogeneity that exists on the surface as noted by the very high value (equal to 11.0457) of the order of dependence of k on D_f.

Affinity, K values, as indicated earlier, are of interest to practicing biosensorists. For the binding and dissociation of 400–1000 nM Dengue virus envelope protein in solution to heparin immobilized on a sensor chip surface, Figure 9.5b shows the increase in the affinity, $K(= k/k_d)$ with an increase in the fractal dimension ratio (D_f/D_{fd}). For the data in Table 9.2a,b and in Figure 9.5b, the affinity, K is given by:

$$K(= k/k_d) = (13.476 \pm 0.1436)(D_f/D_{fd})^{4.1225 \pm 0.2699} \tag{9.4b}$$

Only four data points are available. The availability of more data points would lead to a more reliable and better fit. The affinity, K is very sensitive to the ratio of fractal

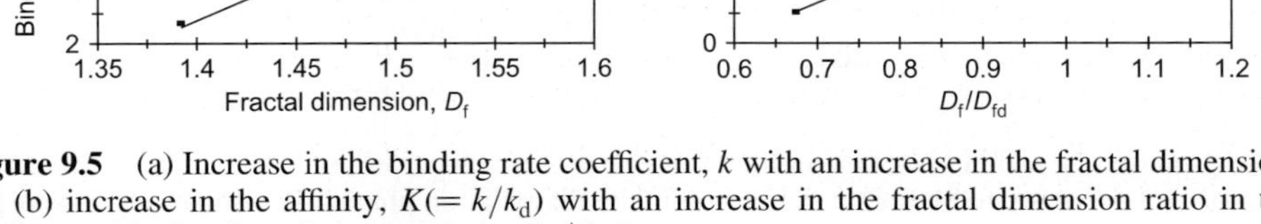

Figure 9.5 (a) Increase in the binding rate coefficient, k with an increase in the fractal dimension, D_f; (b) increase in the affinity, $K(= k/k_d)$ with an increase in the fractal dimension ratio in the binding and in the dissociation phase $(= D_f/D_{fd})$.

dimensions in the binding and in the dissociation phase as noted by the very high order (equal to 4.1225) of dependence of the affinity, K on (D_f/D_{fd}). The above predictive equation for the affinity, K is useful in that it permits one to manipulate the affinity, K in desired directions. For example, and as indicated earlier, if one would prefer to increase the affinity values, then one needs to increase the ratio, D_f/D_{fd}. This may require, as indicated earlier, a little ingenuity in selectively increasing the fractal dimension for the binding phase, D_f whilst simultaneously increasing only to a small or to a lesser extent the fractal dimension for the dissociation phase, D_{fd}.

For the binding of 400–1000 nM Dengue virus envelope protein in solution to heparin immobilized on a sensor chip surface, Figure 9.6a shows the increase in the binding rate coefficient, k with an increase in the Dengue virus envelope protein concentration in solution. For the data in Table 9.2a,b and in Figure 9.6a, the binding rate coefficient, k is given by:

$$k = (0.000176 \pm 0.000043)\,[\text{Dengue virus envelope protein}]^{1.6271 \pm 0.3034} \qquad (9.5a)$$

Only four data points are available. The fit is quite good. The availability of more data points would lead to a more reliable and better fit. The binding rate coefficient for a single-fractal analysis, k is quite sensitive to the Dengue virus envelope protein concentration in solution since it exhibits an order of dependence that lies between first and second

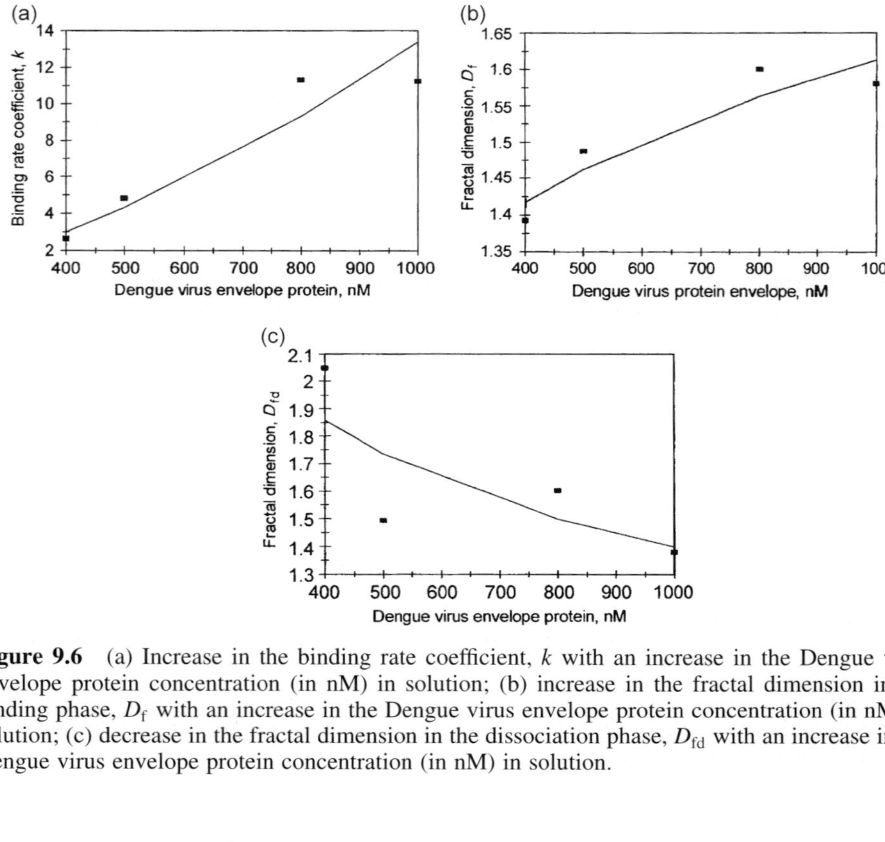

Figure 9.6 (a) Increase in the binding rate coefficient, k with an increase in the Dengue virus envelope protein concentration (in nM) in solution; (b) increase in the fractal dimension in the binding phase, D_f with an increase in the Dengue virus envelope protein concentration (in nM) in solution; (c) decrease in the fractal dimension in the dissociation phase, D_{fd} with an increase in the Dengue virus envelope protein concentration (in nM) in solution.

(equal to 1.6271) order. The non-integer order of dependence exhibited reinforces the fractal nature of the system.

For the binding of 400–1000 nM Dengue virus envelope protein in solution to heparin immobilized on a sensor chip surface, Figure 9.6b shows the increase in the fractal dimension, D_f with an increase in the Dengue virus envelope protein concentration in solution. For the data presented in Table 9.2a,b and in Figure 9.6b, the fractal dimension, D_f is given by:

$$D_f = (0.6076 \pm 0.0172)\,[\text{Dengue virus envelope protein}]^{0.1414 \pm 0.0383} \quad (9.5b)$$

Only four data points are available. The fit is quite reasonable. The availability of more data points would lead to a more reliable and better fit. The fractal dimension, D_f is only slightly dependent on the Dengue virus envelope protein concentration in solution since it exhibits an order of dependence that is equal to 0.1414.

For the binding of 400–1000 nM Dengue virus envelope protein in solution to heparin immobilized on a sensor chip surface, Figure 9.6c shows the decrease in the fractal dimension for dissociation, D_{fd} with an increase in the Dengue virus envelope protein concentration in solution. For the data in Table 9.2a,b and in Figure 9.6c, the fractal dimension, D_{fd} is given by:

$$D_{fd} = (11.897 \pm 1.716)\,[\text{Dengue virus envelope protein}]^{-0.3098 \pm 0.0383} \quad (9.5c)$$

Only four data points are available. The fit is reasonable. There is scatter in the data. Note the decrease in the fractal dimension for dissociation, D_{fd} with an increase in the Dengue virus envelope protein concentration in solution. The availability of more data points would lead to a more reliable and better fit. The fractal dimension for dissociation, D_{fd} is only slightly dependent on the Dengue virus envelope protein concentration in solution since it exhibits an order of dependence that is equal to -0.3098.

For the binding and dissociation of 400–1000 nM Dengue virus envelope protein concentration in solution to heparin immobilized on a sensor chip surface, Figure 9.7a shows the increase in the fractal dimension ratio, D_f/D_{fd} as the Dengue virus envelope protein concentration in solution increases. In the 400–1000 nM Dengue virus envelope

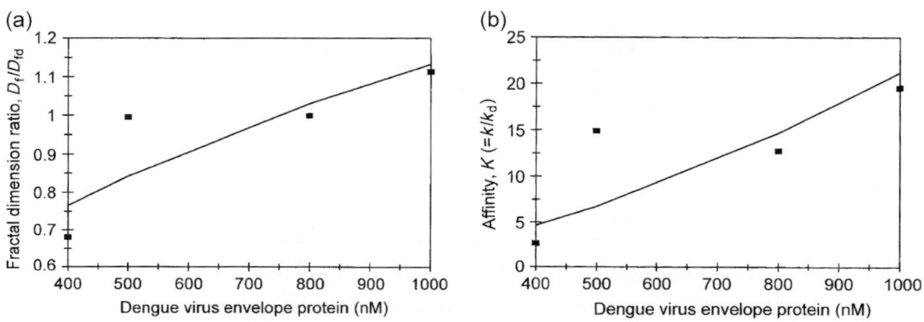

Figure 9.7 (a) Increase in the fractal dimension ratio, D_f/D_{fd} with an increase in the Dengue virus envelope protein concentration (in nM) in solution; (b) increase in the affinity, $K(= k/k_d)$ with an increase in the Dengue virus envelope protein concentration (in nM) in solution.

protein concentration range in solution, the ratio D_f/D_{fd} is given by:

$$D_f/D_{fd} = (0.05873 \pm 0.00927)\,[\text{Dengue virus envelope protein}]^{0.4285 \pm 0.2019} \quad (9.6a)$$

The fit is quite good. There is some scatter in the data. The availability of more data points would lead to a more reliable and better fit. The fractal dimension ratio, D_f/D_{fd} is only mildly sensitive to the Dengue virus envelope protein concentration in solution as noted by the less than one-half order (equal to 0.4285) of dependence exhibited by D_f/D_{fd} on the Dengue virus envelope protein concentration in solution.

For the binding and dissociation of 400–1000 nM Dengue virus envelope protein concentration in solution to heparin immobilized on a sensor chip surface, Figure 9.7b shows the increase in the affinity, K as the Dengue virus envelope protein concentration in solution increases. In the 400–1000 nM Dengue virus envelope protein concentration range in solution, the affinity, $K(= k/k_d)$ is given by:

$$K(= k/k_d) = (0.00242 \pm 0.00242)\,[\text{Dengue virus envelope protein}]^{1.674 \pm 0.9514} \quad (9.6b)$$

The fit is quite good. The availability of more data points would lead to a more reliable and better fit. The affinity, K exhibits an order of dependence that lies between $1\frac{1}{2}$ and 2 (equal to 1.6474) on the Dengue virus envelope protein concentration in solution. The non-integer order of dependence exhibited once again reinforces the fractal nature of the system.

9.4 CONCLUSIONS

A fractal analysis is presented for the binding and dissociation of heparin binding proteins in solution to covalently immobilized heparin on a SPR biochip using a preformed albumin–heparin conjugate (Zhang *et al.*, 2002). The binding and dissociation of aFGF and of Dengue virus envelope protein was analyzed. A single- as well as a dual-fractal analysis was used to adequately model the binding and the dissociation kinetics. The regression analysis was performed using Corel Quattro Pro (1997). A dual-fractal analysis was used only when a single-fractal analysis was inadequate to describe either the binding or the dissociation kinetics.

A single- as well as a dual-fractal analysis was used to model the binding and the dissociation kinetics for aFGF. However, a single-fractal analysis was adequate to describe the binding and the dissociation kinetics for the Dengue virus envelope protein in solution. This indicates a change in the mechanism of binding and dissociation for the different aFGF concentrations in solution. For example, at the lowest aFGF concentration (13.3 nM) in solution, a single-fractal analysis is adequate to describe the binding kinetics. At the higher aFGF concentrations (33.3–133 nM) in solution a dual-fractal analysis is required to adequately describe the binding kinetics.

The fractal analysis provides values of the binding and dissociation rate coefficients as well as the degree of heterogeneity (fractal dimension values) present on the biosensor chip surface. Predictive relations are presented for the binding and the dissociation rate coefficients as well as for the fractal dimension in the binding phase as a function of the analyte concentration in solution. In general, the predictive equations indicate that the binding rate coefficient(s) or the fractal dimension values exhibit an increase in their respective values with an increase in the analyte (aFGF or Dengue virus envelope protein) concentration in solution. For the data presented, the predictive equations for the binding rate coefficient exhibit a higher order of dependence on the analyte concentration than the fractal dimension.

For the binding and dissociation of aFGF in solution to heparin immobilized on a sensor chip surface predictive equations are also presented for (a) the binding rate coefficient, k_2 as a function of the fractal dimension, D_{f2}, (b) the dissociation rate coefficient, k_{d1} as a function of the fractal dimension, D_{fd1}, and (c) the affinity, K as a function of the ratio of the fractal dimensions in the binding and in the dissociation phase. The binding rate coefficient, k_2 and the dissociation rate coefficient, k_{d1} are very sensitive to the degree of heterogeneity present on the biosensor chip surface as noted by the high order of dependence exhibited on the fractal dimensions present in the binding (equal to 5.956) and dissociation (equal to 5.6489) phases. The affinity K is quite sensitive to the ratio of fractal dimension ratios as noted by the order of dependence equal to 1.295.

For the binding and dissociation of the Dengue virus envelope protein in solution to heparin immobilized on a sensor chip surface (a) the binding rate coefficient, k is extremely sensitive to the degree of heterogeneity present on the sensor chip surface exhibiting an order of dependence equal to 11.0457, and (b) the affinity, K is also very sensitive to the ratio of fractal dimensions present in the binding and in the dissociation phase exhibiting an order of dependence equal to 4.1225.

Practicing biosensorists need to pay more attention to the nature and the degree of heterogeneity present on the sensor chip surface. The quantitative relationships developed for the binding and the dissociation rate coefficients, and the affinity values as a function of the degree of heterogeneity present on the sensor chip surface will permit one to manipulate these parameters in desired directions. This type of flexibility with biosensors will permit one to increase biosensor performance parameters such as sensitivity, selectivity, stability, and other biosensor parameters of interest.

REFERENCES

Capila, I, MJ Hernaiz, YD Mo, TR Mealy, B Campos, RJ Linhardt and BA Seaton, Annexin V-heparin oligosaccharide complex suggests heparin sulfate-mediated assembly on cell surfaces. *Structure*, 2001, **9**, 57–64.

Capila, I and RJ Linhardt, Heparin–protein interactions, *Angewandte Chemie*, 2002, **41**, 390–412.

Chen, Y, T Maguire, RE Hileman, JR Fromm, JD Esko, RJ Linhardt and RM Marks, *Nature Medicine*, 1997, **3**, 866–871.

Corel Quattro Pro 8.0 (1997). Corel Corporation Limited, Ottawa, Canada.

Dong, J, C Peters-Libeu, KH Weisbarger, BW Segelle, B Rupp, I Capila, MJ Hernaiz, LA Le Brun and RJ Linhardt, *Biochemistry*, 2001, **40**, 2826–2834.

Faham, S, RE Hileman, JR Fromm, RJ Linhardt and DC Rees, *Science*, 1996, **271**, 1116–1120.

Havlin, S, *The Fractal Approach to Heterogeneous Chemistry: Surfaces, Colloids, Polymers*, pp. 251–169, Wiley, New York, 1989.

Lindahl, U, Biosynthesis of heparin and related polysaccharides, in *Chemical and Biological Properties, Clinical Applications*, (eds DA Lane and U Lindahl), CRC Press, Boca Raton, FL, 1989, pp. 159–189.

Mach, D, DB Volkin, CJ Burke, CR Middaugh, RJ Linhardt, J Fromm, D Loganathan and L Mattsson, *Biochemistry*, 1993, **32**, 5480–5489.

Marks, RM, H Lu, R Sundaresan, T Toida, A Suzuki, T Imanari, MJ Hernaiz and RJ Linhardt, Probing the interaction of dengue virus envelope protein with heparin: assessment of glycosaminoglycon-derived inhibtors. *Journal of Medicinal Chemistry*, 2001, **44**, 2178–2187.

Nagata, K and H Handa, *Real-time Analysis of Biomolecular Interactions*. Springer, New York, 2000.

Pfeifer, P and M Obert, *The Fractal Approach to Heterogeneous Chemistry: Surfaces, Colloids, Polymers*, pp. 11–43, Wiley, New York, 1989.

Ramakrishnan, A and A Sadana, *Automedica*, 2001, **20**(314), 313–340.

Sadana, A, *Analytical Biochemistry*, 2001, **29**(1), 34–47.

Sadana, A and T Vo-Dinh, *Biotechnology & Applied Biochemistry*, 2001, **33**, 17–31.

Swaitzky, D, *Medical and Microbiological Immunology*, 1996, **184**, 155–161.

Van Delden, CJ, JP Lens, RP Kooyman, GH Engbers and J Feijen, *Biomaterials*, 1997, **18**, 845–852.

Weiss, GH, *Fractals in Science*, Springer, Berlin, 1994.

Zhang, F, M Fath, R Marks and RJ Linhardt, A highly stable covalent conjugated heparin biochip for heparin-protein interaction studies. *Analytical Biochemistry*, 2002, **304**, 271–273.

– 10 –

Fractal Analysis of Binding and Dissociation Kinetics of Thrombin on Biosensor Surfaces

10.1 INTRODUCTION

The prevention of clot formation due to thrombin is an important area of investigation. Elg *et al.* (1997) and Stone and Tapparelli (1995) emphasize the importance of both the equilibrium constant of the complex between thrombin and inhibitor, and the association rate for complex formation. The reversible equilibrium between thrombin and a thrombin inhibitor may be described by:

$$\text{thrombin} + \text{thrombin inhibitor} \leftrightarrow \text{thrombin·thrombin inhibitor complex} \quad (10.1)$$

The affinity is given by:

$$K_A = [\text{thrombin·thrombin inhibitor complex}]/[\text{thrombin}][\text{thrombin inhibitor}] \quad (10.2a)$$

and the equilibrium dissociation rate constant, K_i is given by:

$$K_i = [\text{thrombin}][\text{thrombin inhibitor}]/[\text{thrombin inhibitor complex}] \quad (10.2b)$$

Lower K_i values lead to stronger affinity values. Deinum *et al.* (2002) indicate that analyte (for example, thrombin) inhibitors selected as drug candidates are often chosen on the basis of K_i values. These authors have recently presented the binding and dissociation kinetics of three potent inhibitors, Inogatran, Melagatran, and CH-248 in solution to human α-thrombin immobilized on a surface plasmon resonance (SPR) biosensor. The structures of these three high-affinity thrombin inhibitors are given in Figure 10.1.

Karlsson *et al.* (2000) had previously use the SPR biosensor technology to identify thrombin inhibitors from a compound library. In this chapter we present a fractal analysis method to determine the binding and the dissociation rate coefficients when the binding rate of the analyte in solution is limited by diffusion to the receptor immobilized on the biosensor surface.

The BIACORE biosensor based on the SPR principle is being increasingly used nowadays in a wide variety of areas as an important screening tool (in the present case to identify thrombin inhibitors) to monitor biomolecular reactions in real time. It is significant to obtain kinetic and dissociation rate coefficients for the thrombin inhibitor

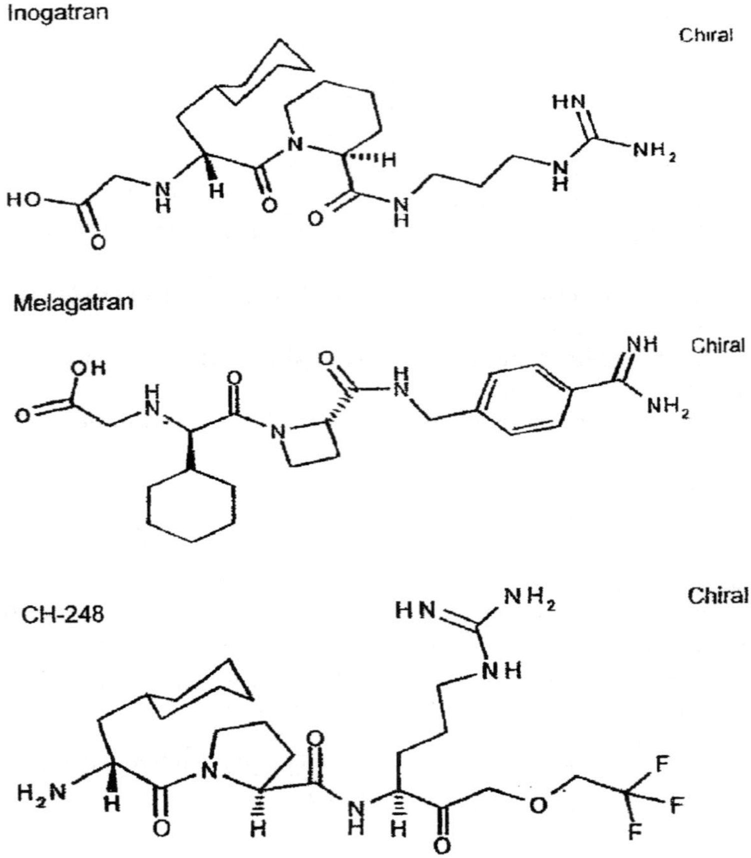

Figure 10.1 Three high-affinity thrombin inhibitors (Deinum *et al.*, 2002): Inogatran, Melagatran, and CH-248.

interactions in order to obtain better physical insights. An analysis of the interaction, wherein human α-thrombin is immobilized on the sensor chip of a SPR biosensor and the corresponding analyte is in solution, is an initial step in this direction.

The SPR biosensor protocol analyzes the binding (and dissociation wherever applicable) kinetic curves using classical saturation models involving analyte–receptor binding using 1:1, 1:2, etc. ratios, generally under diffusion-free and assuming that the receptors are homogeneously distributed over the sensor surface. Computer programs and software that come with the equipment provide values of the binding (and the dissociation) rate coefficients (Biacore, 2002). Though a careful analysis and experimental protocol may eliminate or minimize the influence of diffusional limitations; realistically speaking, it is more appropriate to include a heterogeneous distribution of receptors on the sensor surface. It is for this reason that the computer programs and the software indicated above, and provided by the manufacturers (Biacore, 2002) are not used in the present analysis. Instead, an alternate analysis is used that incorporates theoretically in the kinetic model the heterogeneous distribution of receptors on the sensing surface. This is a more realistic approach to the real-life situation. This would become more significant if the degree of heterogeneity of the receptors on the surface affects the binding and the dissociation rate coefficients to a large degree.

The system by its design is heterogeneous (for example, the receptors immobilized on the biosensor surface may exhibit some heterogeneity; that is, surface roughness), and often other factors like mass transport limitations (unless they are carefully eliminated or minimized) play a significant role and further complicate the design (especially the kinetics aspects) of the assay or the correct interpretation of the assay results. One possible way of accounting for the presence of heterogeneity that exists on the surface is by using fractals. A characteristic feature of fractals is the self-similarity at different levels of scale. Fractals exhibit dilatational symmetry. Fractals are particularly useful for this type of analysis because they help characterize the heterogeneity that exists on the surface by a lumped parameter, the fractal dimension.

In this chapter we use fractals to analyze the binding and dissociation of Inogatran, Melagatran, and CH-248 in solution to human α-thrombin immobilized on a SPR biosensor surface (Deinum *et al.*, 2002). The data has been analyzed before by the software that comes along with the SPR biosensor (Deinum *et al.*, 2002). But, in the analysis to be presented in this chapter we include (and as indicated above) the involvement of heterogeneity on the receptor surface. This aspect has been neglected in the previous study (Deinum *et al.*, 2002). We assume that the surface heterogeneity is responsible, along with other factors, for the complex character of the thrombin–thrombin inhibitor interactions. Binding and dissociation rate coefficients, affinity values, as well as fractal dimension values for the binding and dissociation phases are provided for the thrombin–thrombin inhibitor system. As indicated before the fractal analysis is offered to help improve the understanding. We do not imply that this is better than the original (SPR-based software) analysis (Deinum *et al.*, 2002).

10.2 THEORY

Havlin (1989) has reviewed and analyzed the diffusion of reactants towards fractal surfaces. The details of the theory and the equations involved for the binding and the dissociation phases for analyte–receptor (in our case, thrombin–thrombin inhibitor) are available (Ramakrishnan and Sadana, 2001; Sadana and Vo-Dinh, 2001). Here we will attempt to apply these equations to the thrombin–thrombin inhibitor interaction.

The basic idea is to fold everything into the lumped parameter, the fractal dimension, D_f. This lumped parameter takes care of the chemical, orientational, and geometrical details. It is also assumed that the kinetic rate depends only on the geometric (static) parameter, D_f. It would also be useful to indicate the low and high limits in angstroms of the fractal character to be presented.

10.2.1 Single-fractal analysis

Binding rate coefficient

Havlin (1989) indicates that the diffusion of a particle (analyte) from a homogeneous solution to a solid surface (e.g. receptor-coated surface) on which it reacts to form

a product (analyte–receptor) complex is given by:

$$(\text{Analyte}\cdot\text{Receptor}) \sim \begin{cases} t^{(3-D_{\mathrm{f,bind}})/2} = t^{p} & (t < t_{\mathrm{c}}) \\ t^{1/2} & (t > t_{\mathrm{c}}) \end{cases} \tag{10.3a}$$

where the analyte–receptor represents the association complex formed on the surface. Here $p = -b$, and D_{f} is the fractal dimension of the surface. Havlin (1989) states that the crossover value may be determined by $r_{\mathrm{c}}^2 \sim t_{\mathrm{c}}$. Above the characteristic length, r_{c}, the self-similarity of the surface is lost and the surface may be considered homogeneous. Eq. (10.3a) indicates that the concentration of the product [analyte–receptor] on a solid fractal surface scales at short and intermediate times as analyte–receptor $\sim t^{p}$ with the coefficient $p = (3 - D_{\mathrm{f}})/2$ at short time scales and $p = 1/2$ at intermediate time scales. Note that D_{f}, $D_{\mathrm{f,assoc}}$, and $D_{\mathrm{f,bind}}$ are used interchangeably. This equation is associated with the short-term diffusional properties of a random walk on a fractal surface. Note that, in perfectly stirred kinetics on a regular (nonfractal) structure (or surface), k_1 is a constant, that is, it is independent of time. In other words, the limit of regular structures (or surfaces) and the absence of diffusion-limited kinetics leads to k_1 being independent of time. In all other situations, one would expect a scaling behavior given by $k_1 \sim k' t^{-b}$ with $-b = p < 0$. Also, the appearance of the coefficient, p different from $p = 0$ is the consequence of two different phenomena, that is, the heterogeneity (fractality) of the surface and the imperfect mixing (diffusion-limited) condition. Finally, for a homogeneous surface where D_{f} is equal to 2, and when only diffusional limitations are present, $p = 1/2$ as it should be. Another way of looking at the $p = 1/2$ case (where $D_{\mathrm{f,bind}}$ is equal to two) is that the analyte in solution views the fractal object, in our case the receptor-coated biosensor surface, from a 'large distance'. In essence, in the association process, the diffusion of the analyte from the solution to the receptor surface creates a depletion layer of width $(Đt)^{\frac{1}{2}}$ where $Đ$ is the diffusion constant. This gives rise to the fractal power law, $(\text{Analyte}\cdot\text{Receptor}) \sim t^{(3-D_{\mathrm{f,bind}})/2}$.

The values of the parameters k (association rate coefficient), p, and D_{f} in Eq. (10.3a) may be obtained for analyte–receptor binding kinetics data. This may be done by a regression analysis using, for example, Corel Quattro Pro (1997) along with Eq. (10.3a) where $(\text{analyte}\cdot\text{receptor}) = kt^{p}$. The fractal dimension may be obtained from the parameter p. Higher values of the fractal dimension would indicate higher degrees of 'disorder' or heterogeneity or inhomogeneity on the surface.

Another way of looking at the diffusive process is that it inherently involves fluctuations at the molecular level that may be described by a random walk (Weiss, 1994). This author indicates that the kinetics of transport on disordered (or heterogeneous) media itself needs to be described by a random-walk model. When both of these are present, that is the diffusion phenomena as well as a fractal surface, then one needs to analyze the interplay of both of these fluctuations. In essence, the disorder on the surface (or a higher fractal dimension, D_{f}) tends to slow down the motion of a particle (analyte in our case) moving in such a medium. Basically, according to Weiss (Weiss, 1994) the particle (random walker analyte) is trapped in regions in space, as it oscillates for a long time before resuming its motion.

Havlin (1989) indicates that the crossover value may be determined by $r_{\mathrm{c}}^2 \sim t_{\mathrm{c}}$. Above the characteristic length, r_{c}, the self-similarity of the surface of the surface is lost.

Above t_c, the surface may be considered homogeneous, and 'regular' diffusion is now present. One may consider the analysis to be presented as an intermediate 'heuristic' approach in that in the future one may also be able to develop an autonomous (and not time-dependent) model of diffusion-limited kinetics in disordered media. No evidence of fractality is presented.

Dissociation rate coefficient

The diffusion of the dissociated particle (receptor or analyte) from the solid surface (e.g. analyte–receptor complex coated surface) into solution may be given, as a first approximation by:

$$\begin{aligned}(\text{Analyte}\cdot\text{Receptor}) &\sim -t^{(3-D_{\mathrm{f,diss}})/2} \qquad (t > t_{\mathrm{diss}}) \\ &= -k_{\mathrm{diss}}t^{(3-D_{\mathrm{f,diss}})/2}\end{aligned} \tag{10.3b}$$

Here $D_{\mathrm{f,diss}}$ is the fractal dimension of the surface for the dissociation step. t_{diss} represents the start of the dissociation step. This corresponds to the highest concentration of the analyte–receptor complex on the surface. Henceforth, its concentration only decreases. $D_{\mathrm{f,assoc}}$ may or may not be equal to $D_{\mathrm{f,diss}}$. k_{d} and k_{diss}, and $\mathrm{D_{f,d}}$ and $\mathrm{D_{f,diss}}$ are used interchangeably in the chapter.

10.2.2 Dual-fractal analysis

Binding rate coefficient

The single-fractal analysis we have just presented is extended to include two fractal dimensions. At present, the time ($t = t_1$) at which the first fractal dimension 'changes' to the second fractal dimension is arbitrary and empirical. For the most part it is dictated by the data analyzed and the experience gained by handling a single-fractal analysis. The r^2 (regression coefficient) value obtained is also used to determine if a single-fractal analysis is sufficient, or one needs to use a dual-fractal analysis to provide an adequate fit. Only if the r^2 value is less than 0.97 for a single-fractal analysis, do we use a dual-fractal model. In this case, the analyte–receptor complex is given by:

$$(\text{Analyte}\cdot\text{Receptor}) \sim \begin{cases} t^{(3-D_{\mathrm{f1,bind}})/2} = t^{p_1} & (t < t_1) \\ t^{(3-D_{\mathrm{f2,bind}})/2} = t^{p_2} & (t_1 < t < t_2 = t_c) \\ t^{1/2} & (t > t_c) \end{cases} \tag{10.3c}$$

In analyte–receptor association the analyte–receptor associates with the active site on the surface and the product is released. In this sense the catalytic surface exhibits an unchanging fractal surface to the reactant in the absence of fouling and other complications. In the case of analyte–receptor association the biosensor surface exhibits a changing fractal surface to the analyte in solution. This occurs since as each association (binding) reaction takes place, smaller and smaller amounts of 'association' sites or receptors are available on the biosensor surface to which the analyte may bind. Furthermore, as the reaction proceeds, there is an increasing degree of heterogeneity on the biosensor surface for some reaction systems. This is manifested by two degrees of heterogeneity, or two fractal dimensions on the biosensor surface. In the theoretical limit one might envisage a temporal fractal dimension wherein there is a continuous change in the degree of heterogeneity on the surface; though of course, such situations would be very rare, if at all.

Note that, at present, the dual-fractal analysis does not have a basis at the molecular level. This represents two different levels of heterogeneity on the biosensor surface. But, in some examples presented, a single-fractal analysis is clearly inadequate to model the data. Only in these cases does one resort to a dual-fractal analysis.

Dissociation rate coefficient

In this case the dissociation rate coefficient is given by:

$$(\text{Analyte}\cdot\text{Receptor}) \sim \begin{cases} -t^{(3-D_{\mathrm{f1,diss}})/2} & (t_{\mathrm{diss}} < t < t_{\mathrm{d1}}) \\ -t^{(3-D_{\mathrm{f2,diss}})/2} & (t_{\mathrm{d1}} < t < t_{\mathrm{d2}}) \end{cases} \qquad (10.3\text{d})$$

Here $D_{\mathrm{f,diss}}$ is the fractal dimension of the surface for the dissociation step. t_{diss} represents the start of the dissociation step. This corresponds to the highest concentration of the analyte·receptor on the surface. Henceforth, its concentration only decreases. $D_{\mathrm{f,bind}}$ or $D_{\mathrm{f,assoc}}$ may or may not be equal to $D_{\mathrm{f,diss}}$.

10.3 RESULTS

A fractal analysis will be applied to the data obtained for thrombin–thrombin inhibitor complex binding and dissociation taken from the literature (Deinum *et al.*, 2002). The fractal analysis is only one possible approach to analyzing the diffusion-limited binding kinetics assumed to be present in the systems analyzed.

Understandably, alternate expressions for fitting the data that include saturation, first-order reaction, and no diffusion limitations are available, but these expressions are apparently deficient in describing the heterogeneity that inherently exists on the surface. Another advantage of this technique is that the analyte–receptor binding (as well as the dissociation reaction) is a complex reaction, and the fractal analysis via the fractal dimension and the rate coefficient provides a useful lumped parameter(s) analysis of the diffusion-limited reaction occurring on a heterogeneous surface.

In the classical situation to demonstrate fractality, one should make a log–log plot, and one should definitely have a large amount of data. It may be useful to compare the fit to some other forms, such as an exponential form or one involving saturation. At present, we do not present any independent proof or physical evidence of fractals in the examples presented. It is a convenient means (since it provides a lumped parameter) to make the degree of heterogeneity that exists on the surface more quantitative. Thus, there is some arbitrariness in the fractal model to be presented. One might justifiably argue that appropriate modeling may be achieved by using a Langmuirian or other approach. The Langmuirian approach has a major drawback because it does not allow for or accommodate the heterogeneity that exists on the surface.

Deinum *et al.* (2002) have analyzed the binding of different low-molecular mass, active-site-directed thrombin inhibitors (299–575 Da) to human α-thrombin using a BIACORE SPR biosensor. Figure 10.2a shows the binding of a potent thrombin inhibitor, 192 nM CH-248, H-(R)-Cha-Pro-Arg[$CH_2OCH_2CF_3$] (520 Da) in solution at 6 °C to human α-thrombin immobilized on a biosensor chip surface. A single-fractal analysis is adequate to describe the binding as well as the dissociation kinetics.

The values of the binding and the dissociation rate coefficients are given in Table 10.1a. The values of the fractal dimensions for the binding and the dissociation phase are presented in Table 10.1b. These values of the binding and the dissociation rate coefficient(s), and the fractal dimensions for the binding and the dissociation phase presented in Table 10.1 were obtained using Sigmaplot (1993). The binding and the dissociation rate coefficient values presented in Table 10.1a and 10.1b are within 95% confidence limits. For example, for the binding of 192 nM CH-248 in solution at 6 °C to immobilized human α-thrombin on a sensor chip surface, the rate coefficient, k is given by 0.2688 ± 0.0251. The 95% confidence limit indicates that the k value lies between 0.2437 and 0.2939.

Figure 10.2b shows the binding of 24 nM CH-248 in solution at 6 °C to human α-thrombin immobilized on a biosensor chip surface. Once again, a single-fractal analysis is adequate to describe the binding and the dissociation kinetics. The values of (a) the binding rate coefficient, k, and the dissociation rate coefficient, k_d are given in Table 10.1a, and the values of the fractal dimensions for the binding phase, D_f and for the dissociation phase, D_{fd} are given in Table 10.1b. Apparently, at 6 °C, as one increases the CH-248 inhibitor concentration in solution from 24 to 192 nM, (a) the values of the fractal dimension in either the binding or the dissociation phase increase, and (b) the binding and the dissociation rate coefficient values increase. An eight-fold increase in the CH-248 concentration in solution from 24 to192 nM leads to (a) an increase in the value of the fractal dimension in the binding phase by 16.56% from 1.0 to 1.1656, (b) an increase in the value of the fractal dimension in the dissociation phase by 15.2% from 1.6766 to 1.9312, (c) an increase in the binding rate coefficient, k value by a factor of 2.37 from 0.1136 to 0.2688, and (d) an increase in the dissociation rate coefficient, k_d value by a factor of 4.47 from 0.0324 to 0.1448. In this case, the dissociation rate coefficient is more sensitive than the binding rate coefficient to the eight-fold increase in the CH-248 concentration in solution.

The affinity, K defined as equal to k/k_d is equal to 1.856 and 3.506 at 192 and 24 nM CH-248, respectively. Apparently, a decrease in the CH-248 thrombin concentration in solution by a factor of eight leads to an increase in the affinity value by a factor of 1.89. Thus, if high affinity values are required for CH-248, then the lower CH-248 (24 nM) concentration should be used.

Figure 10.2c shows the binding of 192 nM CH-248 in solution at 15 °C to human α-thrombin immobilized on a sensor chip surface. A dual-fractal analysis is required to adequately describe the binding kinetics. A single-fractal analysis is adequate to describe the dissociation kinetics. The values of the binding and the dissociation rate coefficients are presented in Table 10.1a. The values of the fractal dimensions in the binding and in the dissociation phase are given in Table 10.1b. Note that an increase in the temperature from 6 to 15 °C leads to a change in the binding mechanism for the 192 nM CH-248 concentration, since a dual-fractal analysis is required at 15 °C to describe the binding kinetics, whereas a single-fractal analysis is required at 6 °C to describe the binding kinetics.

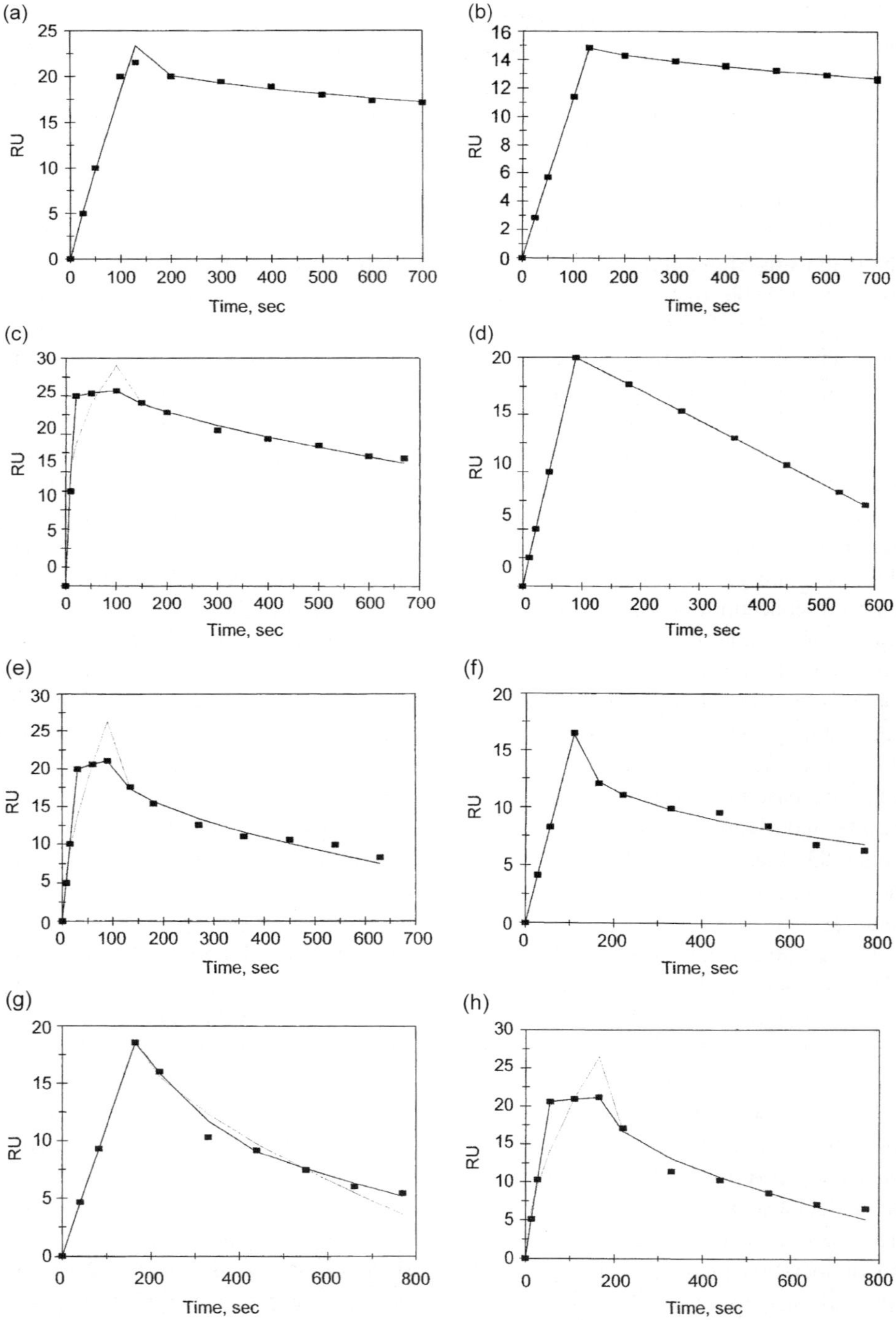

Figure 10.2 Binding of a potent thrombin inhibitor, CH-248, H-(R)-Cha-Pro-Arg [CH2 OCH2CF3] (520 Da) in solution at different concentrations and at different temperatures to human α-thrombin immobilized on a sensor chip surface: (a) 192 nM, 6 °C; (b) 24 nM, 6 °C; (c) 192 nM, 15 °C; (d) 24 nM, 15 °C; (e) 192 nM, 25 °C; (f) 24 nM, 25 °C; (g) 192 nM, 30 °C; (h) 24 nM, 30 °C.

Table 10.1a

Rate coefficients for the binding and the dissociation phase for CH-248 in solution to thrombin immobilized on a sensor chip surface (Deinum *et al.*, 2002)

Analyte in solution/ Receptor on surface	Temperature, °C	k	k_1	k_2	k_d	k_{d1}	k_{d2}
192 nM CH-248/ immobilized thrombin	6	0.2688 ± 0.0251	na	na	0.1448 ± 0.0116	na	na
24 nM CH-248/ immobilized thrombin	6	0.1136 ± 0.0.0006	na	na	0.0324 ± 0.0019	na	na
192 nM CH-248/ immobilized thrombin	15	8.298 ± 2.588	5.557 ± 1.073	23.74 ± 1.073	0.111 ± 0.0085	na	na
24 nM CH-248/ immobilized thrombin	15	0.2222 ± 0.0	na	na	0.0262 ± 0.00002	na	na
192 nM CH-248/ immobilized thrombin	25	1.9225 ± 0.619	0.6666 ± 0	16.97 ± 0.058	0.5452 ± 0.043	na	na
24 nM CH-248/ immobilized thrombin	25	0.1494 ± 0.00007	na	na	1.1499 ± 0.0734	na	na
192 nM CH-248/ immobilized thrombin	30	1.443 ± 0.476	0.3794 ± 0.0003	18.627 ± 0.020	0.5374 ± 0.07	na	na
24 nM CH-248/ immobilized thrombin	30	0.1137 ± 0.00002	na	na	0.2031 ± 0.0397	0.08669 ± 0.0223	0.8289 ± 0.020

Table 10.1b

Fractal dimensions for the binding and the dissociation phase for CH-248 in solution to thrombin immobilized on a sensor chip surface (Deinum *et al.*, 2002)

Analyte in solution/ Receptor on surface	Temperature, °C	D_f	D_{f1}	D_{f2}	D_{fd}	D_{fd1}	D_{fd2}
192 nM CH-248/ immobilized thrombin	6	1.1656 ± 0.1386	na	na	1.9312 ± 0.0890	na	na
24 nM CH-248/ immobilized thrombin	6	1.0 ± 0.0008	na	na	1.6766 ± 0.0682	na	na
192 nM CH-248/ immobilized thrombin	15	2.456 ± 0.309	2.1684 ± 0.5552	22.9661 ± 0.0054	1.5986 ± 0.0674	na	na
24 nM CH-248/ immobilized thrombin	15	1.0 ± 2.2E-15	na	na	1.0012 ± 0.0011	na	na
192 nM CH-248/ immobilized thrombin	25	1.8380 ± 0.2766	3.0 ± 0	2.9040 ± 0.0087	1.9780 ± 0.0684	na	na
24 nM CH-248/ immobilized thrombin	25	2.0 ± 0.001	na	na	2.3456 ± 0.0556	na	na
192 nM CH-248/ immobilized thrombin	30	1.8602 ± 0.2824	1.0074 ± 0.0018	2.9507 ± 0.0027	1.9416 ± 0.124	na	na
24 nM CH-248/ immobilized thrombin	30	1.0042 ± 0.00036	na	na	1.6574 ± 0.1812	0.08669 ± 0.0223	0.8289 ± 0.020

Figure 10.2d shows the binding of 24 nM CH-248 in solution at 15 °C to α-human thrombin immobilized on a sensor chip surface. A single-fractal analysis is adequate to describe the binding and the dissociation kinetics. It is of interest to compare the binding and the dissociation rate coefficient values for 24 nM CH-248 at 6 and at 15 °C. In both cases, a single-fractal analysis is adequate to describe the binding as well as the dissociation kinetics. An increase in the temperature from 6 to 15 °C, leads to (a) an increase in the binding rate coefficient, k value by a factor of 1.96 from 0.1136 to 0.2222, and (b) to a decrease in the dissociation rate coefficient, k_d value by 19.1% from 0.0324 to 0.0262.

For the 24 nM CH-248 concentration in solution, the affinity, K value is 8.48. It is of interest to note that for the 24 nM CH-248 concentration in solution, an increase in the temperature from 6 to 15 °C, leads to an increase in the affinity, K value by a factor of 2.42 from 3.506 to 8.48.

Figure 10.2e shows the binding of 192 nM CH-248 in solution at 25 °C to α-human thrombin immobilized on a sensor chip surface. A dual-fractal analysis is adequate to describe the binding kinetics. A single-fractal analysis is adequate to describe the dissociation kinetics.

Figure 10.2f shows the binding of 24 nM CH-248 in solution at 25 °C to α-human thrombin immobilized on a sensor chip surface. A single-fractal analysis is adequate to describe the binding and the dissociation kinetics. The affinity, K value is equal to 0.13. Note the sharp drop in the affinity value by a factor of 65.23 from a value of 8.48 to 0.13 as the temperature increases from 15 to 25 °C.

Figure 10.2g shows the binding of 192 nM CH-248 in solution at 30 °C to α-human thrombin immobilized on a sensor chip surface. A dual-fractal analysis is adequate to describe the binding kinetics. A single-fractal analysis is adequate to describe the dissociation kinetics.

Figure 10.2h shows the binding of 24 nM CH-248 in solution at 25 °C to α-human thrombin immobilized on a sensor chip surface. A single-fractal analysis is adequate to describe the binding and the dissociation kinetics. The affinity, K value is equal to 0.559. Note the increase in the affinity value by a factor of 4.3 for the 24 nM CH-248 concentration from a value of 0.13 to 0.559 as the temperature increases from 25 to 30 °C.

Table 10.2 shows the affinity values obtained for CH-248 at 6, 15, 25, and 30 °C for its binding in solution to human α-thrombin immobilized on a sensor chip surface. The highest affinity value (equal to 8.48) is at 15 °C, and the lowest value (equal to 0.13) is at 25 °C.

Table 10.2

Affinity values for a potent thrombin inhibitor, 24 nM CH-248 (analyte) in solution to human α-thrombin (receptor) immobilized on a BIACORE sensor chip surface

Temperature, °C	Affinity, $K(= k/k_d)$
5	3.51
15	8.48
25	0.13
30	0.559

Figure 10.3a shows the binding and dissociation of 1.8 nM melagatran in solution at 6 °C to thrombin immobilized on a SPR sensor chip surface. A single-fractal analysis is adequate to describe both the binding and the dissociation kinetics. The values of (a) the binding rate coefficient, k for a single fractal analysis, and (b) the dissociation rate coefficient, k_d for a single-fractal analysis are given in Table 10.3a. The values of the fractal dimensions for the binding and the dissociation phase are given in Table 10.3b. The affinity, K value is equal to 2.766.

Figure 10.3b shows the binding and dissociation of 7.3 nM melagatran in solution at 6 °C to thrombin immobilized on a SPR sensor chip surface. Once again, a single-fractal analysis is adequate to describe both the binding and the dissociation kinetics. The affinity, K value is equal to 8.25. It is of interest to note that an increase in the melagatran concentration in solution by a factor of about 4 from 1.8 to 7.3 nM leads to an increase in the binding rate coefficient, k by a factor of 2.65 from a value of 0.0647 to 0.1716. Also, for this same change in melagatran concentration from 1.8 to 7.3 nM there is an increase in the affinity, K value by a factor of 2.98 from 2.766 to 8.25.

Figure 10.3c shows the binding and dissociation of 29.1 nM melagatran in solution at 6 °C to thrombin immobilized on a SPR sensor chip surface. A dual-fractal analysis is required to adequately describe the binding kinetics. The dissociation kinetics is adequately described by a single-fractal analysis. The values of (a) the binding rate coefficient, k for a single fractal analysis, and (b) the binding rate coefficients, k_1 and k_2 for

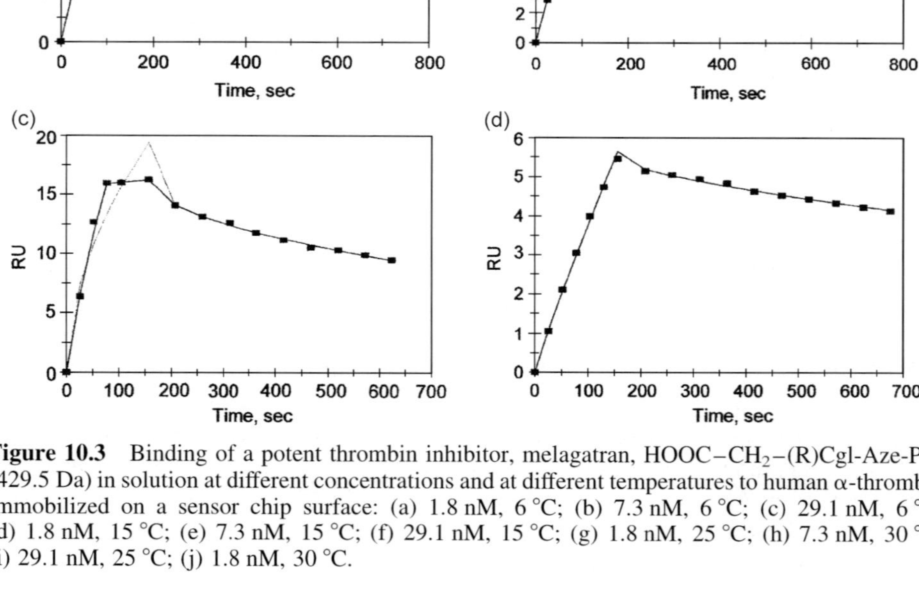

Figure 10.3 Binding of a potent thrombin inhibitor, melagatran, HOOC–CH_2–(R)Cgl-Aze-Pab (429.5 Da) in solution at different concentrations and at different temperatures to human α-thrombin immobilized on a sensor chip surface: (a) 1.8 nM, 6 °C; (b) 7.3 nM, 6 °C; (c) 29.1 nM, 6 °C; (d) 1.8 nM, 15 °C; (e) 7.3 nM, 15 °C; (f) 29.1 nM, 15 °C; (g) 1.8 nM, 25 °C; (h) 7.3 nM, 30 °C; (i) 29.1 nM, 25 °C; (j) 1.8 nM, 30 °C.

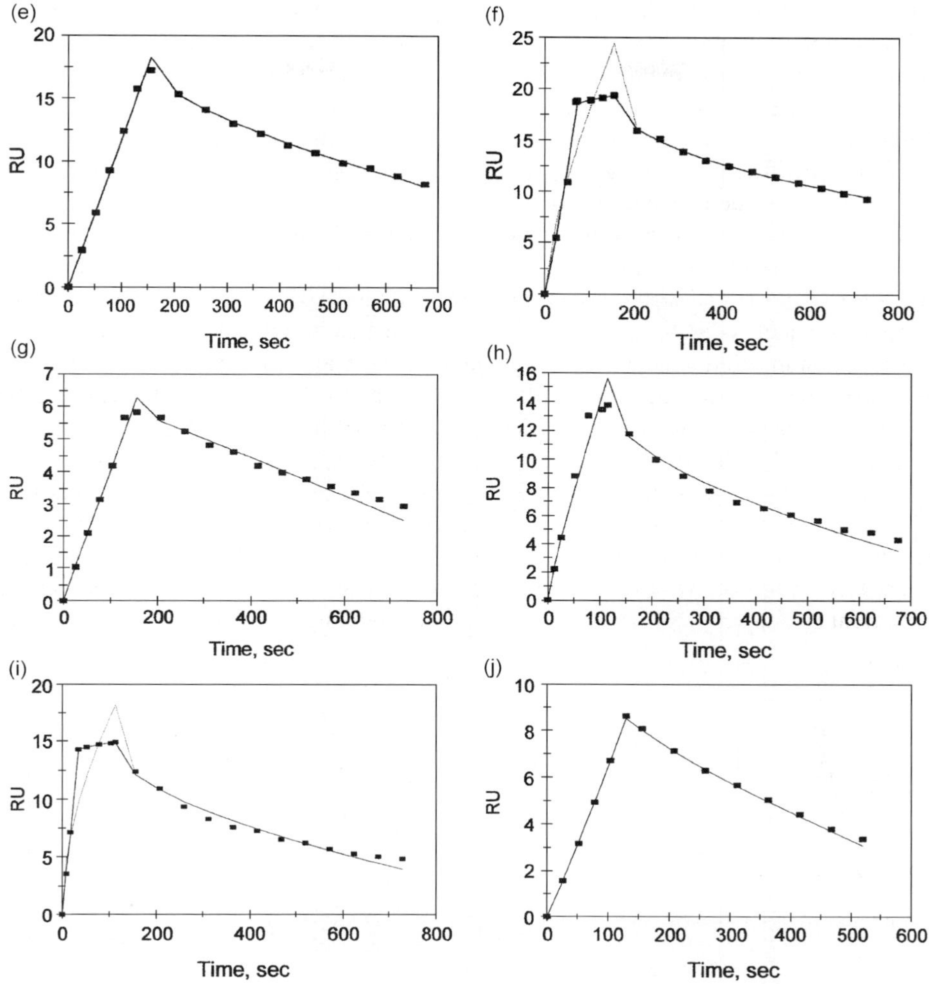

Figure 10.3 Continued.

a dual-fractal analysis are given in Table 10.3a. The values of (a) the fractal dimension, D_f for a single-fractal analysis, and (b) the fractal dimensions, D_{f1} and D_{f2} for a dual-fractal analysis are given in Table 10.3b.

There is apparently a change in the binding mechanism at the higher (29.1 nM) concentration in solution when compared with the lower (1.8 and 7.3 nM) concentrations in solution. This is because at the lower concentrations a single-fractal analysis is adequate to describe the binding kinetics, and at the higher concentration a dual-fractal analysis is required to describe the binding kinetics. Apparently, at the higher melagatran concentration in solution there is a saturation of the active (receptor) sites on the surface that leads to a change in the binding mechanism. Subsequently, a dual-fractal analysis is required to adequately describe the binding kinetics.

Table 10.3a

Rate coefficients for the binding and the dissociation phase for melagatran in solution to thrombin immobilized on a sensor chip surface (Deinum *et al.*, 2002)

Analyte in solution/ Receptor on surface	Temperature, °C	k	k_1	k_2	k_d	k_{d1}	k_{d2}
1.8 nM melagatran/ immobilized thrombin	6	0.06473 ± 0.0041	na	na	0.0234 ± 0.0032	na	na
7.3 nM melagatran/ immobilized thrombin	6	0.1716 ± 0.012	na	na	0.021 ± 0.002	na	na
29.1 nM melagatran/ immobilized thrombin	6	1.3248 ± 0.2872	0.3958 ± 0.037	14.08 ± 0.065	0.2753 ± 0.0074	na	na
1.8 nM melagatran/ immobilized thrombin	15	0.0533 ± 0.00169	na	na	0.0187 ± 0.0016	na	na
7.3 nM melagatran/ immobilized thrombin	15	0.1083 ± 0.0044	na	na	0.136 ± 0.0031	na	na
29.1 nM melagatran/ immobilized thrombin	15	0.6848 ± 0.1862	0.1006 ± 0.0100	14.7 ± 0.003	0.515 ± 0.0141	na	na
1.8 nM melagatran/ immobilized thrombin	25	0.04155 ± 0.0023	na	na	0.0045 ± 0.0007	na	na
7.3 nM melagatran/ immobilized thrombin	25	0.2554 ± 0.0335	na	na	0.2496 ± 0.0169	na	na
29.1 nM melagatran/ immobilized thrombin	25	1.4857 ± 0.4294	0.6285 ± 0.1152	12.645 ± 0.023	0.3171 ± 0.022	na	na
1.8 nM melagatran/ immobilized thrombin	30	0.0492 ± 0.00087	na	na	0.0429 ± 0.0062	0.08669 ± 0.0223	0.8289 ± 0.020

Table 10.3b

Fractal dimensions for the binding and the dissociation phase for melagatran in solution to thrombin immobilized on a sensor chip surface (Deinum *et al.*, 2002)

Analyte in solution/Receptor on surface	Temperature, °C	D_f	D_{f1}	D_{f2}	D_{fd}	D_{fd1}	D_{fd2}
1.8 nM melagatran/immobilized thrombin	6	1.276 ± 0.0935	na	na	1.762 ± 0.0746	na	na
7.3 nM melagatran/immobilized thrombin	6	1.253 ± 0.099	na	na	1.2290 ± 0.0738	na	na
29.1 nM melagatran/immobilized thrombin	6	1.9368 ± 0.2848	1.2822 ± 0.227	2.9434 ± 0.018	1.9536 ± 0.0258	na	na
1.8 nM melagatran/immobilized thrombin	15	1.153 ± 0.0423	na	na	1.6392 ± 0.0778	na	na
7.3 nM melagatran/immobilized thrombin	15	0.9692 ± 0.0534	na	na	1.6522 ± 0.021	na	na
29.1 nM melagatran/immobilized thrombin	15	1.5846 ± 0.324	0.5696 ± 0.2324	2.892 ± 0.0058	2.069 ± 0.0228	na	na
1.8 nM melagatran/immobilized thrombin	25	1.0132 ± 0.0734	na	na	0.9176 ± 0.1212	na	na
7.3 nM melagatran/immobilized thrombin	25	1.2626 ± 0.129	na	na	1.8276 ± 0.0518	na	na
29.1 nM melagatran/immobilized thrombin	25	1.9432 ± 0.2162	1.3368 ± 0.2472	2.931 ± 0.0059	1.8994 ± 0.0534	na	na
1.8 nM melagatran/immobilized thrombin	30	0.8826 ± 0.0274	na	na	1.3710 ± 0.1118	na	na

Figure 10.3d shows the binding and dissociation of 1.8 nM melagatran in solution at 15 °C to thrombin immobilized on a sensor chip surface. A single-fractal analysis is adequate to describe both the binding and the dissociation kinetics. The values of (a) the binding rate coefficient, k for a single-fractal analysis, and (b) the dissociation rate coefficient, k_d for a single-fractal analysis are given in Table 10.3a. The values of the fractal dimension for binding and for the dissociation phase are given in Table 10.3b. The affinity, K value is equal to 2.842.

Figure 10.3e shows the binding and dissociation of 7.3 nM melagatran in solution at 15 °C to thrombin immobilized on a SPR sensor chip surface. Once again, a single-fractal analysis is adequate to describe both the binding and the dissociation kinetics. It is of interest to note that an increase in the melagatran concentration in solution by a factor of about 4 from 1.8 to 7.3 nM leads to an increase in the binding rate coefficient, k by a factor of 2.03 from a value of 0.0647 to 0.1716. The affinity, K value is equal to 0.796. In this case, an increase in the melagatran concentration in solution by a factor of about 4 from 1.8 to 7.3 nM leads to a decrease in the affinity, K value by a factor of 3.57.

Figure 10.3f shows the binding and dissociation of 29.1 nM melagatran in solution at 15 °C to thrombin immobilized on a SPR sensor chip surface. A dual-fractal analysis is required to adequately describe the binding kinetics. The dissociation kinetics is adequately described by a single-fractal analysis. The values of (a) the binding rate coefficient, k for a single fractal analysis, and (b) the binding rate coefficients, k_1 and k_2 for a dual-fractal analysis are given in Table 10.3a. The values of (a) the fractal dimension, D_f for a single-fractal analysis, and (b) the fractal dimensions, D_{f1} and D_{f2} for a dual-fractal analysis are given in Table 10.3b.

As noted previously for 6 °C, here to at 15 °C there is apparently a change in the binding mechanism at the higher (29.1 nM) concentration in solution when compared with the lower (1.8 and 7.3 nM) concentration in solution. This is because at the lower concentrations a single-fractal analysis is adequate to describe the binding kinetics, and at the higher concentration a dual-fractal analysis is required to describe the binding kinetics. Apparently, and as indicated earlier, the higher melagatran concentration in solution (29.1 nM) tends to saturate the active (receptor) sites on the sensor chip surface, which eventually leads to a change in the binding mechanism.

Figure 10.3g shows the binding and dissociation of 1.8 nM melagatran in solution at 25 °C to thrombin immobilized on a sensor chip surface. A single-fractal analysis is adequate to describe both the binding and the dissociation kinetics. The values of (a) the binding rate coefficient, k for a single-fractal analysis, and (b) the dissociation rate coefficient, k_d for a single-fractal analysis are given in Table 10.3a. The values of the fractal dimension for binding and for the dissociation phase are given in Table 10.3b. The affinity, K value is equal to 9.27.

Figure 10.3h shows the binding and dissociation of 7.3 nM melagatran in solution at 25 °C to thrombin immobilized on a SPR sensor chip surface. Once again, a single-fractal analysis is adequate to describe both the binding and the dissociation kinetics. The affinity, K value is equal to 1.023. Once again, and as noted previously for 6 and 15 °C, an increase in the melagatran concentration in solution by a factor of about 4 from 1.8 to 7.3 nM leads to an increase in the binding rate coefficient, k by a factor of 6.15 from a value of 0.0416 to 0.2554. The dissociation rate coefficient, k_d also increases by a factor of 55.7 from a value of 0.00448 to 0.2496.

Figure 10.3i shows the binding and dissociation of 29.1 nM melagatran in solution at 25 °C to thrombin immobilized on a SPR sensor chip surface. A dual-fractal analysis is required to adequately describe the binding kinetics. The dissociation kinetics is adequately described by a single-fractal analysis. The values of (a) the binding rate coefficient, k for a single fractal analysis, and (b) the binding rate coefficients, k_1 and k_2 for a dual-fractal analysis are given in Table 10.3a. The values of (a) the fractal dimension, D_f for a single-fractal analysis, and (b) the fractal dimensions, D_{f1} and D_{f2} for a dual-fractal analysis are given in Table 10.3b.

As noted previously for 6 °C and at 15 °C, here to at 25 °C there is apparently a change in the binding mechanism at the higher (29.1 nM) concentration in solution when compared with the lower (1.8 and 7.3 nM) concentration in solution. This is because at the lower concentrations a single-fractal analysis is adequate to describe the binding kinetics, and at the higher concentration a dual-fractal analysis is required to describe the binding kinetics. Apparently, once again, the higher melagatran concentration in solution (29.1 nM) tends to saturate the active (receptor) sites on the sensor chip surface, which eventually leads to a change in the binding mechanism.

Table 10.4 shows the affinity, K value at different temperatures and at different melagatran concentrations, when a single-fractal analysis is adequate to describe both the binding as well as the dissociation kinetics. For the data presented in Table 10.4 the highest affinity, K value is obtained at 1.8 nM melagatran concentration in solution at 25 °C. There is a wide range of affinities exhibited (greater than an order of magnitude) with the lowest being 0.796 (7.3 nM, 15 °C) and the highest being 9.27 (1.8 nM, 25 °C).

Figure 10.3j shows the binding and dissociation of 1.8 nM melagatran in solution at 30 °C to thrombin immobilized on a sensor chip surface. A single-fractal analysis is adequate to describe both the binding and the dissociation kinetics. The values of (a) the binding rate coefficient, k, for a single-fractal analysis, and (b) the dissociation rate coefficient, k_d, for a single-fractal analysis are given in Table 10.3a. The values of the fractal dimension for the binding and for the dissociation phase are given in Table 10.3b. The affinity value is equal to 1.143.

Figure 10.4 shows the increase in the affinity, K value with an increase in the ratio of the fractal dimension in the binding and in the dissociation phase (D_f/D_{fd}). For the data

Table 10.4

Affinity, K values for melagatran at different temperatures and concentrations (in nM)

Temperature, °C	Melagatran concentration, nM	Affinity, $K(= k/k_d)$
6	1.8	2.766
6	7.3	8.25
15	1.8	2.842
15	7.3	0.796
25	1.8	9.27
25	7.3	1.023
30	1.8	1.143

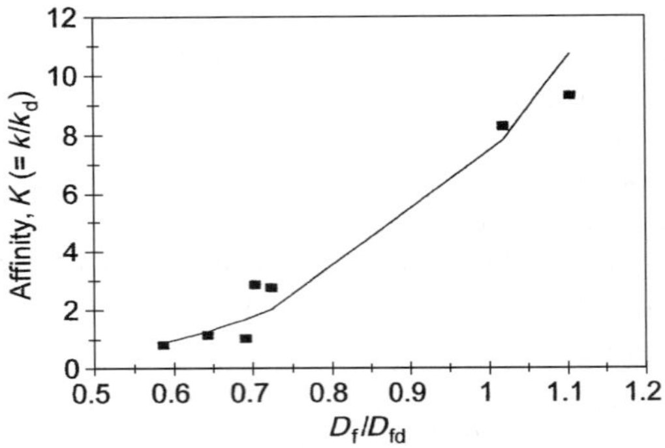

Figure 10.4 Increase in the affinity, K with an increase in the fractal dimension ratio, D_f/D_{fd}.

presented in Table 10.4 for melagatran, the affinity, K is given by:

$$K = (7.237 \pm 2.99)(D_f/D_{fd})^{3.956 \pm 0.5956} \tag{10.4}$$

The fit is very good. The affinity, K is very sensitive to the ratio of the fractal dimensions in the binding and in the dissociation phase, as noted by the close to fourth order dependence. The non-integer dependence reinforces the fractal nature of the system.

10.4 CONCLUSIONS

A fractal analysis of the binding and dissociation of potent thrombin inhibitors in solution to human α-thrombin immobilized on a SPR sensor chip surface provides a quantitative indication of the state of disorder or the degree of heterogeneity on the biosensor surface. The analysis of both the binding as well as the dissociation steps provides a more complete picture of the reaction occurring on the surface besides providing a value of the affinity, K. This is the ratio of the rate coefficient for binding, k, and for dissociation, k_d, steps. This is important, as indicated by Deinum *et al.* (2002), that analyte inhibitors (for example, for thrombin) selected as drug candidates are often chosen on the basis of $K_i (= 1/K)$ values. In our case, the lowest K_i value for 24 nM CH-248 occurs at 15 °C, and is equal to 0.118. Similarly, the lowest K_i value for melagatran is equal to 0.108. This occurs at 1.8 nM melagatran and at 25 °C. There is very little difference between these two values, and analysis of other or more data is required to more clearly define and distinguish which inhibitor needs to be selected as a possible drug candidate.

The fractal dimension value provides a quantitative measure of the degree of heterogeneity that exists on the sensor chip surface for the thrombin inhibitor–thrombin interactions. The degree of heterogeneity for the binding and dissociation phases is, in general, different. Both types of examples are presented wherein either a single- or a dual-fractal analysis is required to describe the binding and/or dissociation kinetics. The dual-fractal analysis is used only when the single-fractal analysis did not provide an adequate fit (sum of least squares less than 0.98). This was done by regression provided by Quattro Pro 8.0 (Corel Quattro Pro, 1997).

In accord with the prefactor analysis for fractal aggregates (Sorenson and Roberts, 1997), quantitative (predictive) expressions for the affinity, K are developed as a function of the ratio of fractal dimensions, D_f/D_{fd} for melagatran. The affinity, K is very sensitive to the ratio of the degree of heterogeneity present in the binding and in the dissociation phase. Since enzyme inhibitors (in our case, thrombin inhibitors) are selected on the basis of lower K_i (higher affinity, K) values, it would be advisable to select the inhibitor with as high a ratio for D_f/D_{fd} as possible. Ideally, one would like to increase the D_f value without changing (or with only small changes of) the D_{fd} value. This may not be possible, and may require considerable expertise from the experimentalist.

The fractal dimension for the binding or the dissociation phase is not a typical independent variable, such as the analyte (thrombin inhibitor) concentration in solution or the receptor (human α-thrombin) on the sensor chip surface that may be directly manipulated. It is estimated from Eqs. (10.3a), (10.3b), (10.3c), and (10.3d), as the case may be, and one may consider it as a derived variable. The predictive relationship (for melagatran) presented for the affinity as a function of the degree of heterogeneity that exists on the sensor chip surface (fractal dimension value) provides a means by which the affinity may be manipulated by changing the degree of heterogeneity that exists on the surface.

To the best of this author's knowledge this is the first study that links the affinity, K or the inhibitor constant, K_i to the degree of heterogeneity on the sensor chip surface. This may be of considerable assistance in helping to select a thrombin inhibitor or an inhibitor for another analyte. It provides a fresh avenue for manipulating these affinity, K (or the inhibitor constant, K_i) in desired directions. Even though the analysis is presented for reactions occurring on a sensor chip surface, it does provide insights into reactions occurring in a physiological environment. More such studies are required that (a) clearly determine the dynamics of these types of interactions (such as thrombin and thrombin inhibitor), (b) help determine the affinity, K or inhibitor constant, K_i values, and (c) provide a means by which they may be manipulated in desired directions, for example by changing the degree of heterogeneity on the surface.

REFERENCES

Biacore, AB, BIAevaluation, 3.2 software, Uppsala, Sweden, 2002.

Corel Quattro Pro, Corel Corporation Limited, Ottawa, Canada, 1997.

Deinum, J, L Gustafsson, E Gyzander, M Kullman-Magnusson, A Edstrom and R Karlson, *Analytical Biochemistry*, 2002, **300**, 152–162.

Elg, M, D Gustafsson and J Deinum, *Thrombolyitc Homeostasis*, 1997, **78**, 1286–1292.

Havlin, S, *The Fractal Approach to Heterogeneous Chemistry: Surfaces, Colloids, Polymers*, (ed. D. Avnir), Wiley, New York, 1989, pp. 251–269.

Karlsson, R, M Kullman-Magnusson, MD Hamalainen, A Remaus, K Andersson, P Borg, E Gyzander and J Deinum, *Analytical Biochemistry*, 2000, **278**, 1–13.

Ramakrishnan, A and A Sadana, *Automedica*, 2001, **20**(3/4), 313–340.

Sadana, A and T Vo-Dinh, *Biotechnology & Applied Biochemistry*, 2001, **33**, 17–31.

Sigmaplot, *Scientific Graphic Software, User's Manual*. Jandel Scientific, San Rafael, CA, 1993.

Sorenson, CM and GC Roberts, The prefactor of fractal aggregates. *Journal of Colloid and Interface Science*, 1997, **186**, 447–452.

Stone, SR and C Tapparelli, *Journal of Enzyme Inhibitors*, 1995, **9**, 3–15.

Weiss, GH, *Fractals in Science*. Springer, Berlin, 1994.

– 11 –

Fractal Analysis of Interleukin Binding and Dissociation Kinetics on Biosensor Surfaces

11.1 INTRODUCTION

Biosensors, particularly, the surface plasmon resonance (SPR) biosensor is recently being increasingly used to understand and describe biomolecular interactions. For example, the SPR biosensor has been used to detect triplex formation of modified oligonucleotides (Bates *et al.*, 2002). These authors have used the Biacore biosensor to detect specific triple helix formation at the murine c-myc P2 promoter target. They reported on the relative binding of the triplex forming oligonucleotide with DNA, RNA, or modified backbones that have increased resistance to nucleases. In this chapter we apply the fractal approach to analyze the binding of interleukins to biosensor surfaces.

Interleukin 5 (IL5) is a single chain major cytokine involved in the activation, proliferation, and maturation of eosinophils (Karlen *et al.*, 1998; Sanderson, 1998). Scibek *et al.* (2002) indicate that it is a disulfide-linked homo dimeric protein. It is composed of two glycosylated monomeric units. Each of these monomeric units exhibits structural and sequence homology to other class I cytokines. These authors further indicate that IL5 exerts its hematopoietic growth factor function by binding to a membrane bound receptor. This receptor consists of two subunits, α and β_c. Two subunits are required: one to initiate the binding (α), and the other to initiate the signaling (β_c). Thus, the interactions and the kinetics of formation of the IL5-α-β_c is of importance; especially since there are gaps in the understanding of the molecular mechanism involved. Devos *et al.* (1993); Johannson *et al.* (1995) indicate that the stoichiometry involved in the binding of IL5 to the α subunit is 1:1. Much less is known about the quantitative nature of the binding of the β_c subunit to the α-IL5 complex.

Scibek *et al.* (2002) have used the biosensor approach to analyze the binding kinetics of a soluble β_c ectodomain (β_c) immobilized on an SPR surface to either wild type (wt) IL5:α complex or to a single chain IL5:α complex in solution. Wild type IL5 denotes the native, disulfide-linked dimeric form expressed by Sf9 (*Spodoptera frugiperda*) cells. These authors indicate that the biosensor approach has been used to determine and to analyze the binding kinetics of different cytokine-receptor complexes. Furthermore, this biosensor technique has been used to analyze other interactions such as protein–protein, protein–DNA, protein–sugar, and protein–lipid interactions (Nagata and Handa, 2000).

The system by its very design is heterogeneous (for example, the receptors immobilized on the biosensor surface may exhibit some heterogeneity, that is, surface roughness), and often other factors such as mass transport limitations (unless they are carefully eliminated or minimized) play a significant role and further complicate the design (especially its kinetic aspects) of the assay or the correct interpretation of the assay results. One possible way of monitoring for the presence of diffusional limitations and the heterogeneity that exists on the surface is by using fractals. A characteristic feature of fractals is self-similarity at different levels of scale. Fractals exhibit dilatational symmetry. Fractals are disordered systems, and the disorder is described by nonintegral dimensions (Pfeifer and Obert, 1989). Fractals have previously been used to analyze the binding and dissociation kinetics of a wide variety of analyte–receptor systems (Sadana, 2001). Fractals are particularly useful for this type of analysis because they help characterize the heterogeneity that exists on the surface by a lumped parameter, the fractal dimension.

In this chapter we provide an alternate analysis for the binding and dissociation of wild type IL5:IL5Rα complex in solution to sβ_c immobilized on a Biacore 3000 optical biosensor surface (Scibek *et al.*, 2002). Binding and dissociation rate coefficients, as well as fractal dimension values for the binding and the dissociation phases will be provided wherever applicable. We offer the fractal analysis as an alternate analysis to help improve the understanding; we do not imply that this is better than the original (SPR-based software) analysis.

11.2 THEORY

Havlin (1989) has reviewed and analyzed the diffusion of reactants towards fractal surfaces. The details of the theory and the equations involved for the binding and the dissociation phases for analyte–receptor (in our case, interleukin-β_c ectodomain) are available (Ramakrishnan and Sadana, 2001; Sadana and Vo-Dinh, 2001). Here we will attempt to apply these equations to the interleukin-β_c ectodomain interaction.

The basic idea is to fold everything into the lumped parameter, the fractal dimension, D_f. This lumped parameter takes care of the chemical, orientational, and geometrical details. It is also assumed that the kinetic rate depends only on the geometric (static) parameter, D_f. It would also be useful to indicate the low and high limits in angstroms of the fractal character to be presented.

11.2.1 Single-fractal analysis

Binding rate coefficient

Havlin (1989) indicates that the diffusion of a particle (analyte) from a homogeneous solution to a solid surface (e.g. receptor-coated surface) on which it reacts to form a product (analyte–receptor) complex is given by:

$$(\text{Analyte}\cdot\text{Receptor}) \sim \begin{cases} t^{(3-D_{f,\text{bind}})/2} = t^p & (t < t_c) \\ t^{1/2} & (t > t_c) \end{cases} \tag{11.1a}$$

where the analyte–receptor represents the association complex formed on the surface. Here $p = -b$, and D_f is the fractal dimension of the surface. Havlin (1989) states that the crossover value may be determined by $r_c^2 \sim t_c$. Above the characteristic length, r_c, the self-similarity of the surface is lost and the surface may be considered homogeneous. Eq. 11.1a indicates that the concentration of the product [analyte–receptor] on a solid fractal surface scales at short and intermediate times as analyte–receptor $\sim t^p$ with the coefficient $p = (3 - D_f)/2$ at short time scales and $p = 1/2$ at intermediate time scales. Note that D_f, $D_{f,assoc}$, and $D_{f,bind}$ are used interchangeably. This equation is associated with the short-term diffusional properties of a random walk on a fractal surface. Note that, in perfectly stirred kinetics on a regular (nonfractal) structure (or surface), k_1 is a constant, that is, it is independent of time. In other words, the limit of regular structures (or surfaces) and the absence of diffusion-limited kinetics leads to k_1 being independent of time. In all other situations, one would expect a scaling behavior given by $k_1 \sim k' t^{-b}$ with $-b = p < 0$. Also, the appearance of the coefficient, p different from $p = 0$ is the consequence of two different phenomena, that is, the heterogeneity (fractality) of the surface and the imperfect mixing (diffusion-limited) condition. Finally, for a homogeneous surface where D_f is equal to 2, and when only diffusional limitations are present, $p = 1/2$ as it should be. Another way of looking at the $p = 1/2$ case (where $D_{f,bind}$ is equal to two) is that the analyte in solution views the fractal object, in our case the receptor-coated biosensor surface, from a 'large distance'. In essence, in the association process, the diffusion of the analyte from the solution to the receptor surface creates a depletion layer of width $(Ðt)^{1/2}$ where Ð is the diffusion constant. This gives rise to the fractal power law, $(\text{Analyte}\cdot\text{Receptor}) t^{(3-D_{f,bind})/2}$.

The values of the parameters k (association rate coefficient), p, and D_f in Eq. 11.1a may be obtained for analyte–receptor binding kinetics data. This may be done by a regression analysis using, for example, Quattro Pro (1997) along with Eq. 11.1a where $(\text{analyte}\cdot\text{receptor}) = kt^p$. The fractal dimension may be obtained from the parameter p. Higher values of the fractal dimension would indicate higher degrees of 'disorder' or heterogeneity or inhomogeneity on the surface.

Another way of looking at the diffusive process is that it inherently involves fluctuations at the molecular level that may be described by a random walk (Weiss, 1994). This author indicates that the kinetics of transport on disordered (or heterogeneous) media itself needs to be described by a random-walk model. When both of these are present, that is, the diffusion phenomena as well as a fractal surface, then one needs to analyze the interplay of both of these fluctuations. In essence, the disorder on the surface (or a higher fractal dimension, D_f) tends to slow down the motion of a particle (analyte in our case) moving in such a medium. Basically, according to Weiss (1994) the particle (random walker analyte) is trapped in regions in space, as it oscillates for a long time before resuming its motion.

Havlin (1989) indicates that the crossover value may be determined by $r_c^2 \sim t_c$. Above the characteristic length, r_c, the self-similarity of the surface of the surface is lost. Above t_c, the surface may be considered homogeneous, and 'regular' diffusion is now present. One may consider the analysis to be presented as an intermediate 'heuristic' approach in that in the future one may also be able to develop an autonomous (and not time-dependent) model of diffusion-limited kinetics in disordered media. No evidence of fractality is presented.

Dissociation rate coefficient

The diffusion of the dissociated particle (receptor or analyte) from the solid surface (e.g. analyte–receptor complex coated surface) into solution may be given, as a first approximation by:

$$\begin{aligned}(\text{Analyte}\cdot\text{Receptor}) &\sim -t^{(3-D_{\mathrm{f,diss}})/2} \qquad (t > t_{\mathrm{diss}}) \\ &= -k_{\mathrm{diss}} t^{(3-D_{\mathrm{f,diss}})/2}\end{aligned} \tag{11.1b}$$

Here $D_{\mathrm{f,diss}}$ is the fractal dimension of the surface for the dissociation step. t_{diss} represents the start of the dissociation step. This corresponds to the highest concentration of the analyte–receptor complex on the surface. Henceforth, its concentration only decreases. $D_{\mathrm{f,assoc}}$ may or may not be equal to $D_{\mathrm{f,diss}}$. k_{d} and k_{diss}, and $D_{\mathrm{f,d}}$ and $D_{\mathrm{f,diss}}$ are used interchangeably in the chapter.

11.2.2 Dual-fractal analysis

Binding rate coefficient

The single-fractal analysis we have just presented is extended to include two fractal dimensions. At present, the time ($t = t_1$) at which the first fractal dimension 'changes' to the second fractal dimension is arbitrary and empirical. For the most part it is dictated by the data analyzed and the experience gained by handling a single-fractal analysis. The r^2 (regression coefficient) value obtained is also used to determine if a single-fractal analysis is sufficient or one needs to use a dual-fractal analysis to provide an adequate fit. Only if the r^2 value is less than 0.97 for a single-fractal analysis, do we use a dual-fractal model. In this case, the analyte–receptor complex is given by:

$$(\text{Analyte}\cdot\text{Receptor}) \sim \begin{cases} t^{(3-D_{\mathrm{f1,bind}})/2} = t^{p_1} & (t < t_1) \\ t^{(3-D_{\mathrm{f2,bind}})/2} = t^{p_2} & (t_1 < t < t_2 = t_{\mathrm{c}}) \\ t^{1/2} & (t > t_{\mathrm{c}}) \end{cases} \tag{11.1c}$$

In analyte–receptor association the analyte–receptor associates with the active site on the surface and the product is released. In this sense the catalytic surface exhibits an unchanging fractal surface to the reactant in the absence of fouling and other complications. In the case of analyte–receptor association the biosensor surface exhibits a changing fractal surface to the analyte in solution. This occurs since as each association (binding) reaction takes place, smaller and smaller amounts of 'association' sites or receptors are available on the biosensor surface to which the analyte may bind. Furthermore, as the reaction proceeds, there is an increasing degree of heterogeneity on the biosensor surface for some reaction systems. This is manifested by two degrees of heterogeneity or two fractal dimensions on the biosensor surface. In the theoretical limit one might envisage a temporal fractal dimension wherein there is a continuous change in the degree of heterogeneity on the surface; though of course, such situations would be very rare, if at all.

Note that, at present, the dual-fractal analysis does not have a basis at the molecular level. This represents two different levels of heterogeneity on the biosensor surface.

But, in some examples presented, a single-fractal analysis is clearly inadequate to model the data. Only in these cases does one resort to a dual-fractal analysis.

Dissociation rate coefficient

In this case the dissociation rate coefficient is given by:

$$(\text{Analyte}\cdot\text{Receptor}) \sim \begin{cases} -t^{(3-D_{f1,diss})/2} & (t_{diss} < t < t_{d1}) \\ -t^{(3-D_{f2,diss})/2} & (t_{d1} < t < t_{d2}) \end{cases} \tag{11.1d}$$

Here $D_{f,diss}$ is the fractal dimension of the surface for the dissociation step. t_{diss} represents the start of the dissociation step. This corresponds to the highest concentration of the analyte·receptor on the surface. Henceforth, its concentration only decreases. $D_{f,bind}$ or $D_{f,assoc}$ may or may not be equal to $D_{f,diss}$.

11.3 RESULTS

A fractal analysis will be applied to the binding and dissociation of scIL5:IL5Rα complex in solution to sβ_c immobilized on a Biacore optical biosensor surface (Scibek *et al.*, 2002). IL5Rα is the IL5 receptor α subunit, and scIL5 is the single-chain interleukin 5, a major cytokine. As indicated earlier, the interactions of signaling proteins with the IL5 receptor are dependent on the assembly of the IL5-α-β_c complex. Scibek *et al.* (2002) noted a dependence of sβ_c, the soluble extracellular domain of receptor subunit, β_c, on the pre-assembly of an IL5:IL5Rα complex. β_c also exhibited a weaker interaction of sβ_c with either IL5 or IL5Rα alone. Furthermore, the fractal analysis is only one possible approach to analyze the diffusion-limited binding kinetics assumed to be present in the system analyzed. Scibek *et al.* (2002) have used the software to obtain values of the k_{on}, k_{off}, and the affinity, K_d values.

Figure 11.1a shows the binding of 0.56 μM wtIL5:IL5Rα complex in solution to sβ_c immobilized on an SPR Biacore 3000 optical biosensor chip surface (Scibek *et al.*, 2002). These authors have used a nonlinear least squares fitting method provided by the Biaevaluation 3.0 software (Biacore, Uppsala, Sweden). Individual rate coefficient values for the binding and the dissociation phases were obtained separately. As indicated earlier we will re-analyze the data using fractal analysis. A dual-fractal analysis is required to adequately describe the binding kinetics. A single-fractal analysis is, however, sufficient to describe the dissociation kinetics. The value of the binding and the dissociation rate coefficient(s), and the fractal dimension values for the binding and the dissociation phases presented in Table 11.1b were obtained from a regression analysis using eqs. 11.1a and 11.1c for the binding phase, and eqs. 11.1b and 11.1d for the dissociation phase.

The binding and dissociation rate coefficients presented in Tables 11.1a and 11.1b are within 95% confidence limits. For example, for the binding of 0.56 μM wtIL5:IL5Rα complex in solution to sβ_c immobilized on a sensor chip surface, the binding rate coefficient, k_1 value is 2.166 ± 0.357. The 95% confidence limit indicates that the k_1 value lies between 1.809 and 2.523.

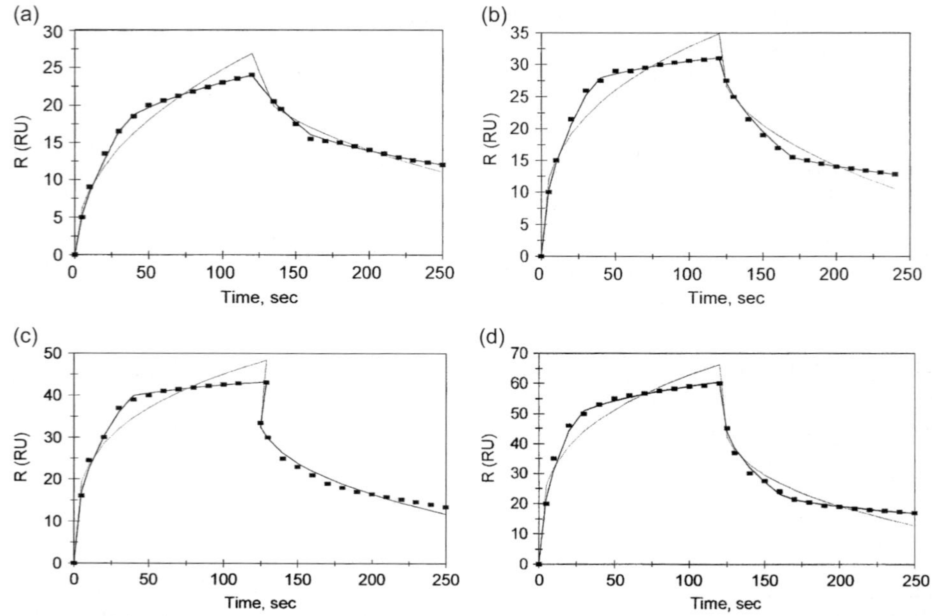

Figure 11.1 Binding and dissociation of different concentrations (in μM) of wtIL5:IL5α complex in solution to soluble β_c ectodomain immobilized on an optical biosensor surface (Scibek *et al.*, 2002): (a) 0.56, (b) 1.12, (c) 2.25, (d) 4.50.

Figure 11.1b shows the binding and dissociation of 1.12 μM wtIL5:IL5Rα complex in solution to sβ_c immobilized on a sensor chip surface. A dual-fractal analysis is required to adequately describe the binding kinetics. The dissociation kinetics is adequately described a single-fractal analysis. The values of (a) the binding rate coefficient, *k*, and (b) the

Table 11.1a

Binding and dissociation rate coefficients for wtIL5:IL5Rα complex in solution to soluble β_c ectodomain immobilized on a Biacore 3000 optical biosensor surface (Scibek *et al.*, 2002)

Analyte in solution/ receptor on surface	k	k_1	k_2	k_d	k_{d1}	k_{d2}
0.56 μM wtIL5: IL5Rα complex/sβ_c	3.5402 ± 0.874	2.166 ± 0.357	14.604 ± 0.235	5.9584 ± 0.315	na	na
1.12 μM wtIL5: IL5Rα complex/sβ_c	7.520 ± 1.292	5.303 ± 0.683	19.640 ± 0.070	5.827 ± 0.207	na	na
2.25 μM wtIL5: IL5Rα complex/sβ_c	13.373 ± 0.302	8.011 ± 0.009	49.982 ± 0.039	5.299 ± 0.620	4.011 ± 0.312	20.962 ± 0.118
4.50 μM wtIL5: IL5Rα complex/sβ_c	21.953 ± 4.143	13.738 ± 3.135	46.479 ± 0.271	5.3414 ± 0.419	na	na

Table 11.1b

Fractal dimensions for the binding and dissociation phase for wtIL5:IL5Rα complex in solution to soluble β_c ectodomain immobilized on a Biacore 3000 optical biosensor surface (Scibek *et al.*, 2002)

Analyte in solution/ receptor on surface	D_f	D_{f1}	D_{f2}	D_{fd}	D_{fd1}	D_{fd2}
0.56 μM wtIL5: IL5Rα complex/sβ_c	2.0662 ± 0.092	1.573 ± 0.126	2.751 ± 0.024	2.397 ± 0.042	na	na
1.12 μM wtIL5: IL5Rα complex/sβ_c	2.2706 ± 0.066	1.918 ± 0.100	2.735 ± 0.005	2.274 ± 0.028	na	na
2.25 μM wtIL5: IL5Rα complex/sβ_c	2.3602 ± 0.075	1.777 ± 0.140	2.980 ± 0.002	2.045 ± 0.089	1.852 ± 0.087	2.673 ± 0.027
4.50 μM wtIL5: IL5Rα complex/sβ_c	2.514 ± 0.072	1.995 ± 0.243	2.880 ± 0.089	1.981 ± 0.091	na	na

binding rate coefficient, k_1 and k_2 for a dual-fractal analysis, and (c) the dissociation rate coefficient, k_d for a single-fractal analysis are given in Table 11.1a. The values of (a) fractal dimension for the binding phase, D_f, for a single-fractal analysis (b) the fractal dimensions for the binding phase, k_1 and k_2 for a dual-fractal analysis, and (c) the fractal dimension for the dissociation phase, D_{fd} for a single-fractal analysis, are given in Table 11.1b. Apparently, a higher concentration of wtIL5:IL5Rα concentration in solution leads to higher values of the binding rate coefficients, k_1 and k_2. An increase in the wtIL5:IL5Rα concentration in solution by a factor of two from 0.56 to 1.12 μM leads to an increase in (a) the binding rate coefficient, k_1 by a factor of 2.45 from a value of 2.166 to 5.303, and in (b) the binding rate coefficient, k_2 by a factor of 1.344 from a value of 14.604 to 19.640.

Figure 11.1c shows the binding and dissociation of 2.25 μM wtIL5:IL5Rα complex in solution to sβ_c immobilized on a sensor chip surface. A dual-fractal analysis is required to adequately describe the binding and the dissociation kinetics. The values of (a) the binding rate coefficient, k, for a single-fractal analysis, and (b) the binding rate coefficient, k_1 and k_2 for a dual-fractal analysis, (c) the dissociation rate coefficient, k_d for a single-fractal analysis, and (d) the dissociation rate coefficient, k_{d1} and k_{d2} for a dual-fractal analysis are given in Table 11.1a. The values of (a) the fractal dimension for the binding phase, D_f, for a single-fractal analysis (b) the fractal dimensions for the binding phase, D_{f1} and D_{f2} for a dual-fractal analysis, (c) the fractal dimension for the dissociation phase, D_{fd} for a single-fractal analysis, and (d) the fractal dimensions for the dissociation phase, D_{fd1} and D_{fd2} for a dual-fractal analysis are given in Table 11.1b. Apparently, a higher concentration of wtIL5:IL5Rα concentration in solution leads to higher values of the binding rate coefficients, $k_1$and k_2. An increase in the wtIL5:IL5Rα concentration in solution by a factor of four from 0.56 to 2.25 μM leads to an increase in (a) the binding rate coefficient, k_1 by a factor of 3.70 from a value of 2.166 to 8.011, and in (b) the binding rate coefficient, k_2 by a factor of 3.42 from a value of 14.604 to 49.982.

Figure 11.1d shows the binding and dissociation of 4.50 μM wtIL5:IL5Rα complex in solution to sβ_c immobilized on a sensor chip surface. Once again, a dual-fractal analysis is required to adequately describe the binding kinetics. The dissociation kinetics is adequately described by a single-fractal analysis. The values of (a) the binding rate coefficient, k, and (b) the binding rate coefficient, k_1 and k_2 for a dual-fractal analysis, and (c) the dissociation rate coefficient, k_d for a single-fractal analysis, are given in Table 11.1a. The values of (a) the fractal dimension for the binding phase, D_f, for a single-fractal analysis (b) the fractal dimensions for the binding phase, D_{f1} and D_{f2} for a dual-fractal analysis, and (c) the fractal dimension for the dissociation phase, D_{fd} for a single-fractal analysis are given in Table 11.1b. Apparently, a higher concentration of wtIL5:IL5Rα concentration in solution still leads to higher values of the binding rate coefficients, $k_1$and k_2. An increase in the wtIL5:IL5Rα concentration in solution by a factor of eight from 0.56 to 4.50 μM leads to an increase in (a) the binding rate coefficient, k_1 by a factor of 6.34 from a value of 2.166 to 13.788, and in (b) the binding rate coefficient, k_2 by a factor of 3.18 from a value of 14.604 to 46.479. No explanation is offered at present to describe the 7% decrease in the binding rate coefficient, k_1 value from 49.982 to 46.479 as the wtIL5:IL5Rα complex concentration in solution increases by a factor of two from 2.25 to 4.5 μM.

Table 11.1a and Figure 11.2a show that the binding rate coefficient, k_1 increases linearly as the wtIL5:IL5Rα complex concentration in solution increases from 0.56 to 4.5 μM. In this concentration range the binding rate coefficient, k_1 is given by:

$$k_1 = (4.0133 \pm 0.693)[\text{wtIL5} : \text{IL5R}\alpha \text{ complex}]^{0.8567 \pm 0.1025} \quad (11.2a)$$

The binding rate coefficient, k_1 exhibits a slightly less than first-order dependence on the wtIL5:IL5Rα complex in solution. The fit is very good. More data points would more firmly establish the above predictive equation for the binding rate coefficient, k_1 as a function of the analyte concentration in solution.

Table 11.1a and Figure 11.2b show that the binding rate coefficient, k_2 increases as the wtIL5:IL5Rα complex concentration in solution increases from 0.56 to 4.5 μM. In this concentration range the binding rate coefficient, k_2 is given by:

$$k_2 = (21.31 \pm 7.173)[\text{wtIL5} : \text{IL5R}\alpha \text{ complex}]^{0.6345 \pm 0.1867} \quad (11.2b)$$

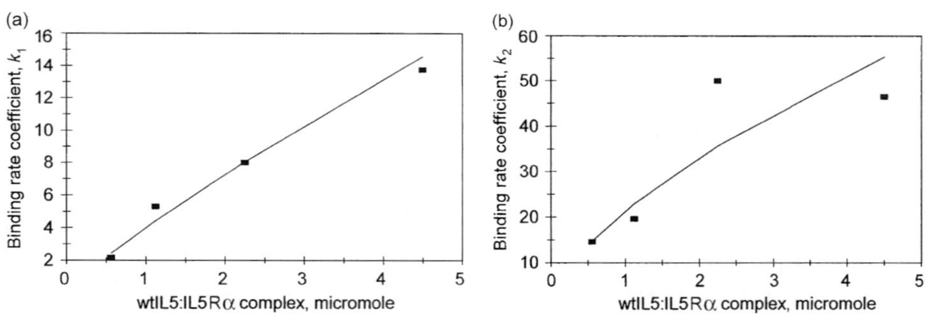

Figure 11.2 (a) Increase in the binding rate coefficient, k_1 with an increase in the wtIL5:IL5Rα complex in solution. (b) Increase in the binding rate coefficient, k_2 with an increase in the wtIL5:IL5Rα complex in solution.

The binding rate coefficient, k_2 exhibits a less than first-order dependence on the wtIL5:IL5Rα complex in solution. The fit is reasonable. More data points would more firmly establish the above predictive equation. The noninteger order of dependence exhibited by k_2 on the wtIL5:IL5Rα complex in solution reinforces the fractal nature of the system. The binding rate coefficient, k_1 exhibits a slightly higher order of dependence on wtIL5:IL5Rα complex concentration (equal to 0.8567) than the binding rate coefficient, k_2 (equal to 0.6345).

Tables 11.1a and 11.1b and Figure 11.3a show that the binding rate coefficient, k_1 increases as the fractal dimension, D_{f1} on the surface increases. For the data given in Tables 11.1a and 11.1b the binding rate coefficient, k_1 is given by:

$$k_1 = (0.1333 \pm 0.0829)D_{f1}^{6.416 \pm 2.665} \quad (11.3a)$$

The binding rate coefficient, k_1 is very sensitive to the degree of heterogeneity that exists on the surface as noted by the very high order of dependence of k_1 on D_{f1}. The fit is quite reasonable. More data points would more firmly establish the above predictive equation for the binding rate coefficient, k_1 as a function of the fractal dimension, D_{f1}.

Tables 11.1a and 11.1b and Figure 11.3b show that the binding rate coefficient, k_2 increases as the fractal dimension, D_{f2} on the surface increases. For the data given in Tables 11.1a and 11.1b the binding rate coefficient, k_2 is given by:

$$k_2 = (1.2 \times 10^{-5} \pm 0.4 \times 10^{-5})D_{f2}^{14.06 \pm 4.122} \quad (11.3b)$$

The binding rate coefficient, k_2 is extremely sensitive to the degree of heterogeneity that exists on the surface as noted by the very high order of dependence of k_2 on D_{f2}. The fit is quite reasonable. More data points would more firmly establish the above predictive equation for the binding rate coefficient, k_2 as a function of the fractal dimension, D_{f2}. The binding rate coefficient, k_2 is more sensitive (order of dependence equal to 14.06) on the degree of heterogeneity that exists on the surface than k_1 (order of dependence equal to 6.416).

Affinity, K_1 ($= k_1/k_d$ or k_1/k_{d1}) is of interest to practicing biosensorists. It would be interest to plot the affinity, K_1 as function of the ratio of the fractal dimensions, D_f/D_{fd} or D_{f1}/D_{fd1}. Figure 11.4a shows the increase in the affinity with an increase in the fractal

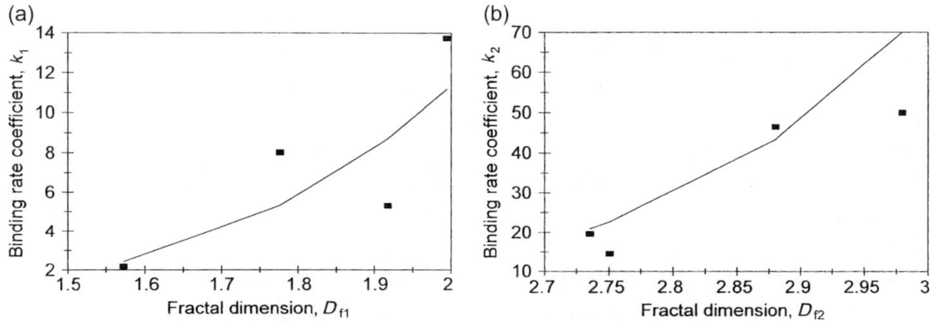

Figure 11.3 (a) Increase in the binding rate coefficient, k_1 with an increase in the fractal dimension, D_{f1}. (b) Increase in the binding rate coefficient, k_2 with an increase in the fractal dimension, D_{f2}.

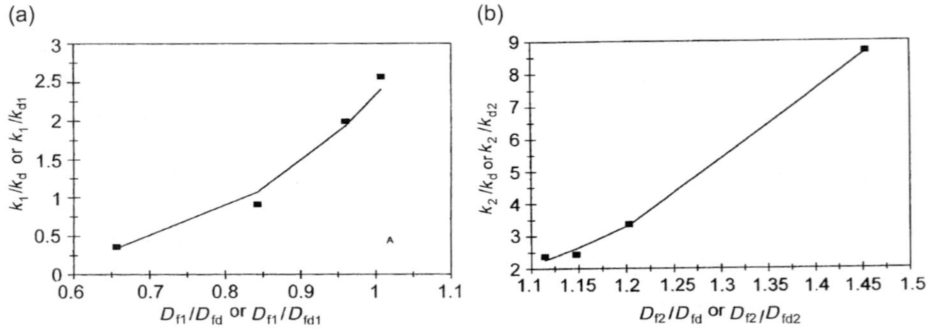

Figure 11.4 (a) Increase in the affinity, K_1 ($= k_1/k_d$ or k_1/k_{d1}) with an increase in the ratio of the fractal dimensions in the binding and in the dissociation phase (D_{f1}/D_{fd} or D_{f1}/D_{fd1}). (b) Increase in the affinity, K_2 ($= k_2/k_d$ or k_2/k_{d1}) with an increase in the ratio of the fractal dimensions in the binding and in the dissociation phase (D_f/D_{fd}) or D_{f2}/D_{fd2}).

dimension ratio. For the data presented in Tables 11.1a and 11.1b, the affinity, K_1 is given by:

$$K_1 = (2.335 \pm 0.336)(D_{f1}/D_{fd} \text{ or } D_{f1}/D_{fd1})^{4.561 \pm 0.405} \quad (11.4a)$$

The fit is very good considering that the independent variable (plotted on the x-axis) is either D_{f1}/D_{fd} or D_{f1}/D_{fd1}. The availability of more data points would more firmly establish the relation. The affinity, K_1 is very sensitive to the ratio of the fractal dimensions in the binding phase and in the dissociation phase as noted by the very high order of dependence (equal to 4.561).

Figure 11.4b shows the increase in the affinity, K_2 with an increase in the fractal dimension ratio. For the data presented in Tables 11.1a and 11.1b, the affinity, K_2 is given by:

$$K_2 = (1.3084 \pm 0.0818)(D_{f2}/D_{fd} \text{ or } D_{f2}/D_{fd2})^{5.055 \pm 0.293} \quad (11.4b)$$

Once again, the fit is very good. The availability of more data points would more firmly establish the relation. The affinity, K_2 is very sensitive to the ratio of the fractal dimensions in the binding phase and in the dissociation phase as noted by the very high order of dependence (equal to 5.055). The affinity, K_2 is slightly more sensitive (order of dependence equal to 5.055) to the ratio of the fractal dimensions in the binding and in the dissociation phase than the affinity, K_1 (order of dependence equal to 4.561).

Scibek *et al.* (2002) used a 1:1 binding model to fit their data. They indicate that since they obtained a poor fit of their global fit analysis of their sensorgrams ($\chi^2 = 0.398$), they fitted their association (binding) and dissociation phases separately. In our analysis presented above we have used the fractal analysis to analyze both the binding (association) and dissociation phases together. Since another data set is available under the 'same' conditions, we will use the fractal analysis again. This will provide indications of the consistency of our numbers, if the values of the binding and dissociation rate coefficients, affinities, and fractal dimensions in the binding and dissociation phases obtained are reasonably close to each other.

Figure 11.5a shows the binding of 0.56 μM wtIL5:IL5Rα complex in solution to sβ_c immobilized on an SPR Biacore 3000 optical biosensor chip surface (Scibek *et al.*, 2002). A dual-fractal analysis is required to adequately describe the binding and the dissociation kinetics. The value of the binding and the dissociation rate coefficient(s), and the fractal dimension values for the binding and the dissociation phases presented in Table 11.2b were obtained from a regression analysis using eqs. 11.1a and 11.1c for the binding phase, and eqs. 11.1b and 11.1d for the dissociation phase.

Figure 11.5b shows the binding and dissociation of 1.12 μM wtIL5:IL5Rα complex in solution to sβ_c immobilized on a sensor chip surface. A dual-fractal analysis is required to adequately describe the binding and the dissociation kinetics. The values of (a) the binding rate coefficient, k, and (b) the binding rate coefficient, k_1 and k_2 for a dual-fractal analysis, and (c) the dissociation rate coefficient, k_d for a single-fractal analysis are given in Table 11.2a. The values of (a) fractal dimension for the binding phase, D_f, for a single-fractal analysis (b) the fractal dimensions for the binding phase, D_{f1} and D_{f2} for a dual-fractal analysis, and (c) the fractal dimension for the dissociation phase, D_{fd} for a single-fractal analysis, are given in Table 11.2b. Apparently, a higher concentration of wtIL5:IL5Rα concentration in solution leads to higher values of the binding rate coefficients, k_1 and k_2. An increase in the wtIL5:IL5Rα concentration in solution by a factor of two from 0.56 to 1.12 μM leads to an increase in (a) the binding rate coefficient, k_1 by a factor of 2.36 from a value of 1.966 to 4.639, in (b) the binding rate coefficient, k_2 by a factor of 2.4 from a value of 8.173 to 19.60, in (c) the dissociation rate coefficient, k_{d1}

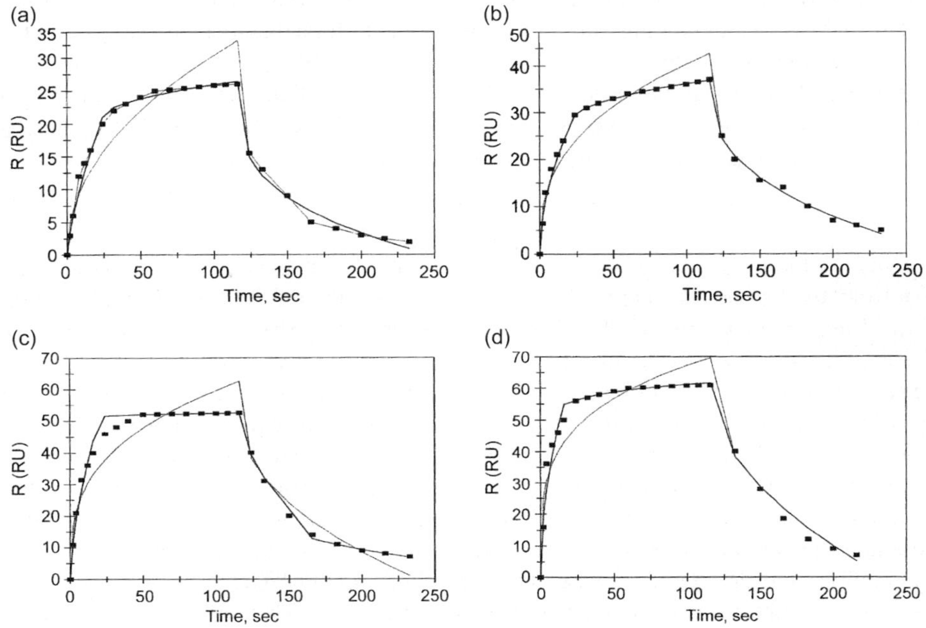

Figure 11.5 Binding and dissociation of different concentrations (in μM) of wtIL5:IL5α complex in solution to soluble β_c ectodomain immobilized on an optical biosensor surface (Scibek *et al.*, 2002): (a) 0.56, (b) 1.12, (c) 2.25, (d) 4.50.

Table 11.2a

Binding and dissociation rate coefficients for wtIL5:IL5Rα complex in solution to soluble β_c ectodomain immobilized on a Biacore optical biosensor surface (Scibek *et al.*, 2002)

Analyte in solution/ receptor on surface	k	k_1	k_2	k_d	k_{d1}	k_{d2}
0.56 μM wtIL5: IL5Rα complex/sβ_c	3.03 ± 0.379	1.966 ± 0.163	8.173 ± 0.065	0.988 ± 0.103	0.400 ± 0.027	2.175 ± 0.311
1.12 μM wtIL5: IL5Rα complex/sβ_c	7.088 ± 0.839	4.639 ± 0.03	19.60 ± 0.052	1.993 ± 0.261	0.620 ± 0.03	7.568 ± 0.02
2.25 μM wtIL5: IL5Rα complex/sβ_c	12.27 ± 1.27	8.537 ± 0.552	31.33 ± 0.069	6.089 ± 0.326	na	na
4.50 μM wtIL5: IL5Rα complex/sβ_c	16.23 ± 0.079	9.594 ± 1.26	35.53 ± 0.334	11.20 ± 0.978	8.613 ± 0.616	24.58 ± 0.095

by a factor of 1.55 from a value of 0.40 to 0.62, and in (d) the dissociation rate coefficient, k_{d2} by a factor of 3.47 from a value of 2.175 to 7.568.

Figure 11.5c shows the binding and dissociation of 2.25 μM wtIL5:IL5Rα complex in solution to sβ_c immobilized on a sensor chip surface. A dual-fractal analysis is required to adequately describe the binding kinetics. The dissociation kinetics is adequately described by a single-fractal analysis. The values of (a) the binding rate coefficient, k, for a single-fractal analysis (b) the binding rate coefficient, k_1 and k_2 for a dual-fractal analysis, and (c) the dissociation rate coefficient, k_d for a single-fractal analysis, are given in Table 11.2a. The values of (a) fractal dimension for the binding phase, D_f, for a single-fractal analysis (b) the fractal dimensions for the binding phase, D_{f1} and D_{f2} for a dual-fractal analysis, and (c) the fractal dimension for the dissociation phase, D_{fd} for a single-fractal analysis are

Table 11.2b

Fractal dimensions for the binding and dissociation phase for wtIL5:IL5Rα complex in solution to sβ_c immobilized on a Biacore 3000 optical biosensor surface (Scibek *et al.*, 2002)

Analyte in solution/ receptor on surface	D_f	D_{f1}	D_{f2}	D_{fd}	D_{fd1}	D_{fd2}
0.56 μM wtIL5: IL5Rα complex/sβ_c	2.0880 ± 0.0696	1.75 ± 0.094	2.551 ± 0.015	1.942 ± 0.083	1.382 ± 0.131	2.299 ± 0.031
1.12 μM wtIL5: IL5Rα complex/sβ_c	2.3336 ± 0.066	2.00 ± 0.06	2.81 ± 0.008	2.026 ± 0.069	1.761 ± 0.061	2.633 ± 0.007
2.25 μM wtIL5: IL5Rα complex/sβ_c	2.435 ± 0.058	2.154 ± 0.074	2.868 ± 0.008	2.328 ± 0.029	na	na
4.50 μM wtIL5: IL5Rα complex/sβ_c	2.41 ± 0.071	1.982 ± 0.18	2.754 ± 0.014	2.411 ± 0.047	2.314 ± 0.07	2.769 ± 0.008

given in Table 11.2b. Apparently, a higher concentration of wtIL5:IL5Rα concentration in solution leads to higher values of the binding rate coefficients, $k_1$and k_2. An increase in the wtIL5:IL5Rα concentration in solution by a factor of four from 0.56 to 2.25 μM leads to an increase in (a) the binding rate coefficient, k_1 by a factor of 4.342 from a value of 1.966 to 8.537, and in (b) the binding rate coefficient, k_2 by a factor of 3.83 from a value of 8.173 to 31.33.

Figure 11.5d shows the binding and dissociation of 4.50 μM wtIL5:IL5Rα complex in solution to sβ_c immobilized on a sensor chip surface. A dual-fractal analysis is required to adequately describe the binding and the dissociation kinetics. The values of (a) the binding rate coefficient, k, for a single-fractal analysis, and (b) the binding rate coefficient, k_1 and k_2 for a dual-fractal analysis, (c) the dissociation rate coefficient, k_d for a single-fractal analysis, and (d) the dissociation rate coefficient, k_{d1} and k_{d2} for a dual-fractal analysis are given in Table 11.2a. The values of (a) the fractal dimension for the binding phase, D_f, for a single-fractal analysis (b) the fractal dimensions for the binding phase, D_{f1} and D_{f2} for a dual-fractal analysis, (c) the fractal dimension for the dissociation phase, D_{fd} for a single-fractal analysis, and (d) the fractal dimensions for the dissociation phase, D_{fd1} and D_{fd2} for a dual-fractal analysis are given in Table 11.2b. Apparently, a higher concentration of wtIL5:IL5Rα concentration in solution still leads to higher values of the binding rate coefficients, $k_1$and k_2. An increase in the wtIL5:IL5Rα concentration in solution by a factor of eight from 0.56 to 4.50 μM leads to an increase in (a) the binding rate coefficient, k_1 by a factor of 4.88 from a value of 1.966–9.594, and in (b) the binding rate coefficient, k_2 by a factor of 4.35 from a value of 8.173–35.53.

Table 11.2a and Figure 11.6a show that the binding rate coefficient, k_1 increases linearly as the wtIL5:IL5Rα complex concentration in solution increases from 0.56 to 4.5 μM. In this concentration range the binding rate coefficient, k_1 is given by:

$$k_1 = (3.659 \pm 1.1101)[\text{wtIL5 : IL5R}\alpha\ \text{complex}]^{0.7721 \pm 0.1705} \tag{11.5a}$$

The binding rate coefficient, k_1 exhibits a less than first-order dependence on the wtIL5:IL5Rα complex in solution. The fit is good. More data points would more firmly establish the above predictive equation for the binding rate coefficient, k_1 as a function of the analyte concentration in solution.

Table 11.2a and Figure 11.6b show that the binding rate coefficient, k_2 increases as the wtIL5:IL5Rα complex concentration in solution increases from 0.56 to 4.5 μM. In this concentration range the binding rate coefficient, k_2 is given by:

$$k_2 = (14.812 \pm 4.889)[\text{wtIL5 : IL5R}\alpha\ \text{complex}]^{0.6769 \pm 0.1836} \tag{11.5b}$$

The binding rate coefficient, k_2 exhibits a less than first-order dependence on the wtIL5:IL5Rα complex in solution. The fit is reasonable. More data points would more firmly establish the above predictive equation. The noninteger order of dependence exhibited by k_2 on the wtIL5:IL5Rα complex in solution reinforces the fractal nature of the system. The binding rate coefficient, k_1 exhibits a slightly higher order of dependence on wtIL5:IL5Rα complex concentration (equal to 0.7721) than the binding rate coefficient, k_2 (equal to 0.6769).

Affinity, K_1 ($= k_1/k_d$ or k_1/k_{d1}) values are of interest to practicing biosensorists. It would be of interest to plot the affinity, K_1 as function of the ratio of the fractal

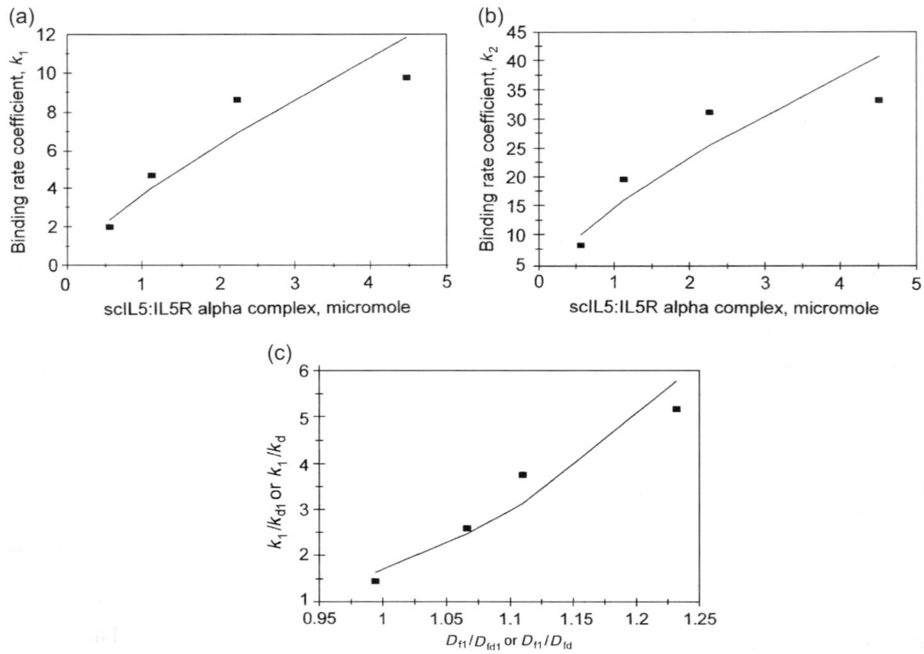

Figure 11.6 (a) Increase in the binding rate coefficient, k_{d1} with an increase in the wtIL5:IL5Rα complex in solution. (b) Increase in the binding rate coefficient, k_2 with an increase in the wtIL5:IL5Rα complex in solution. (c) Increase in the affinity, K_1 (= k_1/k_d or k_1/k_{d1}) with an increase in the ratio of the fractal dimensions in the binding and in the dissociation phase (D_{f1}/D_{fd} or D_{f1}/D_{fd1}).

dimensions, D_{f1}/D_{fd} or D_{f1}/D_{fd1}. Figure 11.6c shows the increase in the affinity with an increase in the fractal dimension ratio. For the data presented in Tables 11.2a and 11.2b, the affinity, K_1 is given by:

$$K_1 = (2.335 \pm 0.336)(D_{f1}/D_{fd} \text{ or } D_{f1}/D_{fd1})^{4.561 \pm 0.405} \tag{11.5c}$$

The fit is very good considering that the independent variable (plotted on the x-axis) is either D_{f1}/D_{fd} or D_{f1}/D_{fd1}. The availability of more data points would more firmly establish the relation. The affinity, K_1 is very sensitive to the ratio of the fractal dimensions in the binding phase and in the dissociation phase as noted by the very high order of dependence (equal to 4.561).

Tables 11.2a and 11.2b and Figure 11.7a show that the dissociation rate coefficient, k_{d1} increases as the fractal dimension, D_{fd1} on the surface increases. For the data given in Tables 11.2a and 11.2b the dissociation rate coefficient, k_{d1} is given by:

$$k_{d1} = (0.04 + 0.09)D_{fd1}^{6.033 \pm 2.249} \tag{11.6a}$$

The dissociation rate coefficient, k_{d1} is very sensitive to the degree of heterogeneity that exists on the surface as noted by the very high order of dependence (equal to 6.033) of k_{d1}

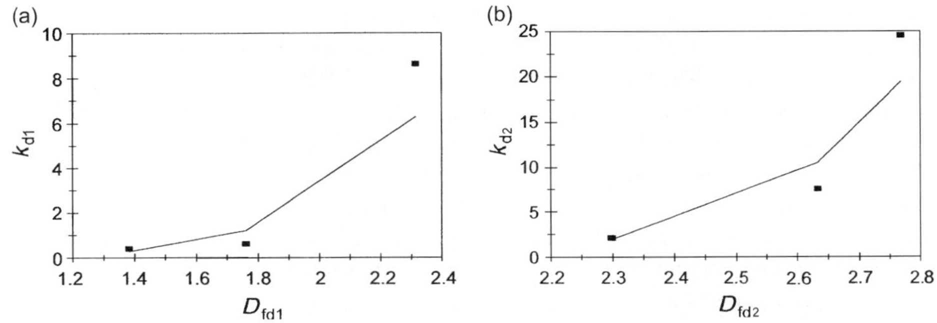

Figure 11.7 (a) Increase in the dissociation rate coefficient, k_{d1} with an increase in the fractal dimension, D_{fd1}. (b) Increase in the dissociation rate coefficient, k_{d2} with an increase in the fractal dimension, D_{fd2}.

on D_{fd1}. Since the error is high, the negative error is not given, as it would yield a negative dissociation rate coefficient. Nevertheless, the fit is quite reasonable. More data points would more firmly establish the above predictive equation for the dissociation rate coefficient, k_{d1} as a function of the fractal dimension, D_{fd1}.

Tables 11.2a and 11.2b and Figure 11.7b show that the dissociation rate coefficient, k_{d2} increases as the fractal dimension, D_{fd2} on the surface increases. For the data given in Tables 11.2a and 11.2b the dissociation rate coefficient, k_{d2} is given by:

$$k_{d2} = (7.6 \times 10^{-5} \pm 3.8 \times 10^{-5}) D_{fd2}^{12.23 \pm 3.03} \tag{11.6b}$$

The dissociation rate coefficient, k_{d2} is extremely sensitive to the degree of heterogeneity that exists on the surface as noted by the very high order of dependence of k_{d2} on D_{fd2} (equal to 12.23). The fit is quite reasonable. More data points would more firmly establish the above predictive equation for the dissociation rate coefficient, k_{d2} as a function of the fractal dimension, D_{fd2}. The dissociation rate coefficient, k_{d2} is more sensitive (order of dependence equal to 12.23) on the degree of heterogeneity that exists on the surface than k_{d1} (order of dependence equal to 6.033).

11.4 CONCLUSIONS

A fractal analysis is presented for the binding and dissociation of wtIL5:IL5Rα complex in solution to soluble β_c immobilized on a Biacore optical biosensor surface (Scibek *et al.*, 2002). The analysis provides a quantitative indication of the state of disorder or the degree of heterogeneity on the biosensor chip surface. The fractal analysis is offered as an alternate way to improve the understanding of the kinetics in the heterogenous case with diffusion-limited reactions occurring on structured surfaces. The analysis of both the binding as well as the dissociation steps provide a more complete picture of the reaction occurring on the surface besides providing a value of the affinity, *K*.

The fractal dimension value provides a quantitative measure of the degree of heterogeneity that exists on the sensor chip surface for the wtIL5:IL5R complex-soluble β_c ectodomain interaction. The degree of heterogeneity for the binding and the dissociation phases is, in general, different. Both types of examples are presented wherein either a single- or a dual-fractal analysis is required to describe the binding and/or dissociation kinetics. The dual-fractal analysis is used only when the single-fractal analysis did not provide an adequate fit (sum of least squares less than 0.98). This was done by regression provided by Quattro Pro 8.0 (Corel Quattro Pro, 1997).

Even though the analysis is presented for reactions occurring on a sensor chip surface, it does provide insights into the reactions occurring in a physiological environment. More such studies are required that determine the dynamics of the interleukin interactions, and provide a means by which they may be manipulated in desired directions, for example, by changing the degree of heterogeneity on the surface.

REFERENCES

Bates, PJ, JF Reddoch, P Hansakul, A Arrow, R Dale and DM Miller, *Analytical Biochemistry*, 2002, **307**, 235–243.

Corel Quattro Pro 8.0, Corel Corporation Limited, Ottawa, Canada, 1997.

Devos, R, Y Ginsez, S Cornelius, A Verhee, J Van der Heyden, M Manneberg, HW Lahm, W Fiers, J Tavernier and G Plaetinck, *Journal of Biological Chemistry*, 1993, **268**, 6581–6587.

Havlin, S, *The Fractal Approach to Heterogeneous Chemistry: Surfaces, Colloids, Polymers*, (ed. D. Avnir), Wiley, New York, 1989, pp. 251–269.

Johannson, K, E Appelbaum, M Doyle, P Hensley, B Zhao, SS Abdel Meguid, P Young, R Cook, S Carr, R Matico, D Cuismano, E Dul, M Angelichio, I Brooks, E Winborne, P McDonnell, T Morton, D Bennett, T Sokolski and D McNulty, *Journal of Biological Chemistry*, 1995, **270**, 9549–9571.

Karlen, S, ML De Boer, RJ Lipscombe, W Lutz, VA Mordvinov and CJ Sanderson, *International Reviews on Immunology*, 1998, **16**, 227–247.

Nagata K and H Handa, *Real-Time Analysis of Biomolecular Interactions*, Springer, New York, 2000.

Pfeifer P and M Obert, *The Fractal Approach to Heterogeneous Chemistry: Surfaces, Colloids, Polymers*, pp. 11–43, Wiley, New York, 1989.

Ramakrishnan, A and A Sadana, *Automedica*, 2001, **20**(3/4), 313–340.

Sadana, A, *Analytical Biochemistry*, 2001, **29**(1), 34–47.

Sadana, A and T Vo-Dinh, *Biotechnology & Applied Biochemistry*, 2001, **33**, 17–31.

Sanderson C, *IL5: From Molecule to Drug Target for Asthma*. Marcel Dekker, New York, 1998.

Scibek, JJ, E Evergren, S Zahn, GA Canzini, D Van Ryk and IM Chaiken, *Analytical Biochemistry*, 2002, **307**, 258–265.

Weiss, GH, *Fractals in Science*. Springer, Berlin, 1994.

– 12 –

Fractal Analysis of Environmental Contaminants Binding Kinetics on Biosensor Surfaces

12.1 INTRODUCTION

Methanol significantly affects plant physiology (Nanomura and Benson, 1992; Shirashi *et al.*, 2 000). Hasunuma *et al.* (2004) have recently developed a hybrid alcohol oxidase–peroxidase biosensor for the real-time quantification of methanol in plants. Previously, biosensors have been developed to detect methanol using alcohol oxidase (EC 1.1.3.13; AOD) and alcohol dehydrogenase (EC 1.1.1; ADH) as an alcohol recognition site (Guilbault *et al.*, 1983; Guilbault and Lubrano, 1974; Miyamoto *et al.*, 1991; Wang *et al.*, 1995; Lubrano *et al.*, 1991). Hasunuma *et al.* (2004) indicate that alcohol oxidase catalyzes the oxidation of short-chain aliphatic alcohols, including methanol. Alcohol dehydrogenase possesses low selectivity for the catalytic oxidation of both aromatic and aliphatic alcohols (Barman, 1969).

Hasunuma *et al.* (2004) have successfully made quantitative, the methanol in low-purity tobacco (*Nicotiana tabacum*) plant extracts with their biosensor. These authors fabricated an amperometric biosensor immobilizing two enzymes and an electronic mediator by the self-assembly technique. Hasunuma *et al.* (2004) self-assembled a mixed monolayer of 4,4′-dithiobutyric acid to which was covalently attached *Hansenula* sp. alcohol oxidase (AOD) and horse radish peroxidase (POD) and 11-ferrocenyl-1-undecanethiol as an electron mediator on an Au (gold) electrode. The authors emphasize that the redox-relay biosensor is simple to use. It is also stable due to the covalent attachment to the surface.

Figure 12.1a shows the electronic relay representation of the AOD/POD/11-FUT electrode. POD is peroxidase (EC 1.11.17). Figure 12.1b shows the three kinds of functional alkanethiols used: ferrocenyl-1-hexanethiol (6-FHT), 11-ferrocenyl-1-undecanethiol (11-FUT), and 2-carboxy-1-propanethiol (3-CPT).

Hasunuma *et al.* (2004) emphasize that they have successfully fabricated a miniature methanol sensor. Figure 12.2 shows the schematic diagram of the microsensor on a 96-well polypropylene assay plate. The authors indicated that the AOD/POD/11-FUT/Au sheet is the working electrode (WE) and the coiled Ag/AgCl wire is the counter electrode (CE).

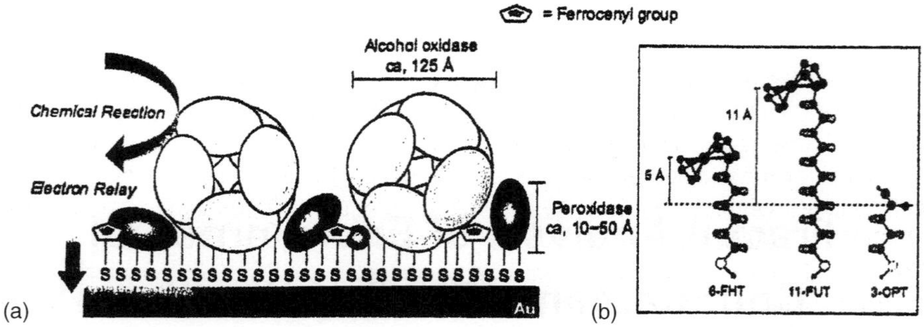

Figure 12.1 (a) Electronic relay schematic representation of the AOD/POD/11-FUT electrode (Hasunuma *et al.*, 2004). (b) Three kinds of functional alkanethiol (Hasunuma *et al.*, 2004): 6-ferrocenyl-1-hexanethiol (6-FHT), 11-ferrocenyl-1-undecanethiol (11-FUT), 3-carboxy-1-propanethiol (3-CPT).

Hasunuma *et al.* (2004) did not provide any values of the binding rate coefficients. Though, they did provide binding curves using their biosensor comprising of the AOD/POD/11-FUT/Au electrode. In this chapter we re-analyze their binding data using a fractal analysis. Rate coefficients are provided for (a) the binding of methanol in plant extract to the AOD/POD/11-FUT/Au electrode, (b) the binding of pectin methylesterase (PME) to the AOD/POD/11-FUT/Au electrode, and (c) the binding of different concentrations of PME to the AOD/POD/11-FUT/Au electrode (Hasunuma *et al.*, 2004). Fractal dimension values are also provided.

In this chapter we also analyze the entrapment (or binding) kinetics of two pollutants, benzene, and toluene to different sol–gel monoliths (Calvo-Munoz *et al.*, 2002). These sol–gel monoliths are either pure or with methyl groups. Finally, we present a fractal analysis for the binding of free 2,4 dintrophenol (DNP) equilibrated with anti-DNP monoclonal antibody in solution to DNA-HSA (human serum albumin) labeled polymethylmethacrylate (PMMA) beads (Carter *et al.*, 2003).

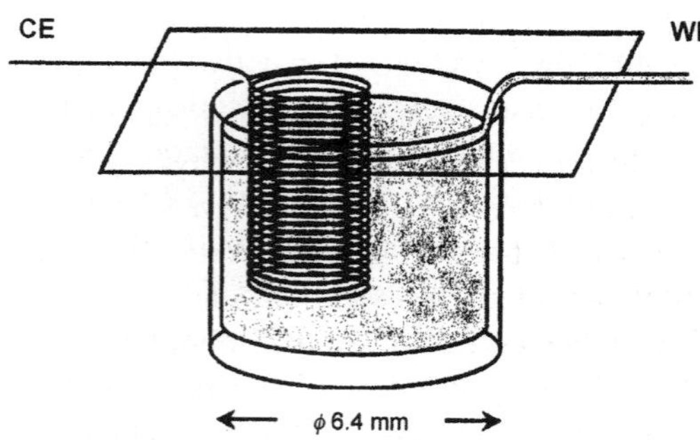

Figure 12.2 Schematic diagram of the microsensor on a 96-well polypropylene assay plate (Hasunuma *et al.*, 2004). WE, working electrode; CE, counter electrode.

12.2 THEORY

Havlin (1989) has reviewed and analyzed the diffusion of reactants towards fractal surfaces. The details of the theory and the equations involved for the binding and the dissociation phases for analyte–receptor (in our case, for example, PME to the AOD/POD/11-FUT/Au electrode interaction) are available (Ramakrishnan and Sadana, 2001; Sadana and Vo-Dinh, 2001). Here we will attempt to apply these equations to the PME-AOD/POD/11-AOD/POD/11-FUT/Au electrode interaction.

The basic idea is to fold everything into the lumped parameter, the fractal dimension, D_f. This lumped parameter takes care of the chemical, orientational, and geometrical details. It is also assumed that the kinetic rate depends only on the geometric (static) parameter, D_f. It would also be useful to indicate the low and high limits in angstroms of the fractal character to be presented.

12.2.1 Single-fractal analysis

Binding rate coefficient

Havlin (1989) indicates that the diffusion of a particle (analyte) from a homogeneous solution to a solid surface (e.g. receptor-coated surface) on which it reacts to form a product (analyte–receptor) complex is given by:

$$(\text{Analyte}\cdot\text{Receptor}) \sim \begin{cases} t^{(3-D_{f,\text{bind}})/2} = t^p & (t < t_c) \\ t^{1/2} & (t > t_c) \end{cases} \tag{12.1a}$$

where the analyte–receptor represents the association complex formed on the surface. Here $p = -b$, and D_f is the fractal dimension of the surface. Havlin (1989) states that the crossover value may be determined by $r_c^2 \sim t_c$. Above the characteristic length, r_c, the self-similarity of the surface is lost and the surface may be considered homogeneous. Eq. 12.1a indicates that the concentration of the product [analyte–receptor] on a solid fractal surface scales at short and intermediate times as analyte–receptor $\sim t^p$ with the coefficient $p = (3 - D_f)/2$ at short time scales and $p = 1/2$ at intermediate time scales. Note that D_f, $D_{f,\text{assoc}}$, and $D_{f,\text{bind}}$ are used interchangeably. This equation is associated with the short-term diffusional properties of a random walk on a fractal surface. Note that, in perfectly stirred kinetics on a regular (nonfractal) structure (or surface), k_1 is a constant, that is, it is independent of time. In other words, the limit of regular structures (or surfaces) and the absence of diffusion-limited kinetics leads to k_1 being independent of time. In all other situations, one would expect a scaling behavior given by $k_1 \sim k' t^{-b}$ with $-b = p < 0$. Also, the appearance of the coefficient, p different from $p = 0$ is the consequence of two different phenomena, that is, the heterogeneity (fractality) of the surface and the imperfect mixing (diffusion-limited) condition. Finally, for a homogeneous surface where D_f is equal to 2, and when only diffusional limitations are present, $p = 1/2$ as it should be. Another way of looking at the $p = 1/2$ case (where $D_{f,\text{bind}}$ is equal to two) is that the analyte in the solution views the fractal object, in our case the receptor-coated biosensor surface, from a 'large distance'. In essence, in the association process, the diffusion of the analyte from the solution to the receptor surface creates a depletion layer of width $(Đt)^{1/2}$ where $Đ$ is the diffusion constant. This gives rise to the fractal power law, (Analyte·Receptor) $\sim t^{(3-D_{f,bind})/2}$.

The values of the parameters k (association or binding rate coefficient), p, and D_f in eq. 12.1a may be obtained for analyte–receptor binding kinetics data. This may be done by a regression analysis using, for example, Quattro Pro (1997) along with eq. 12.1a where (analyte·receptor) $= kt^p$. The fractal dimension may be obtained from the parameter p. Higher values of the fractal dimension would indicate higher degrees of 'disorder' or heterogeneity or inhomogeneity on the surface.

Another way of looking at the diffusive process is that it inherently involves fluctuations at the molecular level that may be described by a random walk (Weiss, 1994). This author indicates that the kinetics of transport on disordered (or heterogeneous) media itself needs to be described by a random-walk model. When both of these are present, that is, the diffusion phenomena as well as a fractal surface, then one needs to analyze the interplay of both of these fluctuations. In essence, the disorder on the surface (or a higher fractal dimension, D_f) tends to slow down the motion of a particle (analyte in our case) moving in such a medium. Basically, according to Weiss (Weiss, 1994) the particle (random walker analyte) is trapped in regions in space, as it oscillates for a long time before resuming its motion.

Havlin (1989) indicates that the crossover value may be determined by $r_c^2 \sim t_c$. Above the characteristic length, r_c, the self-similarity of the surface of the surface is lost. Above t_c, the surface may be considered homogeneous, and 'regular' diffusion is now present. One may consider the analysis to be presented as an intermediate 'heuristic' approach in that in the future one may also be able to develop an autonomous (and not time-dependent) model of diffusion-limited kinetics in disordered media. No evidence of fractality is presented.

12.2.2 Dual-fractal analysis

Binding rate coefficient

The single-fractal analysis we have just presented is extended to include two fractal dimensions. At present, the time ($t = t_1$) at which the first fractal dimension 'changes' to the second fractal dimension is arbitrary and empirical. For the most part it is dictated by the data analyzed and the experience gained by handling a single-fractal analysis. The r^2 (regression coefficient) value obtained is also used to determine if a single-fractal analysis is sufficient, or one needs to use a dual-fractal analysis to provide an adequate fit. Only if the r^2 value is less than 0.97 for a single-fractal analysis, do we use a dual-fractal model. In this case, the analyte–receptor complex is given by:

$$(\text{Analyte}\cdot\text{Receptor}) \sim \begin{cases} t^{(3-D_{f1,bind})/2} = t^{p_1} & (t < t_1) \\ t^{(3-D_{f2,bind})/2} = t^{p_2} & (t_1 < t < t_2 = t_c) \\ t^{1/2} & (t > t_c) \end{cases} \tag{12.1b}$$

In catalytic reactions the reactant binds to the active site on the surface and the product is released. In this sense the catalytic surface exhibits an unchanging fractal surface to the reactant in the absence of fouling and other complications. In the case of analyte–receptor association the biosensor surface exhibits a changing fractal surface to the analyte in solution. This occurs since as each association (binding) reaction takes place, smaller and smaller amounts of 'association' sites or receptors are available on the biosensor surface to which the analyte may bind. Furthermore, as the reaction proceeds, there is an increasing degree of heterogeneity on the biosensor surface for some reaction systems. This is manifested by two degrees of heterogeneity or two fractal dimensions on the biosensor surface. In the theoretical limit one might envisage a temporal fractal dimension wherein there is a continuous change in the degree of heterogeneity on the surface; though of course, such situations would be very rare, if at all.

Note that, at present, the dual-fractal analysis does not have a basis at the molecular level. This represents two different levels of heterogeneity on the biosensor surface. But, in some examples presented, a single-fractal analysis is clearly inadequate to model the data. Only in these cases does one resort to a dual-fractal analysis.

12.3 RESULTS

A fractal analysis is applied to (a) the binding of methanol in plant extract to the AOD/POD/11-FUT/Au electrode, (b) the binding of pectin methylesterase (PME) to the AOD/POD/11-FUT/Au electrode, and (c) the binding of different concentrations of PME to the AOD/POD/11-FUT/Au electrode (Hasunuma *et al.*, 2004).

Figure 12.3a shows the binding of methanol in plant extract to 0.1 M phosphate (pH 6.0) to *Hansenula* sp. alcohol oxidase and horse radish peroxidase and 11-ferrocenyl-1-undecane thiol (11-FUT) on an electrode (Hasunuma *et al.*, 2004). These authors did not provide a kinetic analysis with values of the binding rate coefficients. As indicated earlier, the data is re-analyzed using a fractal analysis. A dual-fractal analysis is required to adequately describe the binding kinetics. The values of the binding rate coefficients and

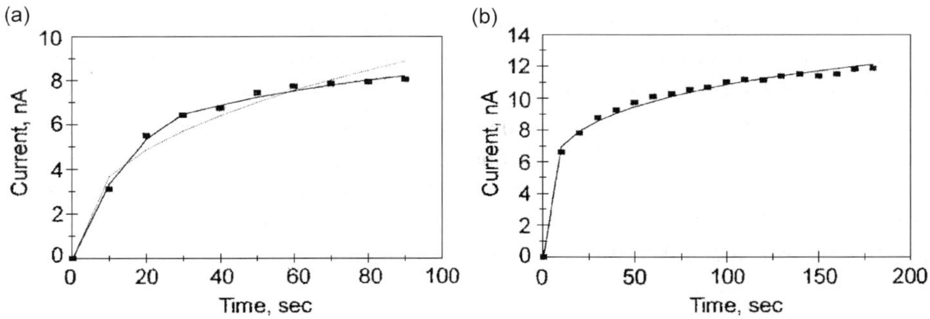

Figure 12.3 (a) Binding of methanol in plant extract in 0.1 M sodium phosphate (pH 6.0) to *Hansenula* sp. alcohol dehydrogenase (AOD), and horse radish peroxidase (POD), and 11-ferrocenyl-1-undecanethiol (11-FUT) on an Au electrode. (b) Binding of pectin methylesterase (PME) in 0.1 M sodium phosphate (pH 7.0) containing 50 mg/ml pectin to the AOD/POD/11-FUT/Au electrode. When only a solid line (—) is used then a single-fractal analysis applies. When both a (···) and a (—) solid line is used, then the (···) line represents a single-fractal analysis, and the (—) line represents a dual-fractal analysis.

the fractal dimensions presented in Table 12.1 for a single- and a dual-fractal analysis were obtained from a regression analysis using eq. 12.1a and b.

The binding rate coefficients presented in Table 12.1 are within 95% confidence limits. For example, for the binding of the methanol in plant extract to the AOD/POD/11-FUT/Au electrode for a dual-fractal analysis, the binding rate coefficient, k_1 is 0.6693 ± 0.0657. The 95% confidence limit indicates that the k_1 value lies between 0.6036 and 0.735. This indicates that the values are precise and significant.

It is of interest to note that as the fractal dimension value increases by 56.5% from a value of D_{f1} equal to 1.6444 to D_{f2} equal to 2.5742, the binding rate coefficient increases by a factor of 4.71 from a value of k_1 equal to 0.6933 to k_2 equal to 3.1350.

Figure 12.3b shows the binding of 50 μM methanol in 0.1 M sodium phosphate (pH 6.0) to the AOD/POD/11-FUT/Au electrode. A single-fractal analysis is adequate to describe the binding kinetics. The value of the binding rate coefficient, k and the fractal dimension, D_f are presented in Table 12.1.

Figure 12.4a shows the binding of 0.25 units/ml PME in 0.1 M sodium phosphate (pH 7.0) containing 50 mg/ml pectin in solution to the AOD/POD/11-FUT/Au electrode. A dual-fractal analysis is required to adequately describe the binding kinetics. The values of (a) the binding rate coefficient, k and the fractal dimension, D_f for a single-fractal analysis, and (b) the binding rate coefficients, k_1 and k_2 and the fractal dimensions, D_{f1} and D_{f2} for a dual-fractal analysis are given in Table 12.1

For a dual-fractal analysis, note once again, that as the fractal dimension increases by 49.4% from a value of D_{f1} equal to 1.3814 to D_{f2} equal to 2.0674, the binding rate coefficient increases by a factor of 4.97 from a value of k_1 equal to 0.0409 to k_2 equal to 0.2034. Once again, an increase in the degree of heterogeneity on the Au electrode surface (increase in the fractal dimension value) leads to an increase in the binding rate coefficient.

Note also that an increase in the PME concentration in solution from 0.25 to 0.50 units/ml leads to an increase in the fractal dimension values (D_{f1} and D_{f2}), and also in the respective binding rate coefficient values (k_1 and k_2).

Figure 12.4b shows the binding of 0.50 units/ml PME in 0.1 M sodium phosphate (pH 7.0) containing 50 mg/ml pectin in solution to the AOD/POD/11-FUT/Au electrode. A dual-fractal analysis is required to adequately describe the binding kinetics. The values of (a) the binding rate coefficient, k and the fractal dimension, D_f for a single-fractal analysis, and (b) the binding rate coefficients, k_1 and k_2 and the fractal dimensions, D_{f1} and D_{f2} for a dual-fractal analysis are given in Table 12.1

For a dual-fractal analysis, note once again, that as the fractal dimension increases by 55.8% from a value of D_{f1} equal to 1.7098 to D_{f2} equal to 2.664 the binding rate coefficient increases by a factor of 3.57 from a value of k_1 equal to 0.1587 to k_2 equal to 0.5673. Once again, an increase in the degree of heterogeneity on the Au electrode surface (increase in the fractal dimension value) leads to an increase in the binding rate coefficient.

Figure 12.4c shows the binding of 1.0 unit/ml PME in 0.1 M sodium phosphate (pH 7.0) containing 50 mg/ml pectin in solution to the AOD/POD/11-FUT/Au electrode. A dual-fractal analysis is required to adequately describe the binding kinetics. The values of (a) the binding rate coefficient, k and the fractal dimension, D_f for a single-fractal analysis, and (b) the binding rate coefficients, k_1 and k_2 and the fractal dimensions, D_{f1} and D_{f2} for a dual-fractal analysis are given in Table 12.1.

Table 12.1

Rate coefficient and fractal dimension values for the binding of (a) methanol in plant extract and 50 μM methanol in 0.1 M sodium phosphate (pH 6.0) to *Hansenula* sp. alcohol oxidase (AOD) and horseradish peroxidase (POD) and 11-ferrocenyl-1-undecane thiol (11-FUT) on an electrode, and (b) pectin methylesterase (PME) in 0.1 M sodium phosphate (pH 7.0) containing 50 mg/ml pectin to AOD/POD/11-FUT/Au electrode

Analyte in solution	k	k_1	k_2	D_f	D_{f1}	D_{f2}
Methanol in plant extract	1.4720 ± 0.166	0.6693 ± 0.0657	3.1350 ± 0.0698	2.2028 ± 0.105	1.6444 ± 0.2382	2.5742 ± 0.0458
50 μM methanol	4.453 ± 0.107	na	na	0.1926 ± 0.0071	na	na
0.25 unit/ml PME	0.0587 ± 0.0054	0.0409 ± 0.0020	0.2034 ± 0.0027	1.5768 ± 0.0426	1.3814 ± 0.0338	2.0674 ± 0.0482
0.50 unit/ml PME	0.1879 ± 0.0168	0.15868 ± 0.011	0.5663 ± 0.0034	1.8290 ± 0.0308	1.7098 ± 0.0326	2.2664 ± 0.01414
1.0 unit/ml PME	0.4295 ± 0.0386	0.2978 ± 0.010	1.3216 ± 0.0451	1.8854 ± 0.0420	1.6868 ± 0.0249	2.3290 ± 0.0464

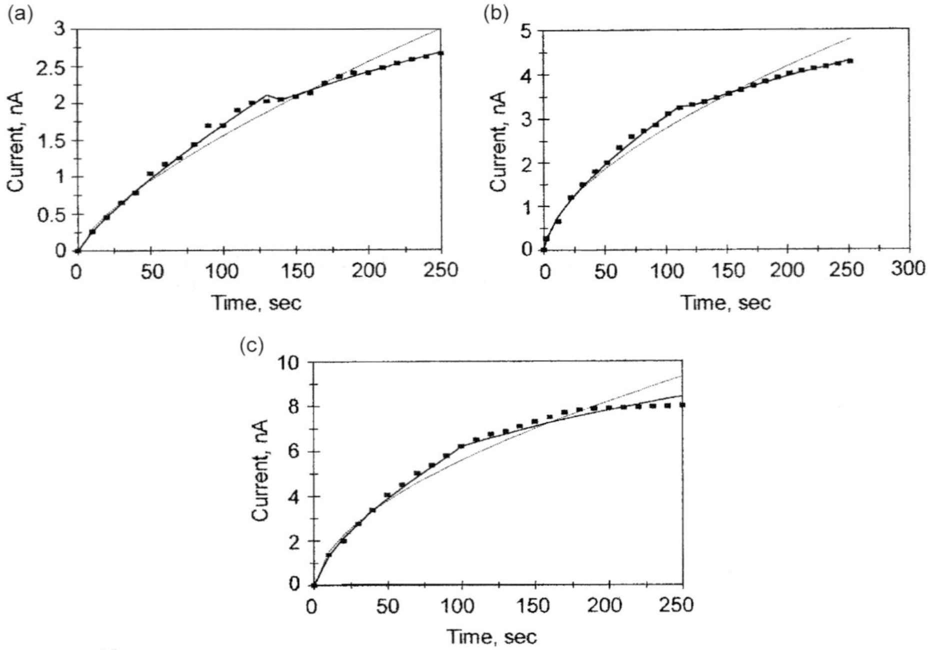

Figure 12.4 Binding of different concentrations (in units/ml) pectin methylesterase (PME) in 0.1 M sodium phosphate buffer (pH 7.0) containing 50 mg/ml pectin to the AOD/POD/11-FUT/Au electrode: (a) 0.25, (b) 0.5, (c) 1.0. When both a (···) and a (—) solid line is used, then the (···) line represents a single-fractal analysis, and the (—) line represents a dual-fractal analysis.

For a dual-fractal analysis, note once again, that as the fractal dimension increases by 38.1% from a value of D_{f1} equal to 1.7098 to D_{f2} equal to 2.3290 the binding rate coefficient increases by a factor of 4.44 from a value of k_1 equal to 0.2978 to k_2 equal to 1.3216. Once again, an increase in the degree of heterogeneity on the Au electrode surface (increase in the fractal dimension value) leads to an increase in the binding rate coefficient.

Table 12.1 and Figure 12.5a show that the binding rate coefficient, k_1 increases as the PME concentration in solution increases from 0.25 to 1.0 units/ml. In this concentration range, the binding rate coefficient, k_1 is given by:

$$k_1 = (0.3347 \pm 0.1112)[\text{PME}]^{1.4320 \pm 0.2925} \tag{12.2a}$$

The binding rate coefficient, k_1 exhibits a slightly less than one and one-half order (equal to 1.432) dependence on the PME concentration in solution. The fit is good. Only three data points are available. The availability of more data points would more firmly establish the above predictive equation for the binding rate coefficient, k_1 as function of PME concentration in solution.

Table 12.1 and Figure 12.5b show that the binding rate coefficient, k_2 increases as the PME concentration in solution increases from 0.25 to 1.0 units/ml. In this concentration

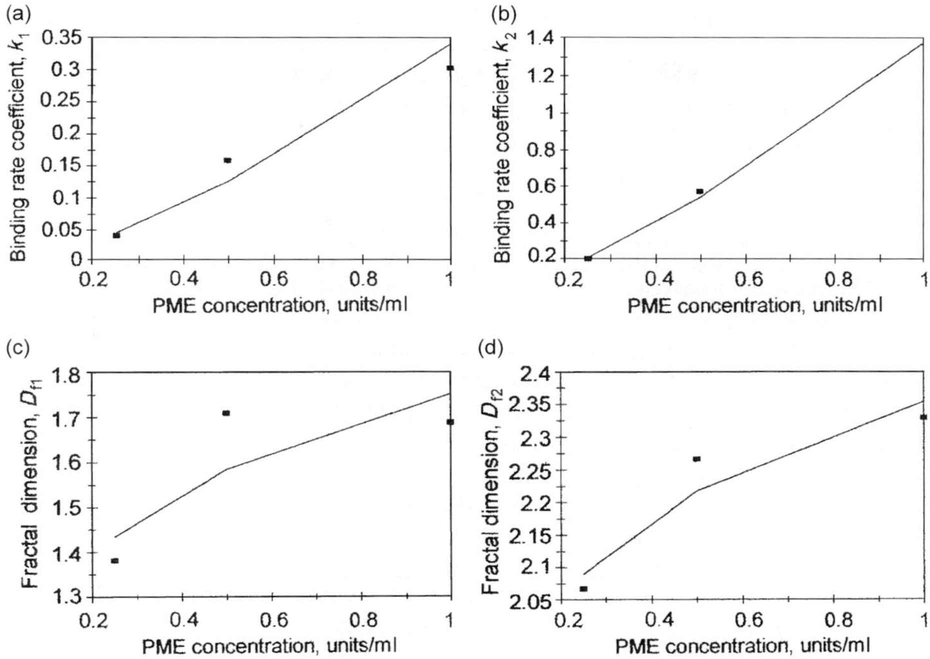

Figure 12.5 (a) Increase in the binding rate coefficient, k_1 with an increase in the PME concentration (in units/ml) in solution. (b) Increase in the binding rate coefficient, k_2 with an increase in the PME concentration (in units/ml) in solution. (c) Increase in the fractal dimension, D_{f1} with an increase in the PME concentration (in units/ml) in solution. (d) Increase in the fractal dimension, D_{f2} with an increase in the PME concentration (in units/ml) in solution.

range, the binding rate coefficient, k_2 is given by:

$$k_2 = (1.3610 \pm 0.1016)[\mathrm{PME}]^{1.3499 \pm 0.0734} \tag{12.2b}$$

The binding rate coefficient, k_2 exhibits a slightly less than one and one-half order (equal to 1.3499) dependence on the PME concentration in solution. The fit is good. Only three data points are available. The availability of more data points would more firmly establish the above predictive equation for the binding rate coefficient, k_2 as function of PME concentration in solution. The binding rate coefficient, k_2 exhibits a slightly lower order of dependence than k_1 on the PME concentration in solution.

For the binding of 0.25–1.0 units/ml PME in solution to the AOD/POD/11-FUT/Au electrode Figure 12.5c shows for a dual-fractal analysis the increase in the fractal dimension, D_{f1} with an increase in the PME concentration in solution. For the data presented in Table 12.1 and in Figure 12.5c, the fractal dimension, D_{f1} is given by:

$$D_{f1} = (1.7517 \pm 0.170)[\mathrm{PME}]^{0.1440 \pm 0.0944} \tag{12.2c}$$

Only three data points are available. The fit is reasonable. The availability of more data points would lead to a better fit. The fractal dimension, D_{f1} is only slightly dependent on the PME concentration in solution since it exhibits an order of dependence equal to 0.1440. Note that, however, and as indicated previously in this book, the fractal dimension is based on a log scale, and even small changes in the fractal dimension may lead to significant changes in the degree of heterogeneity on the electrode surface.

For the binding of 0.25–1.0 units/ml PME in solution to the AOD/POD/11-FUT/Au electrode Figure 12.5d shows for a dual-fractal analysis the increase in the fractal dimension, D_{f2} with an increase in the PME concentration in solution. For the data presented in Table 12.1 and in Figure 12.5d, the fractal dimension, D_{f2} is given by:

$$D_{f2} = (2.354 \pm 0.063)[\mathrm{PME}]^{0.0859 \pm 0.0269} \tag{12.2d}$$

Only three data points are available. The fit is reasonable. The availability of more data points would lead to a better fit. The fractal dimension, D_{f2} is only slightly dependent on the PME concentration in solution since it exhibits an order of dependence equal to 0.0859. For all practical purposes this is negligible. However, one must bear in mind that the fractal dimension, as indicated earlier, is based on a log scale. Thus, even small changes in the fractal dimension could lead to a significant change in the degree of heterogeneity on the Au electrode surface.

Table 12.1 and Figure 12.6 show that the binding rate coefficient, k_2 increases as the fractal dimension, D_{f2} on the Au electrode surface increases. For the data shown in Table 12.1 and in Figure 12.6 the binding rate coefficient, k_2 is given by:

$$k_2 = (2.0 \pm 0.7) \times 10^{-5}\, D_{f2}^{12.74 \pm 1.72} \tag{12.3}$$

The binding rate coefficient, k_2 is very sensitive to the degree of heterogeneity that exists on the surface as noted by the extremely high order of dependence (equal to 12.74) of k_2 on D_{f2}. The fit is reasonable. More data points would more firmly establish the above predictive equation for the binding rate coefficient, k_2 as a function of the fractal dimension, D_{f2}.

Calvo-Munoz *et al.* (2002) have developed a chemical sensor for the aromatic hydrocarbons, benzene, and toluene using the sol–gel process. These authors indicate that benzene, toluene, and xylene are organic air contaminants. Besides, benzene is a carcinogenic (Kaneko *et al.*, 1997). Calvo-Munoz *et al.* (2002) indicate that sensors for these contaminants are required (a) to provide monitoring and capabilities and daily

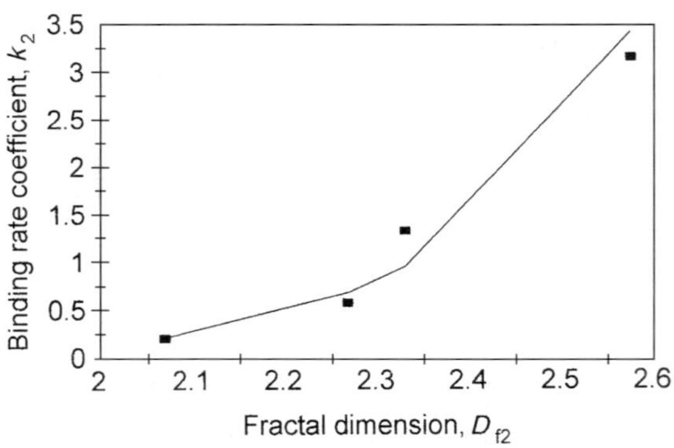

Figure 12.6 Increase in the binding rate coefficient, k_2 with an increase in the fractal dimension, D_{f2}.

average pollution values in big cities, and (b) measurements of peaks of pollution. These authors have presented data for the trapping process of benzene and toluene in porous, hybrid organic–inorganic materials. These materials are prepared by the sol–gel process.

Calvo-Munoz *et al.* (2002) indicate that benzene and toluene can be detected due to their absorbance as they are trapped in the monoliths. The kinetics of trapping of benzene and toluene is followed by noting the changes in absorbance as a function of time. The authors indicate that the absorbance of the pollutants, benzene and toluene was examined at their lowest transitions, 253 and 262 nm, respectively. They analyzed their results on the basis of hydrogen bond interactions between the aromatic cycle and the Si-OH and S-CH_3 groups of the xerogels. They prepared three types of xerogels:

(a) Pure tetramethoxysilane (TMOS)
(b) Hybrid analogue—90% pure TMOS and 10% MeTMOS (methyl groups) (9/1)
(c) Hybrid analogue—80% pure TMOS and 20% MeTMOS (methyl groups) (8/2)

Figure 12.7a shows the kinetics of entrapment (binding) of benzene in a 9/1 TMOS sol–gel. The 9/1 sol–gel is 10% MeTMOS (methyl groups 10%) and 90% pure TMOS. The absorbance change was noted at 253 nm. The 9/1 sol–gel was exposed to 1000 ppm benzene which was diluted in nitrogen. A single-fractal analysis is adequate to describe

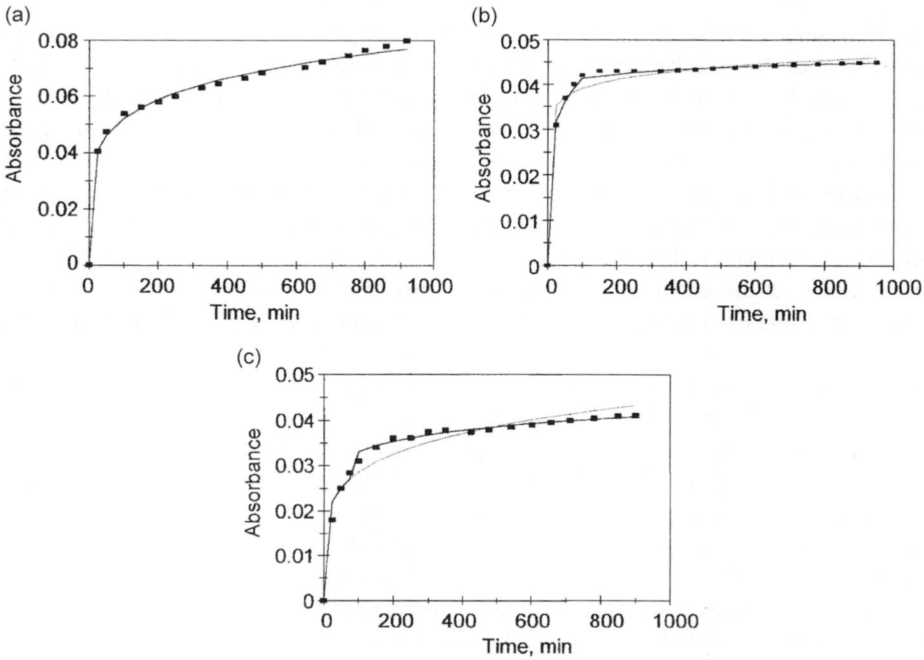

Figure 12.7 Entrapment (or binding) kinetics of benzene to different 9/1 sol–gel monoliths with different amounts of methyl groups. Monoliths are 2.0 ± 0.2 mm thick (Calvo-Munoz *et al.*, 2002): (a) 9/1 TMOS, (b) pure TMOS, (c) 8/2 TMOS. When only a solid line (—) is used then a single-fractal analysis applies. When both a (···) and a (—) solid line is used, then the (···) line represents a single-fractal analysis, and the (—) line represents a dual-fractal analysis.

the binding (or entrapment) kinetics. The values of (a) the binding rate coefficient, k, and the fractal dimension, D_f for a single-fractal analysis are given in Table 12.2.

Figure 12.7b shows the kinetics of entrapment of benzene in a pure TMOS sol–gel. A dual-fractal analysis is required to adequately describe the entrapment (or binding) kinetics. The values of (a) the binding rate coefficient, k, and the fractal dimension, D_f, for a single-fractal analysis, and (b) the binding rate coefficients, k_1 and k_2, and the fractal dimensions, D_{f1} and D_{f2} for a dual-fractal analysis are given in Table 12.2. It is of interest to note that there is a change in the binding (or entrapment) mechanism as one goes from the 9/1 TMOS sol–gel to the pure TMOS sol–gel. A single-fractal analysis is adequate to describe the binding process for the 9/1 TMOS sol–gel, and a dual-fractal analysis is required to adequately describe the entrapment (or binding) process for the pure TMOS sol–gel.

Also, it is of interest to note that for the pure TMOS sol–gel for the entrapment of benzene, where a dual-fractal analysis applies, an increase in the fractal dimension by 11.4% from a value of D_{f1} equal to 2.624 to D_{f2} equal to 2.923 leads to an increase in the binding rate coefficient by a factor of 1.98 from a value of k_1 equal to 0.0174 to k_2 equal to 0.0345. Once again, an increase in the degree of heterogeneity on the sol–gel surface leads to an increase in the binding (or entrapment) rate coefficient.

Figure 12.7c shows the kinetics of entrapment of benzene in a 8/2 TMOS sol–gel. A dual-fractal analysis is required to adequately describe the entrapment (binding) kinetics. The values of (a) the binding rate coefficient, k, and the fractal dimension, D_f, for a single-fractal analysis, and (b) the binding rate coefficients, k_1 and k_2, and the fractal dimensions, D_{f1} and D_{f2} for a dual-fractal analysis are given in Table 12.2. It is of interest to note that for the 8/2 TMOS sol–gel for the entrapment of benzene, an increase in the fractal dimension by 20% from a value of D_{f1} equal to 2.3424 to D_{f2} equal to 2.8114 leads to an increase in the binding rate coefficient by a factor of 3.25 from a value of k_1 equal to 0.00661 to k_2 equal to 0.0215. Once again, an increase in the degree of heterogeneity on the sol–gel surface leads to an increase in the binding (or entrapment) rate coefficient. The thickness of the sol–gel monolith in each of the cases presented in Figure 12.7a–c is 2.0 mm or 2000 μm. Calvo-Munoz *et al.* (2002) also presented data for the effect of sol–gel thickness on the entrapment (or binding) kinetics.

It is of interest to compare the fractal dimensions and binding (entrapment) rate coefficients obtained with a pure TMOS sol–gel and the 8/2 TMOS sol–gel. In both cases, a dual-fractal analysis is required to adequately describe the binding kinetics. Note that as one goes from the pure TMOS silca gel to the 8/2 TMOS silica gel (a) the fractal dimensions decrease, and (b) the corresponding binding rate coefficients, k_1 and k_2 decrease. For example, as one goes from the pure TMOS silca gel to the 8/2 TMOS silica gel, the fractal dimension, D_{f1} decreases by 10.73% from a value of 2.624 to 2.3424. The corresponding binding rate coefficient, k_1 decreases by a 62.4% from a value of 0.0174 to 0.00661. Similarly, the fractal dimension, D_{f2} decreases by 3.82% from a value of 2.923 to 2.8114 as one goes from the pure TMOS silica gel to the 8/2 TMOS silica gel. This leads to a corresponding decrease in the binding rate coefficient, k_2 by 37.7% from a value of 0.0345 to 0.0215. Once again, changes in the degree of heterogeneity or the fractal dimension and in the binding rate coefficient are in the same direction.

Calvo-Munoz *et al.* (2002) also analyzed the influence of 9/1 sol–gel thickness on the entrapment kinetics of benzene. Figure 12.8a shows the entrapment of benzene in a 680 μm thick 9/1 TMOS silica gel. A single-fractal analysis is adequate to describe

Table 12.2

Rate coefficient and fractal dimension values for the binding (trapping process) of benzene and toluene to three different porous hybrid organic–inorganic materials prepared by the sol–gel process (Calvo-Munoz *et al.*, 2002)

Sol gel properties	Analyte	k	k_1	k_2	D_f	D_{f1}	D_{f2}
9/1 2.0 ± 0.02 mm thick; 10% MeTMOS + 90% pure TMOS (10% methyl groups)	Benzene	0.0233 ± 0.0005	na	na	2.6498 ± 0.0107	na	na
Pure TMOS; 2.0 ± 0.02 mm thick; (0% methyl groups)	Benzene	0.0279 ± 0.0013	0.0174 ± 0.00057	0.0345 ± 0.00012	2.8545 ± 0.0198	2.624 ± 0.0473	2.923 ± 0.0047
8/2 2.0 ± 0.02 mm thick; 20% MeTMOS + 80% pure TMOS (20% methyl groups)	Benzene	0.01199 ± 0.00093	0.00661 ± 0.000343	0.0215 ± 0.00025	2.6222 ± 0.03423	2.3424 ± 0.0596	2.8114 ± 0.1886
9/1 680 μm thick; 90% pure TMOS (10% methyl groups)	Benzene	0.09143 ± 0.00042	na	na	2.5678 ± 0.0130	na	na
9/1 1980 μm thick; 90% pure TMOS (10% methyl groups)	Benzene	0.1707 ± 0.0162	0.08774 ± 0.00335	0.3059 ± 0.0014	2.7466 ± 0.0261	2.3760 ± 0.0261	2.9311 ± 0.0039
9/1 2.0 ± 0.02 mm thick; 90% pure TMOS (10% methyl groups)	Benzene	0.01975 ± 0.0011	na	na	2.5936 ± 0.0316	na	na
9/1 2.0 ± 0.02 mm thick; 90% pure TMOS (10% methyl groups)	Toluene	0.03979 ± 0.00081	na	na	2.8258 ± 0.0089	na	na

Absorbance recorded at 252 nm.

the binding or entrapment kinetics. The values of (a) the binding rate coefficient, k and the fractal dimension, D_f for a single-fractal analysis are given in Table 12.2.

Figure 12.8b shows the binding of benzene to a 1980 μm thick 9/1 TMOS sol–gel. In this case, a dual-fractal analysis is required to describe the binding kinetics. The values of (a) the binding rate coefficient, k and the fractal dimension, D_f for a single-fractal analysis, and (b) the binding rate coefficients, k_1 and k_2, and the fractal dimension, D_{f1} and D_{f2} for a dual-fractal analysis are given in Table 12.2. It is of interest to note that as the thickness of the silica gel monolith increases from 680 to 1980 μm, the mechanism of entrapment of benzene changes. This is because a single-fractal analysis is adequate to describe the binding kinetics at the lower (680 μm) thickness, whereas a dual-fractal analysis is required to describe the binding kinetics at the higher (1980 μm) thickness. Understandably, a more complex analysis is required to describe the entrapment (or binding kinetics) at the higher sol–gel monolith thickness. One might expect that for the lower (680 μm) monolith thickness, the degree of diffusional limitations may be less than that exhibited at the higher (1980 μm) monolith thickness.

Figure 12.9a shows the binding of benzene to a 9/1 2.0 ± 0.2 mm thick TMOS sol–gel. A single-fractal analysis is adequate to describe the binding (or entrapment) kinetics. The values of the binding rate coefficient, k and the fractal dimension, D_f for a single-fractal analysis are given in Table 12.2.

Figure 12.9b shows the binding of toluene to a 9/1 2.0 ± 0.2 mm thick TMOS sol–gel. A single-fractal analysis is adequate to describe the binding (or entrapment) kinetics. The values of the binding rate coefficient, k and the fractal dimension, D_f for a single-fractal analysis are given in Table 12.2. It is of interest to note that as one goes from benzene to toluene, Table 12.2 indicates that the fractal dimension increases by 8.95% from a value of D_f equal to 2.5936 to D_f equal to 2.8258. This leads to an increase in the binding (or entrapment) rate coefficient, k by a factor of 2.01 (just a little more than double) from a value of 0.01975 to 0.03979. This is interesting since the molecular weight of benzene is 78 and the molecular weight of toluene is 92. One might anticipate that since the molecular weight of toluene is more than that of benzene, its size would be bigger (especially due to the side group, CH_3), and subsequently its entrapment (or binding) rate

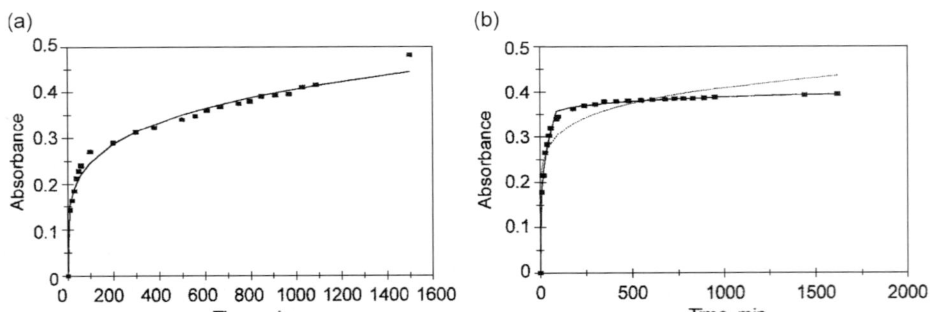

Figure 12.8 Effect of 9/1 sol–gel monolith thickness on the entrapment (or binding) kinetics: (a) 680 μm, (b) 1980 μm. When only a solid line (—) is used then a single-fractal analysis applies. When both a (···) and a (—) solid line is used, then the (···) line represents a single-fractal analysis, and the (—) line represents a dual-fractal analysis.

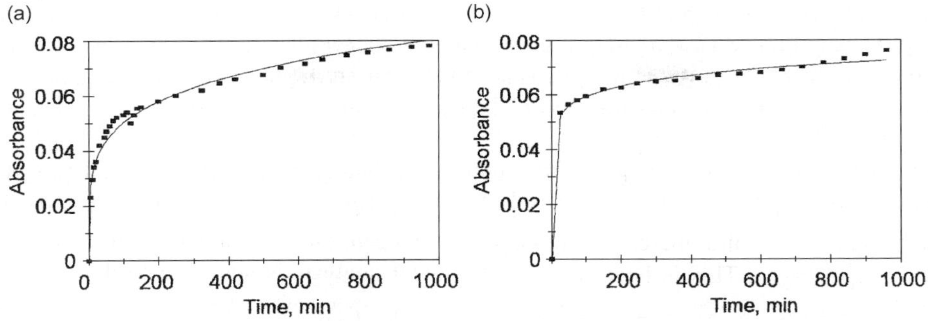

Figure 12.9 Binding (or entrapment) kinetics of different hydrocarbons to a 9/1 sol–gel monolith with a 2.0 ± 0.2 mm thickness (Calvin-Munoz *et al.*, 2002): (a) benzene, (b) toluene.

coefficient is lower. However, note that the fractal dimension for the binding of toluene is higher than that of benzene. Once again, one notes that higher fractal dimensions lead to higher binding (or entrapment) rate coefficients.

Figure 12.10a and Table 12.2 shows that for a single-fractal analysis the binding (or entrapment) rate coefficient, k increases with an increase in the fractal dimension, D_f. For the data presented in Table 12.2 and in Figure 12.10a, the binding rate coefficient, k is given by:

$$k = (7.9 \times 10^{-6} \pm 0.1 \times 10^{-6}) D_f^{8.204 \pm 0.122} \qquad (12.4a)$$

The fit is very good. The R squared value is 0.9997. Only three data points are available. The availability of more data points would lead to a more reliable fit. The binding rate coefficient, k is very sensitive to the degree of heterogeneity (or the fractal dimension, D_f) as noted by the higher than eighth (equal to 8.204) order of dependence on D_f. Note that the data for benzene and toluene are plotted together due to the scarcity of the data points available. Nevertheless, the fit is very good. It would be of interest to entrap (or bind) other

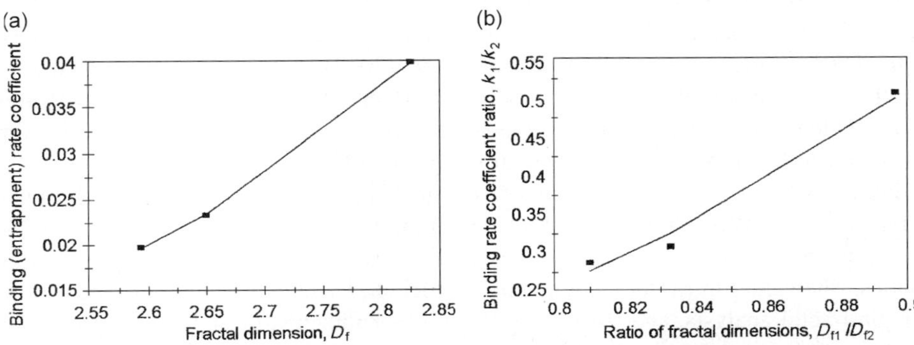

Figure 12.10 (a) Increase in the binding rate coefficient, k with an increase in the fractal dimension, D_f for the entrapment (or binding) of benzene in a TMOS (pure or with methyl groups) monolith sol–gel. (b) Increase in the binding rate coefficient ratio, k_1/k_2 with an increase in the fractal dimension ratio, D_{f1}/D_{f2} for the entrapment (or binding) of benzene in a TMOS (pure or with methyl groups) monolith sol–gel.

hydrocarbons (that are pollution contaminants) such as xylene, and note if they could also be plotted in Figure 12.10a. Also, toluene has a higher molecular weight than benzene, and its binding rate coefficient and fractal dimension are higher than those of benzene, respectively. However, only one set of data points are available, and thus one cannot generalize this result.

Figure 12.10b and Table 12.2 show that for a dual-fractal analysis the ratio, k_1/k_2 increases as the ratio, D_{f1}/D_{f2} increases. In Figure 12.10b and Table 12.2, the ratio k_1/k_2 is given by:

$$k_1/k_2 = (0.9292 \pm 0.0650)(D_{f1}/D_{f2})^{5.7654 \pm 0.9186} \quad (12.4b)$$

The fit is very good. Once again, only three data points are available. The availability of more data points would lead to a more reliable fit. The ratio of binding (or entrapment) rate coefficients is very sensitive to the degree of heterogeneity (or the fractal dimension) as noted by the higher than fifth (equal to 5.7654) order of dependence on the ratio of the fractal dimensions. In this case only the data for benzene are available. It would be of interest to note that if similar trends are also prevalent for toluene, xylene, and other hydrocarbons that are termed as pollutants.

Carter *et al.* (2003) have recently used the KinExA™ instrument to analyze the binding of 2,4 dinitrophenol (DNP). These authors indicate that DNP has been identified for its toxicological properties in animals. Furthermore, even though its effects on humans has not been extensively studied, it is estimated that lethal doses range from 14 to 43 mg/kg (National Research Council, 1982). Cataracts may develop on human contact with DNP (US EPA, 1993). Carter *et al.* (2003) indicate that human exposure may also lead to nausea, dizziness, and headaches.

Carter *et al.* (2003) have used the KinExA™ since they point out that there are errors in using the popular surface plasmon resonance (SPR) instrument in analyzing biomolecular interactions, and in the SPR generated kinetic coefficients. The values generated by the SPR were different from those obtained with the KinExA™ instrument (Blake *et al.*, 1999). The present Carter *et al.* (2003) method for detecting DNP is based on a competitive immunological technique suitable for a fluorescent biosensor format (Blake *et al.*, 2001; Carter *et al.*, 2000). Carter *et al.* (2003) indicate that their KinExA™ instrument comprised of a capillary flow observation cell that contained a microporous screen which enclosed a compact capture antigen-coated bead bed. The beads were PMMA coated with dinitrophenol-human serum albumin (DNP-HSA) conjugate.

Figure 12.11a shows the binding of 0 ng/ml free 2,4 dinitrophenol in phosphate buffered saline (PBS) solution equilibrated with mouse anti-DNP antibody to DNA-HSA labeled beads. A dual-fractal analysis is required to adequately describe the binding kinetics. The values of (a) the binding rate coefficient, k and the fractal dimension, D_f for a single-fractal analysis, and (b) the binding rate coefficients, k_1 and k_2, and the fractal dimensions, D_{f1} and D_{f2} for a dual-fractal analysis are given in Table 12.3.

Figure 12.11b shows the binding of 5 ng/ml free 2,4 dinitrophenol in PBS solution equilibrated with mouse anti-DNP antibody to DNA-HSA labeled beads. A dual-fractal analysis is required to adequately describe the binding kinetics. The values of (a) the binding rate coefficient, k and the fractal dimension, D_f for a single-fractal analysis, and

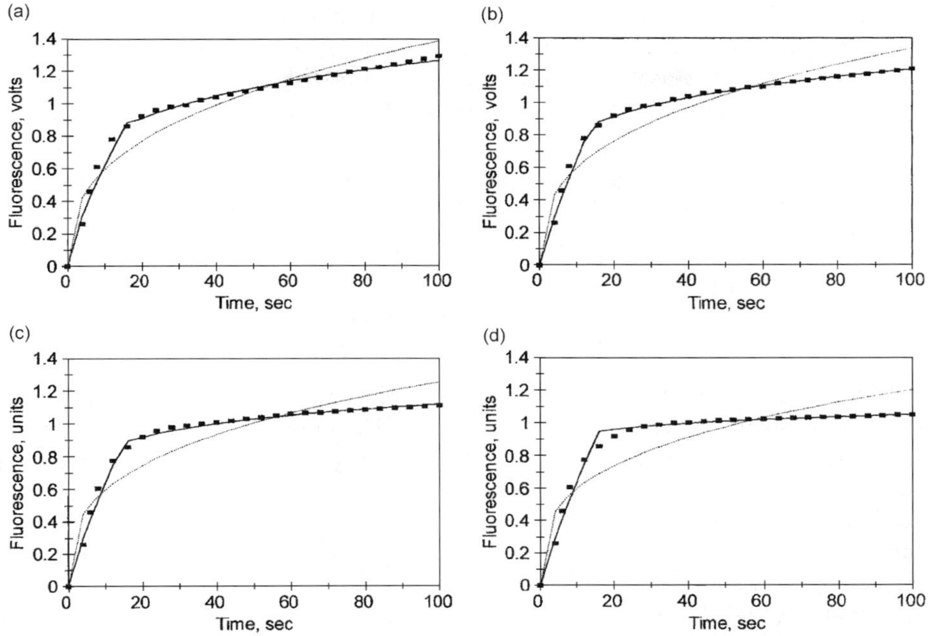

Figure 12.11 Binding of different concentrations (in ng/ml) of free 2,4 dinitrophenol (DNP) in phosphate buffered saline (PBS) equilibrated with anti-DNP monoclonal antibody to DNA-HSA labeled beads (Carter *et al.*, 2003): (a) 0, (b) 5, (c) 10, (d) 100. When only a solid line (—) is used then a single-fractal analysis applies. When both a (· · ·) and a (—) solid line is used, then the (· · ·) line represents a single-fractal analysis, and the (—) line represents a dual-fractal analysis.

(b) the binding rate coefficients, k_1 and k_2, and the fractal dimensions, D_{f1} and D_{f2} for a dual-fractal analysis are given in Table 12.3.

This is a competitive immunoassay, and it is not surprising that as the free 2,4 DNP concentration in solution increases from 0 to 5 ng/ml, the initial binding rate coefficient, k_1 decreases by a factor of 0.841 from a value of k_1 equal to 0.1091 to k_1 equal to 0.0918. It is of interest to note that as one goes from 0 to 5 ng/ml 2,4 DNP in solution the fractal dimension, D_{f1} increases by a factor of 0.885 from a value of D_{f1} equal to 1.4918 to D_{f1} equal to 1.3028. Changes in the degree of heterogeneity (fractal dimension) and in the binding rate coefficient are in the same direction. Also, note that as one goes from 0 to 5 ng/ml 2,4 DNP concentration in solution the fractal dimension, D_{f2} increases by a factor of 1.027 from D_{f2} equal to 2.5864 to D_{f2} equal to 2.6568. This leads to an increase in the binding rate coefficient, k_2 by a factor of 1.124 from a value of k_2 equal to 0.4878 to k_2 equal to 0.5483.

Figure 12.11c shows the binding of 10 ng/ml free 2,4 dinitrophenol in PBS solution equilibrated with mouse anti-DNP antibody to DNA-HSA labeled beads. A dual-fractal analysis is required to adequately describe the binding kinetics. The values of (a) the binding rate coefficient, k and the fractal dimension, D_f for a single-fractal analysis, and (b) the binding rate coefficients, k_1 and k_2, and the fractal dimensions, D_{f1} and D_{f2} for a dual-fractal analysis are given in Table 12.3.

Table 12.3

Binding rate coefficients and fractal dimension values for the binding of different concentrations of free 2,4 dintrophenol (DNP) in phosphate buffered saline (PBS) solution equilibrated with mouse anti-DNP monoclonal antibody to DNA-HSA labeled beads (Carter *et al.*, 2003)

2,4 DNP concentration, ng/ml	k	k_1	k_2	D_f	D_{f1}	D_{f2}
0	0.1091 ± 0.0175	0.1091 ± 0.0175	0.4878 ± 0.0055	0.2676 ± 0.0612	1.4918 ± 0.2168	2.5864 ± 0.0104
5	0.2693 ± 0.0425	0.0918 ± 0.0136	0.5483 ± 0.044	2.3034 ± 0.0652	1.3028 ± 0.2524	2.6568 ± 0.0065
10	0.2895 ± 0.050	0.0918 ± 0.0136	0.6472 ± 0.0086	2.3628 ± 0.0706	1.3028 ± 0.2524	2.7624 ± 0.017
100	0.3041 ± 0.0560	0.0918 ± 0.0136	0.8181 ± 0.0036	2.4020 ± 0.0750	1.3028 ± 0.2524	2.8908 ± 0.0046

Figure 12.11d shows the binding of 100 ng/ml free 2,4 dinitrophenol in PBS solution equilibrated with mouse anti-DNP antibody to DNA-HSA labeled beads. A dual-fractal analysis is required to adequately describe the binding kinetics. The values of (a) the binding rate coefficient, k and the fractal dimension, D_f for a single-fractal analysis, and (b) the binding rate coefficients, k_1 and k_2, and the fractal dimensions, D_{f1} and D_{f2} for a dual-fractal analysis are given in Table 12.3.

Note that in the initial binding region, the binding rate curves for the 5, 10, and 100 ng/ml are almost indistinguishable from each other (Carter *et al.*, 2003). Thus, not surprisingly, the estimated values of the binding rate coefficients, k_1, and the fractal dimension, D_{f1} are identical. The binding rate curves differ in the second regime of binding for these concentrations where, as expected, the binding rate coefficient(s), k_2 and the fractal dimension(s), D_{f2} are different.

Figure 12.12a and Table 12.3 show the increase in the binding rate coefficient, k_2 with an increase in the free 2,4 DNP concentration in solution. In the 0–100 ng/ml 2,4 DNP concentration in solution, the binding rate coefficient, k_2 is given by:

$$k_2 = (0.5614 \pm 0.0669)[2,4\ \mathrm{DNP}]^{0.0641 \pm 0.0204} \tag{12.4c}$$

The fit is quite good. Only four data points are available. The availability of more data points would lead to a more reliable fit. The binding rate coefficient, k_2 is only marginally

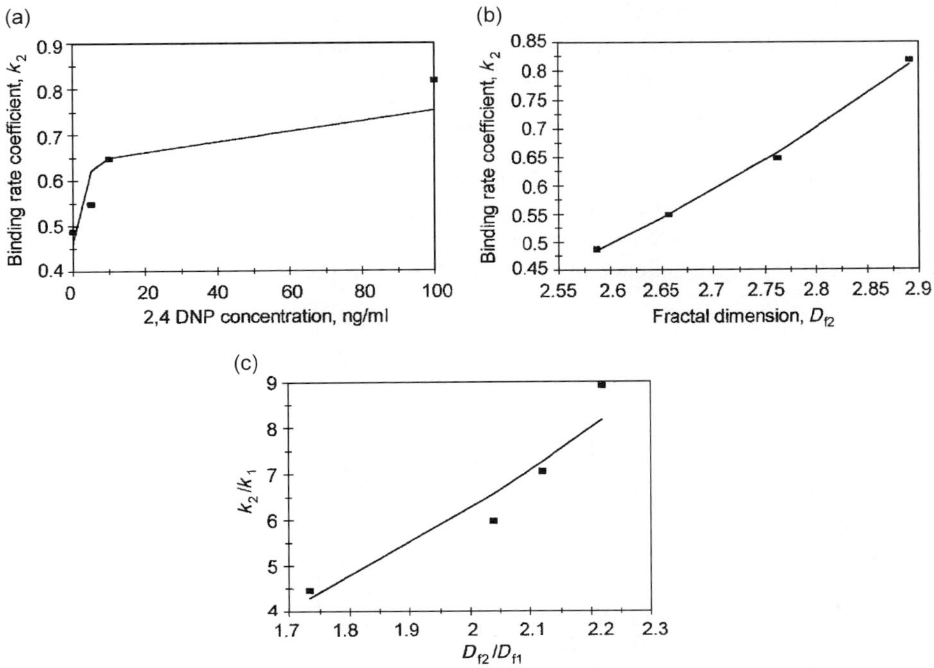

Figure 12.12 (a) Increase in the binding rate coefficient, k_2 with an increase in the free 2,4 DNP concentration in solution. (b) Increase in the binding rate coefficient, k_2 with an increase in the fractal dimension, D_{f2}. (c) Increase in the ratio of the binding rate coefficients, k_2/k_1 with an increase in the ratio of fractal dimensions, D_{f2}/D_{f1}.

sensitive to the free 2,4 DNP concentration in solution as noted by the very low order (equal to 0.0641) of dependence of k_2 on the 2,4 DNP concentration in solution.

Figure 12.12b and Table 12.3 show the increase in the binding rate coefficient, k_2 with an increase in the fractal dimension, D_{f2}. For the data presented in Figure 12.12b and in Table 12.3, the binding rate coefficient, k_2 is given by:

$$k_2 = (0.00597 \pm 0.00009)D_{f2}^{4.625 \pm 0.1588} \tag{12.4d}$$

The fit is very good. Only four data points are available. The availability of more data points would lead to a more reliable fit. The binding rate coefficient, k_2 is very sensitive to the degree of heterogeneity or the fractal dimension, D_{f2} that exists on the PMMA bead bed surface as noted by the very high order (equal to 4.625) of dependence of k_2 on D_{f2}.

Figure 12.12c and Table 12.3 show the increase in the ratio of the binding rate coefficients, k_2/k_1 with an increase in the ratio of the fractal dimensions, D_{f2}/D_{f1}. For the data presented in Figure 12.12c and in Table 12.3, the ratio of the binding rate coefficients, k_2/k_1 is given by:

$$k_2/k_1 = (1.0302 \pm 0.1046)(D_{f2}/D_{f1})^{2.5987 \pm 0.5194} \tag{12.4e}$$

The fit is quite good. Only four data points are available. The availability of more data points would lead to a more reliable fit. The ratio of the binding rate coefficients, k_2/k_1 is sensitive to the ratio of the fractal dimensions, D_{f2}/D_{f1} as noted by the close to two and one-half order (equal to 2.5987) dependence on the ratio, D_{f2}/D_{f1}.

Figure 12.13a and Table 12.3 show the increase in the fractal dimension, D_{f2} with an increase in the free 2,4 DNP concentration in solution. In the 0–100 ng/ml free 2,4 DNP concentration range in solution the fractal dimension, D_{f2} is given by:

$$D_{f2} = (2.6841 \pm 0.0700)[2,4\,\mathrm{DNP}]^{0.0121 \pm 0.0041} \tag{12.4f}$$

The fit is quite good. Only four data points are available. The availability of more data points would lead to a more reliable fit. The fractal dimension, D_{f2} is only marginally sensitive to the free 2,4 DNP concentration in solution as noted by the very low order (equal to 0.0121) of dependence of D_{f2} on the 2,4 DNP concentration in solution.

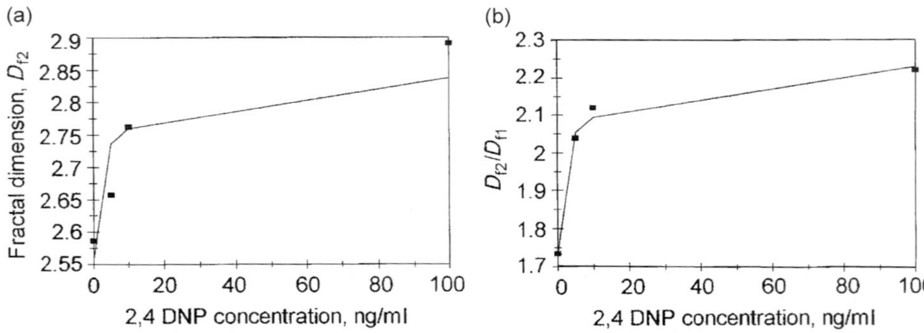

Figure 12.13 (a) Increase in the fractal dimension, D_{f2} with an increase in the free 2,4 DNP concentration in solution. (b) Increase in the ratio of fractal dimensions, D_{f2}/D_{f1} with an increase in the free 2,4 DNP concentration in solution.

Figure 12.13b and Table 12.3 show the increase in the fractal dimension ratio, D_{f2}/D_{f1} with an increase in the free 2,4 DNP concentration in solution. In the 0–100 ng/ml free 2,4 DNP concentration range in solution the ratio of the fractal dimensions, D_{f2}/D_{f1} is given by:

$$D_{f2}/D_{f1} = (1.9664 \pm 0.0216)[2{,}4\text{DNP}]^{0.0272 \pm 0.0016} \tag{12.4g}$$

The fit is quite good. Only four data points are available. The availability of more data points would lead to a more reliable fit. The ratio of the fractal dimensions, D_{f2}/D_{f1} is only marginally sensitive to the free 2,4 DNP concentration in solution as noted by the very low order (equal to 0.0272) of dependence of D_{f2} on the 2,4 DNP concentration in solution. However, it should be noted, and as indicated earlier, that the fractal dimension is on a log scale, and even very small changes in the fractal dimension value can lead to significant changes in the degree of heterogeneity on the PMMA bead surface.

Chomat *et al.* (2002) have recently used an optic fiber to detect toluene in water. They dip-coated a thin dimethoxypolysiloxane layer on the fiber core. These authors used a sensing fiber with an inverted graded-index (IGI) profile of the parabolic shape in the core. This group (Chomat *et al.*, 1999, 2000) had previously indicated this type of fiber permits enhanced sensitivity.

Chomat *et al.* (2002) wanted to investigate which was the best way of exposing the fiber to the analyte (toluene). In the static exposure method the fiber was initially in air. Then, it ws exposed to a solution at rest. In the dynamic exposure method the fiber was initially in flowing water. It was then exposed for some time to a flowing toluene solution.

Figure 12.14a shows the binding of toluene in water to the optical fiber with an IGI profile in the static method and at the beginning of the experiment. A dual-fractal analysis is required to adequately describe the binding and the dissociation kinetics. The values of (a) the binding rate coefficient, k and the fractal dimension, D_f for a single-fractal analysis, (b) the binding rate coefficients, k_1 and k_2 and the fractal dimensions, D_{f1} and D_{f2} for a dual-fractal analysis, (c) the dissociation rate coefficients, k_d, and the fractal dimension, D_{fd} for a single-fractal analysis, and (d) the dissociation rate coefficients, k_{d1} and k_{d2}, and the fractal dimensions, D_{fd1} and D_{fd2} for a dual-fractal analysis are given in Tables 12.4a and 12.4b.

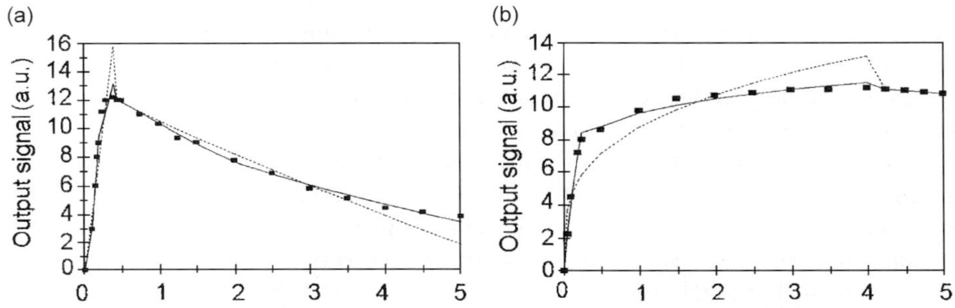

Figure 12.14 (a) Binding of toluene in water to the optical fiber in the static method and at the beginning of the experiment. (b) Binding of toluene in water to the optical fiber in the static method after 600 exposures. (Chomat *et al.*, 2002).

Figure 12.14b shows the binding of toluene in water to the optical fiber with an IGI profile in the static method after 600 exposures. A dual-fractal analysis is required to describe the binding kinetics. A single-fractal analysis is adequate to describe the dissociation kinetics. The values of (a) the binding rate coefficient, k and the fractal dimension, D_f for a single-fractal analysis, (b) the binding rate coefficients, k_1 and k_2 and the fractal dimensions, D_{f1} and D_{f2} for a dual-fractal analysis, and (c) the dissociation rate coefficients, k_d, and the fractal dimension, D_{fd} for a single-fractal analysis are given in Tables 12.4a and 12.4b.

Note that in this case as one goes from the static case at the beginning of the experiment to the static case after 600 exposures, there is an increase in (a) the fractal dimension value, D_{f1} from a value of zero to 1.425, and in (b) the fractal dimension value, D_{f2} from 1.9428 to 2.7510. This leads to a decrease in (a) the binding rate coefficient, k_1 value from 131.11 to 25.159, and in (b) the binding rate coefficient, k_2 value from 21.349 to 9.6053. In these two cases, the changes in the binding rate coefficients (k_1 and k_2) and the changes in the degree of heterogeneity (D_{f1} and D_{f2}) on the fiber optic surface are in opposite directions.

Figure 12.15a shows the binding of toluene in water to the optical fiber in the dynamic mode at 24 °C. A dual-fractal analysis is required to adequately describe the binding kinetics. The values of (a) the binding rate coefficient, k and the fractal dimension, D_f for a single-fractal analysis, and the (b) the binding rate coefficients, k_1 and k_2 and the fractal dimensions, D_{f1} and D_{f2} for a dual-fractal analysis, are given in Tables 12.4a and 12.4b.

Figure 12.15b shows the binding of toluene in water to the optical fiber in the dynamic mode at 24.6 °C. A dual-fractal analysis is required to adequately describe the binding kinetics. The values of (a) the binding rate coefficient, k and the fractal dimension, D_f for a single-fractal analysis, and the (b) the binding rate coefficients, k_1 and k_2 and the fractal dimensions, D_{f1} and D_{f2} for a dual-fractal analysis, are given in Table 12.4a and 12.4b.

Figure 12.15c shows the binding of toluene in water to the optical fiber in the dynamic mode at 26.2 °C. A single-fractal analysis is adequate to describe the binding kinetics. The values of (a) the binding rate coefficient, k and (b) the fractal dimension, D_f for a single-fractal analysis are given in Tables 12.4a and 12.4b.

There is a change in the binding mechanism as one goes from 24 or 24.6 to 26.2 °C, since at 24 and 24.6 °C a dual-fractal analysis is required to describe the binding kinetics, whereas at 26.2 °C a single-fractal analysis is adequate to describe the binding kinetics. Apparently, in this case at the lower temperatures analyzed, the binding is more complex than at the higher temperature analyzed.

Akrajas *et al.* (2002) developed an optical system to detect vapor samples containing 2-propanol. These authors indicate the need for a gas sensor with high sensitivity. This electronic nose has applications in the clinical (Gardner *et al.*, 1997; Grove *et al.*, 1998), environmental (Delpha *et al.*, 2000), and military (Lee *et al.*, 2000) areas. Thin films of four metal-octaethyl porphyrins containing the metal atoms, Mn, Fe, Co, and Ru were used (Akrajas *et al.*, 2002). Langmuir–Blodgett (LB) films of the metallo-octaethyl porphyrins were deposited on a glass substrate. The sensitivity of the porphyrin 2, 3, 7, 8, 12, 13, 17, 18-octaethyl porphine iron (III) chloride (FeOEPCl) towards the saturated vapor of 2-propanol was analyzed. These authors indicate that the optical detection

Table 12.4a

Binding and dissociation rate coefficients for toluene in water using an inverted graded-index (IGI) optical fiber using static or dynamic responses (Chomat *et al.*, 2002)

Type of exposure	Condition of experiment	k	k_1	k_2	k_d	k_{d1}	k_{d2}
Static	At the beginning of the experiment	40.118 ± 8.758	131.11 ± 7.06	21.349 ± 1.854	2.6428 ± 0.602	3.029 ± 0.821	3.4618 ± 0.1238
Static	After 600 exposures	8.7573 ± 2.237	25.159 ± 2.254	9.6053 ± 0.2378	0.40 ± 0	na	na
Dynamic	Water at 24 °C	5.847 ± 2.642	10.624 ± 2.99	9.5065 ± 0.044	na	na	na
Dynamic	Water at 24.6 °C	8.160 ± 1.483	10.675 ± 1.853	10.325 ± 0.243	na	na	na
Dynamic	Water at 26.2 °C	14.344 ± 0.618	na	na	na	na	na

Table 12.4b

Fractal dimensions for the binding and the dissociation phases for toluene in water using an inverted graded-index (IGI) optical fiber using static or dynamic responses (Chomat *et al.*, 2002)

Type of exposure	Condition of experiment	D_f	D_{f1}	D_{f2}	D_{fd}	D_{fd1}	D_{fd2}
Static	At the beginning of the experiment	0.9612 ± 0.3504	0 + 0.2018	1.9428 ± 0.2538	1.1938 ± 0.0828	1.0112 ± 0.152	1.7650 ± 0.0760
Static	After 600 exposures	2.4242 ± 0.0917	1.4250 ± 0.136	2.7510 ± 0.0182	1.0 ± 0	na	na
Dynamic	Water at 24 °C	1.76 ± 0.272	0.1312 + 0.491	2.8882 ± 0.0076	na	na	na
Dynamic	Water at 24.6 °C	2.2886 ± 0.1169	1.5656 ± 0.3264	2.7662 ± 0.0342	na	na	na
Dynamic	Water at 26.2 °C	1.8928 ± 0.053	na	na	na	na	na

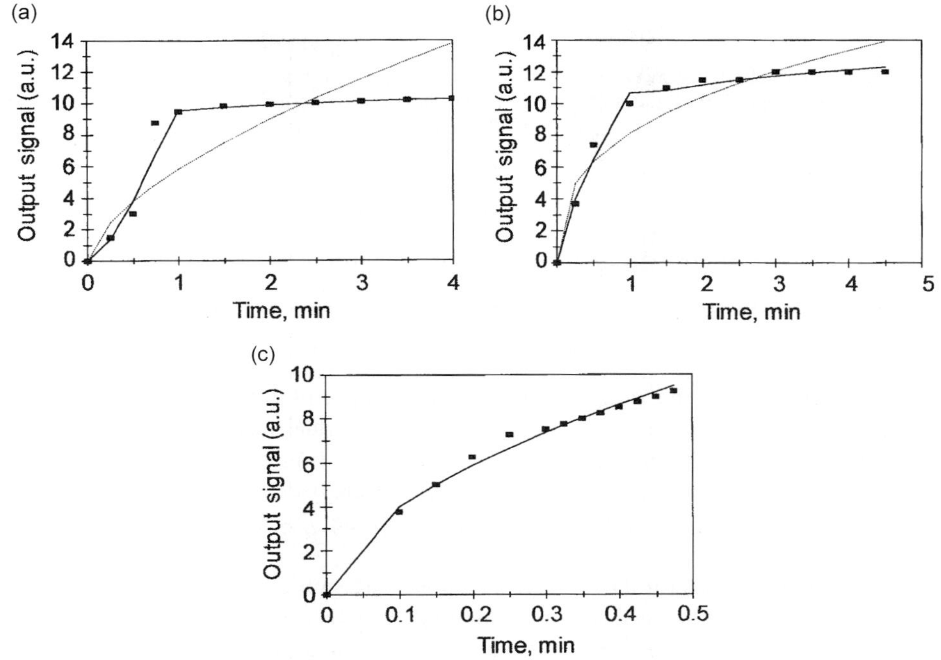

Figure 12.15 Binding of toluene in water to the optical fiber in the dynamic mode (Chomat *et al.*, 2002) at: (a) 24 °C, (b) 24.6 °C, (c) 26.2 °C.

of the gases is made possible due to the change of the optical properties of these thin films which interact with the molecules of the volatile gas such as 2-propanol, ethanol, acetone, and cyclohexane.

Figure 12.16a shows the binding of the saturated vapor of 2-propanol to the thin film deposited on a glass substrate. This is the initial run. A dual-fractal analysis is required to adequately describe the binding kinetics. The values of (a) the binding rate coefficient, k, and the fractal dimension, D_f, for a single-fractal analysis, and (b) the binding rate coefficients, k_1 and k_2 and the fractal dimensions, D_{f1} and D_{f2} for a dual-fractal analysis are given in Table 12.5.

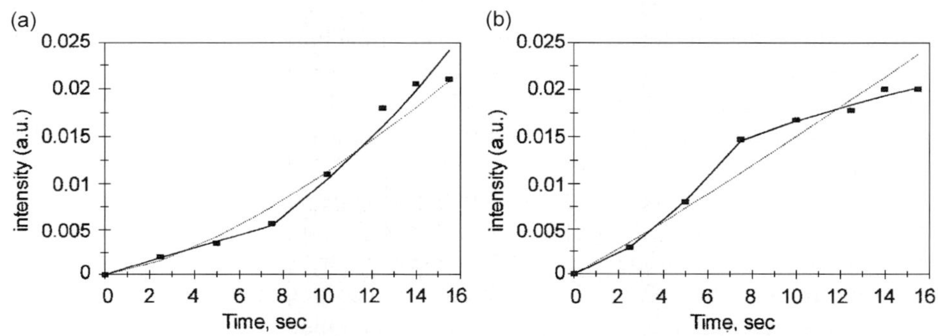

Figure 12.16 Binding of saturated vapor of 2-propanol to the thin film deposited on a glass substrate (Akrajas *et al.*, 2002): (a) initial run, (b) repeated run.

Table 12.5

Binding rate coefficients and fractal dimensions for 2-propanol in the vapor phase to a thin (Langmuir–Blodgett) film of a metalloporphyrin derivative, *n*-octaethyl prophine iron (III) chloride (FeOEPCl) on a glass substrate (Akrajas *et al.*, 2002)

Run	k	k_1	k_2	D_f	D_{f1}	D_{f2}
Initial run	0.000446 ± 0.0001	0.000827 ± 0.00007	0.00137 ± 0.000017	0.1952 + 0.2514	1.1250 ± 0.2028	~0
Repeated run	0.01139 ± 0.00027	0.000795 ± 0.000014	0.00608 ± 0.00017	0.9318 ± 0.2258	0.1138 + 0.2078	2.1240 ± 0.0974

Effect of a repeated run.

Figure 12.16b shows the binding of the saturated vapor of 2-propanol to the thin film deposited on a glass substrate. This is the repeated run. A dual-fractal analysis is required to adequately describe the binding kinetics. The values of (a) the binding rate coefficient, k, and the fractal dimension, D_f, for a single-fractal analysis, and (b) the binding rate coefficients, k_1 and k_2 and the fractal dimensions, D_{f1} and D_{f2} for a dual-fractal analysis are given in Table 12.5.

The binding responses for the two-cycle response are similar in shape. However, there are significant changes in the shape of the binding curve as is evident from Figure 12.16a and b. It is of interest to note that as one goes from the initial run to the repeated run there is a decrease in the fractal dimension, D_{f1} by a factor of 9.885 from a value of 1.1250 to 0.1138, and a corresponding decrease in the binding rate coefficient, k_1, by a factor of 1.04 from a value of 0.000827–0.000795. Also, there is an increase in the fractal dimension, D_{f2} from a value of ~0 to 2.1240. This leads to a corresponding increase in the binding rate coefficient, k_2 by a factor of 44.38 from a value of k_2 equal to 0.000137–0.00608. Note that changes in the fractal dimension (D_{f1} or D_{f2}) and the changes in the binding rate coefficient (k_1 or k_2) are in the same direction.

12.4 CONCLUSIONS

A fractal analysis of the binding of methanol in solution to the AOD/POD/11-FUT/Au electrode provides a quantitative indication of the state of disorder or the degree of heterogeneity on the Au electrode surface. Values of the binding rate coefficient and the fractal dimension for a single- and a dual-fractal analysis are provided. The dual-fractal analysis is used only when the single-fractal analysis did not provide an adequate fit (sum of least squares less than 0.98). This was done by the regression provided by Quattro Pro 8.0.

In accord with the prefactor analysis for fractal aggregates (Sorenson and Roberts, 1997), quantitative (predictive) expressions for the binding rate coefficient, k_2 are developed as a function of the degree of heterogeneity or the fractal dimension, D_{f2} on the Au electrode surface. Predictive equations are also presented for the binding rate coefficients, k_1 and k_2 as a function of the PME concentration in solution. The predictive equation developed helps provide a means by which the binding rate coefficients may be manipulated by changing either the PME concentration in solution or the degree of heterogeneity on the Au surface. This is apparently the first study that links the binding of an environmental contaminant (methanol) to the degree of heterogeneity on the Au electrode surface. In this case, the binding rate coefficient, k_2 is very sensitive to the degree of heterogeneity or the fractal dimension, D_{f2} on the Au electrode surface.

The entrapment (or binding) kinetics of the environmental pollutants, benzene and toluene in sol–gel monoliths (Calvo-Munoz *et al.*, 2002) are also analyzed using fractals. Both pure TMOS sol–gels and MeTMOS (with methyl groups) sol–gels are analyzed. Examples are provided where both a single- or a dual-fractal analysis may be used. In this case too, one notes that, when a single-fractal analysis applies, the binding (or entrapment) rate coefficient, k increases as the fractal dimension, D_f (or the degree of heterogeneity) on the sol–gel surface increases.

It is of interest to note that even though the molecular weight of toluene is higher than that of benzene (its size is bigger due to the side methyl group), its binding (or entrapment) rate coefficient is higher than that of benzene. This is due to the fact that for the case of toluene the fractal dimension, or the degree of heterogeneity, is higher than that of benzene. Once again, emphasizing the influence of the nature of the sol–gel surface on the binding (or entrapment) kinetics. If this is indeed true, then experimentalists who would like to use sol–gel monoliths for biosensor development, should pay more attention to the nature of these sol–gel biosensors.

Carter *et al.* (2003) have used the KinExA™ instrument to analyze the binding kinetics of 2,4 dintrophenol since they claim that using the SPR biosensor leads to errors in the rate coefficients generated. The SPR is a popular biosensor, and has been used frequently. However, there are some assumptions in its usage, such as the absence of (a) diffusional limitations, and (b) heterogeneity of receptors on the sensing surface. Carter *et al.* (2003) have used a competitive immunoassay format, and thus not surprisingly, the binding rate coefficient for the detection of free 2,4 dinitrophenol (DNP) in solution decreases with an increase in the free 2,4 DNP concentration in solution.

More such studies are required that help determine and analyze the influence of the degree of heterogeneity on the binding rate coefficients (a) on the Au electrode surface, (b) on the sol–gel surface, (c) on the PMMA bead surface, and on other biosensor surfaces. The influence of heterogeneity on other relevant biosensor performance parameters on different biosensor surfaces should also be analyzed. Practicing biosensorists apparently need to start paying more attention to the nature of biosensor surface, and how it may influence biosensor performance parameters. This is of particular importance if the biosensor surface very significantly affects the biosensor performance parameters, such as the binding rate coefficient as observed in the present cases.

REFERENCES

Akrajas, MMS and M Yahya, *Sensors & Actuators B*, 2002, **85**, 191–196.

Barman, TE, *Enzyme Handbook*. Springer, New York, 1969.

Blake, II, AR Pavlov, RC and DA Blake, *Analytical Biochemistry*, 1999, **272**(2), 123–134.

Blake, DA, RM Jones, RC Blake, II, AR Pavlov, IA Darwish and H Yu, *Biosensors & Bioelectronics*, 2001, **16**(9–12), 799–809.

Calvo-Munoz, ML, TT Truong and TH Tran-Thi, *Sensors & Actuators B*, 2002, **87**, 173–183.

Carter, RM, RC Blake, II, HP Mayer, AA Echevarria, TD Nguyen and LA Bostanian, *Analytical Letters*, 2000, **33**(3), 405–412.
Carter, RM, RC Blake, II, TD Nguyen and LA Bostanian, *Biosensors & Bioelectronics*, 2003, **18**, 69–72.
Chomat, M, V Berkova, V Matejec, J Ctyroky, I Kasik and H Gagnaire, The detection of refractive index changes by using a sensing fiber with an inverted parabolic-index profile. *Proceedings of SPIE*, 1999, **3860**, 179–189.
Chomat, M, D Berkova, V Matejec, I Kasik, M Hayer, G Kuncova and Z Berka, Detection of hydrocarbons by means of PCS and IGI fibers excited in various ways. *Proceedings of the Conference of EURPTODE*, Lyon, April 16–19, 2000, pp. 141–142.
Chomat, M, D Berkova, V Matejec, I Kasik, G Kuncova and M Hayer, *Sensors & Actuators B*, 2002, **87**, 258–267.
Corel Quattro Pro, 8.0, Corel Corporation, Ottawa, Canada, 1997.
Delpha, C, M Siadat and M Lumbreras, *Sensors & Actuators B*, 2000, **69**, 243–247.
Gardner, JW, M Craven, C Dow and EL Hines, *Measurements in Science and Technology*, 1997, **8**, 120–127.
Grove, WA, ET Zellers and GC Frye, *Analytica Chimica Acta*, 1998, **371**, 131–143.
Guilbault, GG and GJ Lubrano, *Analytica Chimica Acta*, 1974, **69**, 189–194.
Guilbault, GG, B Danielsson, CF Mandenius and K Mosbach, *Analytical Chemistry*, 1983, **55**, 1582–1585.
Hasunuma, T, S Kuwabata, E Fukusaki and A Kobayashi, *Analytical Chemistry*, 2004, **76**, 1500–1506.
Havlin, S, *The Fractal Approach to Heterogeneous Chemistry: Surfaces, Colloids, Polymers*, (ed. D Avnir), Wiley, New York, 1989, pp. 251–269.
Kaneko, T, PY Wang and A Sato, *Journal of Occupational Health*, 1997, **39**, 159.
Lee, DS, HY Jung, JW Lim, M Lee, SW Ban, JS Huh and DD Lee, *Sensors & Actuators B*, 2000, **71**, 565–568.
Lubrano, GJ, MH Faridnia, G Palleschi and GG Guilbault, *Analytical Biochemistry*, 1991, **198**, 97–103.
Miyamoto, S, T Murakami, A Saito and J Kimura, *Biosensors & Bioelectronics*, 1991, **6**, 563–567.
Nanomura, AM and AA Benson, *Proceedings of the National Academy of Sciences*, 1992, **89**, 9794–9798.
National Research Council, *Drinking Water and Health*, **Vol. 4**, National Academy Press, Washington, DC, 1982.
Ramakrishnan, A and A Sadana, *Automedica*, 2001, **20**(3/4), 313–340.
Sadana, A and T Vo-Dinh, *Biotechnology & Applied Biochemistry*, 2001, **33**, 17–31.
Shirashi, T, E Fukusaki, C Miyake, A Yokota and A Kobayashi, *Journal of Bioscience and Bioengineering*, 2000, **89**, 564–568.
Sorenson, CM and GC Roberts, *Journal of Colloid & Interface Science*, 1997, **186**, 447–452.
US EPA, *Integrated Risk Information System (IRIS) on 2,4 Dinitrophenol*. Environmental Criteria and Assessment Office, Office of Health and Environment Assessment, Office of Research and Development, Cincinnati, Ohio, 1993.
Wang, J, Q Chen, M Pedereo and JM Pingarron, *Analytica Chimica Acta*, 1995, **300**, 111–116.
Weiss, GH, *Fractals in Science*. Springer, Berlin, 1994.

– 13 –

Market Size and Economics for Biosensors

13.1 INTRODUCTION

The applications of biosensor and bioelectronic devices may be placed in five categories (Talukder, 2002): agriculture, food analysis, high-throughput screening, medical analysis, and nanobiotechnology. Each of these categories may be further subdivided into different applications, for example, Talukder (2002) indicates that in the category of agriculture one has sensor applications for the detection of herbicide(s) in soil, detection of heavy metals in soil, detection of genetic modifications of foodstuffs, etc. The author provides a thorough in-depth analysis of the market for biosensors, though the report is 3–4 years old.

Zarkoff (2002) indicates that the average annual growth rate (AAGR) for the biosensor market is around 7.5%. Some sectors such as nanobiotechnology have a higher growth rate (8.2%), whereas others such as food monitoring and biosensors used in agriculture have a lower growth rate of 7.2%. This author further indicates that the growth rate for high-throughput screening for the detection of drug targets is 7.5%. However, since presently this sector dominates the biosensor market (77%), Zarkoff (2002) indicates that it is this 7.5% which is also the AAGR of the biosensor market for all biosensor applications. This author further indicates that the market for (a) high-throughput screening and (b) medical analysis for the year 2004 was estimated to be $1.255 billion and $181.4 million, respectively.

Table 13.1 shows the estimated market value for high-throughput screening for the years 2004–2009 (Zarkoff, 2002).

The 8.5% AAGR was presented in Table 13.1 in case there are some reasons, such as (a) strategic new inventions that could significantly impact the biosensor market come into play or (b) a substantial increase in demand for these types of HTS systems to enhance drug discovery.

Zarkoff (2002) emphasizes that medical analysis biosensors is the second largest market, with glucose monitoring for the effective management of diabetes being a major driving force. A gradual increase in obesity of individuals along with increasing stress levels in our daily lives would lead to an increase in the incidence of diabetes, which is already rumored to be reaching epidemic proportions. Thus, apparently, the demand for glucose monitoring systems is not expected to slow down or decrease in the near future. Zarkoff (2002) estimates the medical market for biosensors for the year 2004

Table 13.1

Projected estimated market (in billion dollars) for high-throughput screening systems for drug discovery at an average annual growth rate (AAGR) of 7.5% (adapted from Zarkoff, 2002), and 8.5%, respectively

Year	Estimated market for high-throughput screening systems at 7.5% AAGR $ (in billion)	Estimated market for high-throughput screening systems at 8.5% AAGR $ (in billion)
Base year, 2004	1.255	1.255
2005	1.349	1.362
2006	1.450	1.477
2007	1.559	1.603
2008	1.676	1.739
2009	1.802	1.887

at $200 million. Assuming a conservative 90% of this $200 million for glucose monitoring devices yields $180 million as the estimated market for glucose monitoring systems.

A Freedonia report (2002) on the market for biosensors till the year 2006 estimates that the market will grow at 8.6% annual rate. This report is a couple of years old. Further predictions are provided till the year 2011. A later report by Freedonia (2004) indicates that the demand for chemical sensors will grow at a rate of 8.5% through 2008. This is primarily driven by: (a) falling prices for novel sensors and high-performance sensors and (b) microfabricated innovations. The growth of optical sensors will be the fastest. According to this report, the present chemical sensor market stands at $2.8 billion. Forecasts till the year 2008 and 2013 are provided.

Research and Markets (2004) estimates the biosensor market to grow to $10.8 billion by the year 2007 with an annualized growth rate of 10.4%. The report emphasizes that the medical application market continues to overshadow other application areas. It also points out that the rising rate of obesity in the world is leading to increase in rates of diabetics especially in the industrial world. The Research and Markets report (2004) also mentions other areas of biosensor application that include rapid assay biosensors for drug discovery, detection of biological and chemical agents for the war on terrorism, food monitoring, and environmental applications. As expected, there are differences in the AAGR for the biosensor market estimated by different authors or reports.

Rajan (2002) expects the biosensor market to grow at an annual rate of 6.6% to $1.81 billion by the year 2006. This author emphasizes that advances in photolithographic techniques as well as in microfluidics will go a long way in helping remove some of the hurdles that prevented growth in the biosensor industry. Rajan (2002) further emphasizes the particular use of biosensors to monitor food and crops (especially those that are genetically altered) in Europe. The monitoring of foodstuffs includes spoilage and the detection of toxins.

MedMarket Diligence (2003) has recently indicated the applications and market opportunities for nanotechnology and microelectromechanical systems (MEMS) for current medical applications. The report presents a detailed market and technological assessment for medical applications, with special emphasis on clinical diagnostics.

A very recent London South Bank University report by Chaplin (2004) indicates that the estimated world analytical market is 12 billion British pounds. The author estimates the biosensor market to be 0.1% of this analytical market yielding 12 million British pounds. Using an exchange rate of British £1 = $1.9433 (http://finance.yahoo.com/currency?u) (November 03, 2004; 6:30 pm) yields $2.32 million. Chaplin (2004) emphasizes that this market possesses a significant potential for expansion. Lab_Bell, Inc. (2004), a company that makes fast, inexpensive, and sensitive multi-purpose biosensors for the detection of toxic molecules estimates that the biosensor market for the year 2005 is as follows: agrifood is $237 million, environment applications $237 million, and industrial biotechnology $158 million. The small company (nine employees) presently has plans for a $2 million expansion. One of its products makes quantitative the concentration of herbicides, and evaluates the toxicity of effluents in under 10 min.

Lin and Wang (2005) have very recently presented a theoretical approach to biosensor commercialization strategy. These authors also estimate the biosensor market to grow to $10.8 billion by the year 2007. They expect the biosensor market to mushroom in the coming years. They emphasize that the biosensor market presents both opportunities as well as hurdles that need to be overcome. They also emphasize the importance of predicting the biosensor market, and the need to identify and convert promising biosensor technologies for commercial applications. Furthermore, these authors present alternate commercialization strategies for each specific biosensor application, and outline strategies to help predict the best one.

Glucose monitoring systems come under the fast growing sector of point-of-care (POC) diagnostic testing. Other POC tests include pregnancy testing, hepatitis testing, drugs of abuse screening testing, infectious disease testing, HIV, coagulation testing, and fertility testing (Trimark Publications, 2004). These authors indicate that diagnostic testing is an estimated $21 billion market. Both physicians and patients recognize the better management of health care provided by POC testing, and the demand for these types of products is bound to increase. Trimark Publications (2004) further emphasizes that newer technology platforms and better fabrication techniques would lead to more efficient POC devices. These POC devices range from single-use disposable units for individual home use to moderate-sized instrumented diagnostic systems for clinical and hospital use.

In a more recent report, Fuji-Kerzai (2004) indicates that the worldwide market for biosensors in 2003 was $7.3 billion. In contrast to Zarkoff's (2002) 7.5% AAGR for the biosensor market, Fuji-Kerzai (2004) indicates a growth rate of 10.8%. This represents almost a 44% increase in growth rate between that predicted by Zarkoff (2002) and Fuji-Kerzai (2004). This underscores the 'unreliability' in the projection figures for the growth rate, worldwide market, etc. Using the Fuji-Kerzai (2004) numbers with a $7.3 billion worldwide market for the year 2003, and an annual growth rate of 10.4% leads to a worldwide market for biosensors, in billion dollars of 8.059, 8.897, 9.823,10.84, 11.97, and 13.21, for the years 2004, 2005, 2006, 2007, 2008, and 2009, respectively. This author also indicates the important activities of the different major companies throughout their report that are driving the biosensor market. Using the rule of 72, the biosensor market will double in 6.92 years, if the predicted 10.4% growth rate is maintained. A growth rate of 12% would double the biosensor market in 6 years.

The order of importance of the different sectors in the biosensor market is also slightly different according to Fuji-Kerzai (2004) when compared with the Zarkoff (2002) analysis. Fuji-Kerzai (2004) emphasizes that due to the increasing obesity levels, especially in the industrial world, and the rapid rise in the number of diabetics and the necessity of monitoring glucose levels, the medical area is still the major driving force for the development of biosensors. This is followed by the need for (a) rapid assay biosensors for high-throughput systems to hasten the speed for drug discovery (as required by the pharmaceutical industry), (b) rapid detection biosensors for chemical and biological warfare agents, and finally (c) biosensors to detect food pathogens (food safety) and environmental contaminants. A biosensors and bioelectronics report by Talukder (2002) of the Business Communications Company also indicates the need for biosensors in fermentation and forensic technologies. This report also provides a perspective of public perception and public policy on the regulation of biosensor technologies as they apply to the medical and food industries.

A University of Delaware graduate class, ELEG 667 entitled 'Biosensors and BioMEMS (2004)', indicates that there are at least 50 different types of biosensor systems worldwide. This report was downloaded in the year 2004, and seems like quite an old report. However, in the year 2005, the report estimated the clinical diagnostics market to be around \$8.5 million (with, as expected, 90% comprising the home glucose market). As expected, different individuals have come up with different numbers for the anticipated market for biosensors in different areas.

Kissinger (2004) indicates that the electrochemical sensors market is at \$100 million per year and \$1 billion for all sensors. He indicates that the second most common analyte measured after glucose (which is the first) is lactate. However, lactate is a distant second to glucose, and its market is about two orders of magnitude (about 100 times) lower than that of glucose. The author emphasizes that in spite of the large number of publications in this area, the commercial applications of biosensors face severe challenges. The author classifies biosensors in to three types:

(a) *Single use*. Ninety-nine per cent of the commercial market, easy to use, not very precise or accurate, and the cost versus data rate is very high. For example, the electronics cost approximately, \$50, electrochemical cell and components (\$0.50)
(b) *Intermittent use*. Moderate complexity in use, excellent performance, but high up front cost (approximately \$1000–10,000 per instrument cost), good precision and accuracy, moderate cost versus data rate.
(c) *Continuous use*. Very easy to use, poor performance and accuracy, good precision, very low cost versus data rate.

Finally, Kissinger (2004) indicates that practical issues with regard to biosensor manufacture include inventory time at manufacture (about 2 months), shipping (2–10 days), and finally inventory at point of use (2–3 months). This aspect is often ignored.

The emphasis on the detection of harmful biological and chemical agents for protecting civilians and civilian infrastructure is underscored by the US government's Department of Homeland Security (DHS) budget of \$36.5 billion for the year 2004 (Farrell, 2004). It is expected to increase by about 10.1% to \$40.2 billion for the year 2005. Biosensors are expected to play a major role here. It is of interest to note that, though the DHS does not do any research itself, it does fund other government agencies such as FBI, CIA, EPA, etc.

Miller (2004) recently indicates that an after-effect of the September 11 attacks in the United States followed by the anthrax incidents have prompted the US Congress to approve plans and spend \$500 million for new biosafety space (NIAID, 2003). This biocontainment space is required to develop anti-bioweapon vaccines and drug treatments. The National Institute of Allergy and Infectious Diseases (NIAID, one of the National Institutes of Health in Bethesda, MD) also provided \$7–21 million each to build nine new biosafety level-3 (BSL-3) laboratories around the country. This emphasizes the seriousness of the situation and the perceived need for these new laboratories. Furthermore, Kostel (2004) also indicates that there are presently no rapid diagnostic methods to detect Category A bioweapons such as anthrax, botulism, plague, smallpox, tularemia, and viral hemorrhagic fevers. Vaccines and post-exposure therapy are available to various degrees (yes, no, and limited). Kostel (2004) further adds that the number of biosafety level 4 (BSL-4) laboratories will also increase. These are laboratories which may handle lethal biologicals spread through the air, and for which there is no known cure.

Farrell (2004) further emphasizes the need for rapid deployment, and also that biological agents are more difficult to detect than chemical agents. For the rapid detection of harmful biologicals, Baeumner in Farrell (2004) emphasizes the importance of improving the speed of biosensors. Microfluidics does that, but places a limitation on sample size. Baeumner is trying to increase the sample size. The detection is based on the movement of the reaction products along a test strip by capillary action.

The anthrax scare at a few locations in the United States has also prompted the development of on-site, quick detection tests of anthrax spores. Bohannon (2002) describes a rapid, on-site test that is capable of detecting anthrax spores within minutes. This compares with the pregnancy-like test that requires neither specialized training nor any instrumentation. This, if successfully brought to the market, is a good application of point-of-use detection technology.

Niedbala (2002) indicated a novel technology for the rapid, sensitive, on-site multiplex detection of warfare agents. This author used up-converting phosphor receptors. This is a new technology wherein materials up-convert infrared to ultra violet. The up-converting phosphor technology (UPTm) was demonstrated to detect biological agents such as tularensis, plague, and cholera.

Finally, Richard Mathies of the University of California in Farrell (2004) emphasizes that the information and knowledge gained in developing biosensors for biological agent detection can be applied for the diagnosis of disease (dual-action). By pinpointing the bacterium that causes a disease, the medical personnel can minimize the usage of broad-based antibiotics. Though the etiology of diseases such as systemic lupus erythomatosus (SLE) is not clear, the indiscriminate use of broad-based antibiotics is presumed to be a contributing factor in this complex autoimmune disease.

Diabetes is an autoimmune disease, and is attributed to be the pioneer (disease) for the development of biosensor technology. Individuals affected by diabetes, and who require glucose monitoring, constitute quite a large pool of people. Thus, there are so many major players (companies) in this area of glucose detection diagnostics. It is reasonable to anticipate that biosensing devices are available or will soon become available to facilitate in the early detection of other autoimmune diseases such as rheumatoid arthritis (RA), different forms of Cancer, SLE, etc. At least, this was the general sense gathered by this author whilst attending the International Conference on Immunology held in Stockholm, Sweden, in the summer of 2002, and Montreal, Canada in the summer of 2004. The Singapore Institute of Bioengineering and Nanotechnology (2003) has developed a very sensitive DNA/RNA biosensor for early cancer detection. It uses the principle of hybridization wherein a capture probe is immobilized on the biosensor surface. A DNA/RNA sample hybridizes on the probe surface. The report indicates that the electrical signal generated correlates directly with the extent of hybridization (amount of diseased DNA/RNA).

The autoimmune diseases are slow, insidious, and intractable. Timewise, the earlier one is able to detect these diseases, the earlier one can put the individual on a medical protocol, and help in the management of these diseases. Some of these diseases, like SLE, are difficult to diagnose, since more than one criteria (such as the presence of anti-DNA and creatinine above certain (normal) levels) have to be satisfied before the individual is said to be affected by this disease. It is reasonable to anticipate that in the future, in spite of the low market demand for the detection of autoimmune diseases, biosensors to help detect early markers for these diseases will become a reality, if they have already not done so. For the diagnosis of SLE, the application of microarrays would be a reasonable route to follow. This would permit the simultaneous quantitative detection of the different analytes required and if they are above or below a certain threshold level or range as indicated above for the correct diagnosis of SLE.

Crow and Wohlgemuth (2003) recently indicate that physical insights into the pathogenic mechanisms of disease (for example, SLE) have been attained from the analysis of microarray data of gene expression. Some patterns of gene expression are confirmatory in nature with previous gene expression patterns using other methods, such as increased expression of immune cell surface activation molecules. However, the microarray analysis by Crow and Wohlgemuth (2003) indicates the increased pattern of interferon-induced gene expression in the blood of SLE patients. This study according to these authors underscores the interferon pathway in the hierarchy of gene expression pathways that are involved in systemic autoimmune reactions in general, and SLE, in particular.

In general, it is perhaps safe to say that the immune system is not well understood. The etiology of these autoimmune diseases is definitely not very clear, to say the least, and often there is almost a trial-and-error procedure when prescribing medicines and dosage. There were posters at the International Conference in Montreal held in the summer of 2004, wherein the treatment of refractory arthritis was prescribed, and in some cases, the unfortunate side effects was, for example, tuberculosis. At the International Immunology Conference held in Stockholm in the summer of 2002, one of the invited speakers mentioned that if arthritis has been diagnosed in an individual, it should be 'attacked vigorously'. There should be no procrastination. All of this emphasizes the early detection of these autoimmune diseases, an area where biosensors can be of considerable assistance. The whole scenario is further exacerbated by the (recent) scrutiny and removal of drugs such as Vioxx, Celebrex, and Naproxen (Alleve) (for the treatment of arthritis), and drugs of a similar nature.

The market for the treatment of autoimmune diseases, besides diabetes and perhaps arthritis, is rather small, and thus it is not unusual for pharmaceutical companies not to venture into these detection devices on a 'large scale.' Having said this, it is worthwhile noting that some companies have invested and have marketed detection devices for these and other types of autoimmune diseases. *In lieu* of this, it is imperative that biosensors be used to help detect autoimmune disease at an early stage. More companies need to invest in these areas. Healthy competition would be of considerable assistance in the development of drugs, in general, to treat or at least manage autoimmune diseases. This would greatly impact in bringing the price down for these types of drugs, and make them more available to the general public.

Cancer and other autoimmune diseases go through different stages. It is common knowledge that if a cancer is detected at an early stage, the prognosis is much better than if the cancer is detected at a later ('blast') phase. Roughly, one may characterize the three stages of cancer as: initial (time period typically in years), intermediate (time period typically in months), and blast (time period typically in weeks). Of course, the detection of the different forms of cancer is more and more difficult during the initial stages. This is perhaps true at least for some of the diseases such as arthritis. As expected, the detection of arthritis at an early stage (perhaps by a biosensing method) would greatly facilitate the management of this debilitating disease. Novel methods are required to facilitate the early detection of arthritis and other debilitating autoimmune diseases.

DuBois and Shaw (2004) advocate the use of IR spectroscopy in clinical and diagnostic applications. These authors emphasize that IR spectroscopy provides a 'molecular fingerprint' of a sample. This is the basis of biomedical applications, for example, diseases lead not only to physical symptoms, but also to changes in the chemical composition of the organs, tissues, and fluids. DuBois and Shaw (2004) further emphasize that these changes are the basis of chemical tests and medical imaging techniques.

Eysel *et al.* (1997) initially attempted to extract diagnostic information from biofluid IR spectral patterns for the detection and diagnosis of arthritis. Arthritis, like most autoimmune diseases like SLE and cancer, is difficult to diagnose, and particularly to determine the stage the debilitating disease is in. These authors indicate that the onset of arthritis affects the synovial fluid in a systematic way. Each stage of the arthritis or arthritis variant has a unique molecular imprint. As compared to a normal joint, DuBois and Shaw (2004) indicate that the arthritic joint contains an increased volume of less viscous synovial fluid. The synovial fluid contains hyaluronic acid. This polysaccharide provides the lubricating properties and the high-viscosity properties to the synovial fluid.

DuBois and Shaw (2004) indicate that subtle changes in spectral patterns reflect the different forms of arthritis such as osteoarthritis, rheumatoid arthritis, and spondyloartropathy. These authors indicate the need for fail-safe algorithms that provide an indication of the form of arthritis and the state that it is in based on the spectroscopic signatures of individual fluid specimens. Finally, these authors state that an establishment of links between spectroscopy and diseases will go a long way in the management of these diseases and in lowering health costs.

In a following article in the same journal Mukhopadhay (2004) indicates the increasing application potential for Fourier transform infra red (FTIR) spectrometers. The author quotes Robert Yorkelson of the University of Montana who states, "It is the closest thing we have to an 'everything' detector." Mukhopadhay (2004) emphasizes the emergence of a niche market for portable spectrophotometers after the September 11, 2001 attack. Finally, Mukhopadhay provides cost for portable FTIR spectrometers, which range from a low of $27,000 to a high of $76,000. The weight in kilograms is from a low of 7 to 145 (definitely not portable, as in back-pack form; portable if carried in a car or van). Different sampling accessories are available. Thus, Mukhopadhay (2004) indicates the flexibility of these instruments to detect a wide variety of samples.

Camilleri (2004), CEO of Cambridge (UK)-based Rapid Biosensor Systems Ltd, indicates the development of a prototype biosensor (breath analyzer) to test for tuberculosis (TB). The disease apparently kills 2 million persons each year. A low-cost biosensor that detects TB at an early stage would be of considerable assistance. Camilleri (2004) indicates that the noninvasive biosensor is portable and also durable. Results are obtained within 5 min as compared to present-day blood tests that take 10 days or more. Besides, the biosensor can be used by nonmedics. This should find use in developing countries, for example, India. The biosensor system is based on the collection of a cough sample, followed by the detection of the pathogen. An immunoassay system is used along with fluorometry. The quick isolation of TB-infected patients should help considerably in the prevention of the spread of this re-emergent disease. The particular advantages of this biosensor are the high speed, low cost, and small size. Camilleri (2004) emphasizes that the intention is to develop a low-cost generic biosensor that would rapidly screen for other infectious diseases too. At present, the biosensor is also capable of detecting *Escherichia coli*.

A report from Drexel University (2004) emphasizes that the major market for biosensors may be found when an immediate assay is required (for example during a medical operation, or in a medical clinic where the physician(s) need some quick answers). Mareno (2004) indicates the development of a biosensor within 18 months to detect organisms such as *Legionella pneumophia*. This organism causes Legionnaire's disease, which can cause devastating effects on human health. Researchers at RMIT University in Australia are developing a device that would provide rapid identification of pathogens on-site. This would be a significant improvement in laboratory testing of samples that takes days.

Evans (2004) indicates that recently fluorescent polymer has been used to flag bacterial infections caused by *E. coli*. This technique is also being used for the detection of cholera, gingivitis, and hospital based infections. The principle used for detection is that bacteria bind to cell surface carbohydrates. These carbohydrates may be attached to fluorescent polymers. Evans (2004) indicates that the particular advantage of this method is the rapidity of detection. This technique can detect a few bacteria in 10–15 min, whereas other techniques take days to make a similar detection.

Biosensors can also be effectively used in Process monitoring (Biowise, 2000). These authors emphasize that the use of biosensors permit business decisions to be made at an early stage due to the knowledge gained by utilizing the biosensor(s) appropriately. Information about important variables in a process may be obtained on-site, thus saving the time and transportation costs required to transport a sample to and from a testing laboratory. These quick results facilitate quick changes if need be. Table 13.2 shows some costs provided by Biowise (2000). These numbers are a few years old, and have been updated (doubled) to reflect current prices. Also, the original numbers were in British pounds, and they have been converted to US dollars with an exchange rate of British £1 = \$1.9433 (http://finance.yahoo.com/currency?u) (November 03, 2004; 6:30 pm).

Biowise (2000) emphasizes that the monitoring of 'harmful' chemicals is required due to legislative purposes. For example, hygiene testing is required to safeguard the health of customers and employees. Furthermore, the use of biosensors vis-a-vis laboratory analysis

Table 13.2

Biosensor testing costs per sample for different compounds of interest (adapted from Biowise (2000))

Analyte of interest	Cost per test[a] ($)	Instrumentation cost ($)
Toxicity monitoring	16–20	20,000–200,000 (depending on application)
Biological oxygen demand (BOD)	~2–5	100,000
Hygiene monitoring (microbial contamination)	2–4	80,000–120,000 (luminometer)

[a]Numbers are rounded to the nearest integer.

of compounds such as phenols, dioxins, benzene, algae, and pesticides, etc. would lead to a considerable decrease in cost, even though the initial investment may be high for instrumentation purposes used along with biosensors. The turn-around time is also considerably decreased and, if this is a critical factor, then this is almost priceless, as an example in the early and rapid detection of biological or chemical warfare agents.

Ruzgas *et al.* (2000) attempted to develop a biosensor array to permit the fast monitoring of pollution levels in wastewater and pollution incidents. This was a 36-month long project and was funded at 1.115 million Euros. This is equivalent to $1.476 million (exchange rate: 1 Euro = $1.324). Enzyme-, DNA-, and cell-modified electrodes were used. These electrodes were capable of generating a fast multi-variate response on the interaction of phenols, lignins, heavy metals, and surfactants with the biomolecules. The authors emphasize that their method permits a quick estimate of the wastewater toxicity and composition. Ruzgas *et al.* (2000) further emphasize that an alarm is generated when there is a deviation from a normal or given (set) pattern. One of the biosensor formats that the project intended to generate was screen- and ink-jet printed biosensor arrays.

GenomeCanada (2004) has recently indicated that Virtek in Canada has been awarded a $1.2 million (Canadian) matching grant to develop its fiber optic nucleic acid (FONA) biosensor technology for testing recreational and subsequently drinking water for pathogens. The biosensor is based on a proprietary fluorescence-based fiber optic biosensor and a laser-based detection system. The project is to be able to detect organisms such as *E. Coli*, *Giardia*, and *Crytosporidium*. Virtek uses a DNA-based platform technology to provide genetic tests for (a) food and drinking water and (b) infectious diseases in humans and animals. The platform facilitates a rapid and an accurate analysis of the pathogens in water.

Virtek will collaborate with GAP EnviroMicrobial Service (Waterloo, Canada) for additional financial support, and with research scientists at the University of Toronto, Canada. The company is continuously looking for additional support to help offset the development cost of the biosensor. They indicate that their biosensor permits an accurate analysis that is economic, selective, and easy to use.

Finally, a disease which has come into prominence in recent years is mad cow disease (bovine spongioform encephalopathy, BSE). Biosensors could perhaps be developed and effectively used to detect this infectious disease in animals. An animal with this disease if ingested leads to a rare human brain-wasting disease such as variant Creutzfeldt–Jakob disease (vCJD). There is debate on how many cows need to be tested. For example, Normile (2004) recently indicates that more than 20,000 cows are tested annually in US. This number is much larger in Japan and the European Union (1.2 million and 10.4 million, respectively). Some of these numbers (for testing) correlate with when the disease was first detected in the country or region and how many reported known cases have been accounted for. For example, Normile (2004) indicates that the first known case of BSE in US was reported on December 23, 2003. Japan has had nine cases of BSE reported since 2001, whereas the European Union has had 186,000 reported cases since 1986.

Markus-Moser (CEO of BSE test maker Prionics AG in Switzerland) in Normile (2004) indicates that the appropriate level of testing is, and as it should be, a cost-benefit question. This is exacerbated by the fact that the real risk factor is not clearly understood or defined. However, in the US, the beef industry is a $50 billion industry, and thus the governmental agencies such as the Food & Drug Administration (FDA) do take this threat very seriously.

Normile (2004) further indicates that there are three companies whose detection kit (ELISA) costs range from a low value of $7 to a high value of $25. However, the cost per test is much higher, and is estimated to be as high as $60. This is another large market for the possible use of biosensors, especially since the estimated number of cows that need to be tested annually ranges from a low of 1 million to as high as 20 million over a period of 30 months. This author further indicates that the National World Organization for Animal Health's number for testing is recommended to be 0.01% of the national herd over a period of 30 months. The United Kingdom is more careful with its large number of detected BSE cases. Cows over the age of 30 months are forbidden to enter the food chain.

Depending on each countries' regulations the number of cows that need to be tested may differ, and will increase if (a) there is an outbreak(s) and (b) if there is an increase in the perception of the 'risk factor' (which still needs to be carefully determined) involved. Hueston (Director of the Center of Animal Health and Food Safety at the University of Minnesota) in Normile (2004) correctly provides a sort of balance, and underscores the motive of some testing advocates especially those that have a financial interest in companies that make BSE detection devices. The heart of the problem lies in assessing the appropriate level of risk, which will (a) eventually determine the number of cows that need to be tested annually and (b) thereby define more carefully the market for BSE testing devices such as ELISA and for biosensors.

Biacore (2004) with its SPR biosensor has expanded into the food analysis area. Samples may be analyzed in minutes, which reduces the response time when compared with the traditional methods of testing. Their biosensor can detect vitamins (in health and nutrition food), and veterinary drug residues such as antibiotics and β-agonists in dairy products. Furthermore, their SPR biosensor can detect hazardous natural toxins in food and feed. The Biacore® catalog (2004) provides information on a wide variety of models available. These as mentioned earlier in the book are expensive biosensors. Some of the models available are Biacore C, J, Q, and X. Also in the market are Biacore S51, 2000, and 3000. The different models of the Biacore biosensors may also be used during the different steps involved in drug discovery (Biacore Drug Discovery, 2004). According to the information available, the different models may be used at different steps in drug discovery and development. For example, Biacore® 3000 may be used during target ID and validation, Biacore S51 for secondary screening, etc.

The Agricultural Research Service of the United States Department of Agriculture in its Annual Report (2002) has outlined briefly the problems associated with detecting pathogenic bacteria in food. These include:

Speed. This is essential since processing and distribution systems operate quickly.
High sensitivity. An infectious dose may be as small as one organism.
Selectivity. Only a few bacteria of the total that are present in food are pathogenic. They represent a small fraction of the total benign bacteria present.

The above authors mention the limitations of the traditional microbiological protocols. Thus, the need for the development of rapid detection, inexpensive, easy-to-use biosensors for use by food producers, processors, retailers, and regulatory agencies.

The ARS, USDA Annual Report (2002) emphasizes that over 5 million cases of food borne bacterial diseases occur in the United States every year. This report is 2–3 years old. This number may have increased. The economic impact of these illnesses are significant with regard to (a) time lost at work, (b) medical bills, and (c) costs associated with recall and destruction of contaminated products.

In a recent news item on television (Channel CNBC News; USA, December 28, 2004; 9:30 am) the CEO of Neuogen, Inc. mentioned that the US is considered to be the 'bread basket' for the world. However, this year the food imports were more than the food exports. Thus, the need for the inspection of food that comes into this country. Detection devices, etc are required for bacteria, etc that come along with the food being imported.

Researchers at Georgia Institute of Technology (Georgia Tech, 2004) in Atlanta, GA have spent over 7 years to develop a biosensor that detects pathogen in poultry and other food stuffs. Their biosensor can detect 10^4–10^7 cells/ml in less than 30 min. One may compare this to 72 h required by laboratory screening techniques. The biosensor is based on an antigen (pathogen)–antibody reaction. The biosensor surface has a 'capture' antibody, and a 'reporter' antibody. The primary advantage of this biosensor is that no amplification of the cell counts is required, which is the major cause of the time delay. Furthermore, the biosensor is able to detect the pathogen in the presence of other contaminants. Another distinct advantage is the major reduction in the cost for the biosensor which ranges from \$1000 to \$5000 as compared to the common immunoassay laboratory equipment that may cost in the range of \$12000–\$20000 per instrument. The researchers emphasize that, however, the field trials are necessary to demonstrate the effectiveness of their biosensor.

The researchers also have plans to extend the application of their biosensor for the detection of other pathogens that include *E. Coli* 0157:H7, generic *E. Coli*, *Listeria monocytogenes*, *Campylobacter jejuni*, and *Yersinia enterocolitica*. All these pathogens, the authors indicate, cause stomach illnesses. *Campylobacter* affects more than 2 million persons every year. However, these authors indicate that the disease is usually mild, and rarely life threatening. This biosensor is more sensitive than ELISA (Enzyme-Linked Immunosorbent Assay), less expensive than polymerase chain reaction (PCR) techniques, and has a rugged design that permits on-line usage (Poultry Tech, 2003).

Knecht *et al.* (2003) at the Technische Universitat in Munchen, Germany, report on the development of an immunoassay biosensor to detect antibodies in milk rapidly.

They emphasize the importance of detecting these antibodies since they may increase the danger of bacterial resistance and harm intestinal flora. They are also responsible for allergen reactions. As far as cheese and yoghurt making are concerned antibodies in the milk may also inhibit the fermentation reaction. These authors emphasize that their immunoassay microarray technique leads to a reduction in time for antibody detection when compared with ELISA. Furthermore, their biosensor exhibits a robust design and the system components are of moderate costs.

Applied Nanotech Inc. (2003) has developed a versatile biosensor using carbon nanotubes (CNTs). This company indicates that they have immobilized enzymes into CNTs. Electropolymerization along with conducting polymers were used. They emphasize that their newly developed biosensor is low cost, easy to manufacture, and also versatile. It may be used to detect (a) impurities in air and water, (b) glucose levels in the blood, and (c) chemical and biological warfare agents. Furthermore, this company claims that their biosensor is three times more sensitive than other competing biosensors for the detection of hydrogen peroxide.

Berney (2001) indicates the proposed development of a DNA biosensor to identify genetically modified (GM) crops from nonGM crops. This has become essential according to this author due to EU regulations (EU 258/97 and 1139/980). Validated analytical methods are required that rapidly identify the DNA of genetically modified plant material. Berney (2001) anticipates that the effective monitoring of the plant material would lead to an increase in consumer confidence, followed by a subsequent increase in sales. This is a proposed 3-year project with a cost of $1,889,440 Euros. With an exchange rate of 1 Euro = US $1.324, this works out to US $2.515 million. The project is located at the National University of Ireland at Cork, and has four other partners: Murozone Ltd in UK, Ecole Polytechnique Federale de Lausanne in Switzerland, Sy-Lab Gerate in Austria and the National University of Ireland at Cork.

13.2 BOTTLENECKS, DEVELOPMENT COST, AND FUTURE NEEDS FOR BIOSENSOR DEVELOPMENT

The major resistance apparent for the development of biosensors is the lack of mass markets barring a few exceptional cases such as glucose monitoring for diabetes. With a view to a commercial profit, this places a serious hindrance on investment in biosensor technologies. Walsh (2003) indicates that the development cost of a biosensor may exceed $20 million. It is reasonable to assume that the development cost of a biosensor lies between $20 and 30 million, and the time panel involved is between 7 and 10 years. This significant amount of investment and the losses incurred in the initial (recent) years before a profit occurs are bound to hinder smaller-sized companies from entering the market. Some of the drawbacks pointed out by Walsh (2003) include the reliability to produce a competitive product, and the commercial development of technology required to produce a large number of devices. In all fairness, however, this trend is changing gradually now with the increasing investment by US Governmental agencies such as the National Science Foundation (NSF), Department of Defense (DOD), Department of Energy (DOE), Defense Advanced Research Projects Agency (DARPA), etc. However, this section focuses primarily on the bottlenecks and the development cost of a biosensor.

At the outset, it is perhaps appropriate to indicate the importance of linking marketing intelligence to product development as suggested by Khandelwal (2004). This author indicates that the cost of research is a very small fraction compared to the total expense in bringing a product to the market. Though this author's comments are made with regard to the chemical industry, for all practical purposes they could also be applicable to the development of biosensors.

Walsh (2003) further indicates that some of the drawbacks in biosensor technology include: total integration of the biosensor system, producing inexpensive biosensors in quantity, producing noninvasive biosensors that are self-calibrating, biocompatible biosensors that may be used under *in vivo* circumstances, and reproducible placement of receptors on a biosensor surface.

Medical Technologies (1994) has listed some of the drawbacks or obstacles in biosensor technology that need to be overcome. Most of them are related to the sensing (biomaterial) of the biosensor. They include sterilization (which will inherently destroy, at least, part of the biomaterial), contamination (limits biosensor to a single use), immobilization of the receptor to the biosensor surface (not well understood, especially its impact on analyte–receptor binding and dissociation kinetics; a focus of this book), and uniformity of receptor preparation. Other factors mentioned include selectivity, detection limit, and reproducibility. New nanofabrication devices to produce these miniaturized biosensors are urgently required. Also, emphasis needs to be placed on improving existing transduction technologies and the application of newer transduction technologies. A single use or a disposable biosensor would exhibit considerable potential to enhance the market share. This is especially true since very little is known about the regeneration of biosensors.

Obducat (2004) indicates that biosensor technologies due to cost issues are relying on semiconductor processes that are one or two generations old. This limits their sensitivities and tends to make the detection slower. Besides, the sensor area is limited since patterns cannot be made smaller than 5–50 μm in line widths. This company indicates that its Obducat Micronano process can make structures that are 100 nm and less. The process uses electron beam lithography. The company emphasizes that these smaller structures permit the manufacture of sensors at a low cost that respond faster and also exhibit a high degree of sensitivity.

van Hoof *et al.* (2004) recently indicate that microsystems technology is used presently to make tiny sensors. The principle is similar to that used for integrated circuit production. These authors emphasize the advantages of monolithic integration as compared to side-by-side integration. Figure 13.1 shows the side-by-side and the monolithic integration. These authors indicate that monolithic integration saves volume, electric power, and possibly cost. However, this does come at a price in that different materials and processing

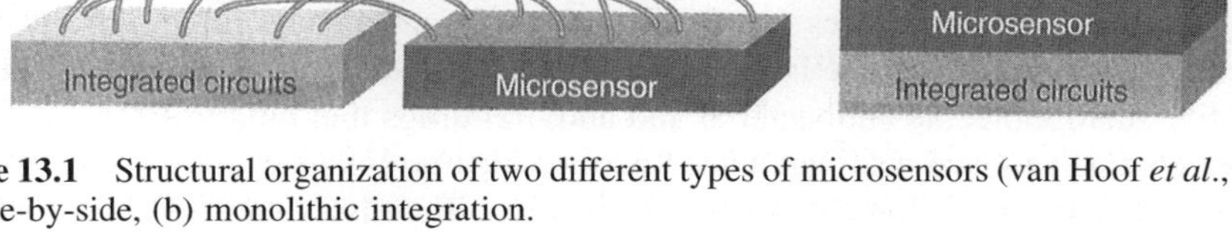

Figure 13.1 Structural organization of two different types of microsensors (van Hoof *et al.*, 2004): (a) side-by-side, (b) monolithic integration.

techniques need to be combined on the same substrate. van Hoof *et al.* (2004) emphasize the two basic tenets of sensor technology: a good sensor material and a viable process technology.

These authors further emphasize that for future smart sensors for personal health, comfort, and safety monitoring (such as glucose monitoring for diabetics), the sensors need to possess power autonomy, be miniaturized as much as possible, and also be disposable. The advantage of being disposable as suggested by van Hoof *et al.* (2004) is that they need not be as durable as present-day sensors.

Biotrace International Plc (2002) (http://www.biotrace.com/content.php?hID + 1&nhID + 21nID = 36), a British company in Bridgend, United Kingdom, specializes in its bioluminescence rapid biological detection system. It has collaborated with Smiths Group's Graseby (another British company that specializes in Detection and Protection Systems) wherein the Biotrace system is fully integrated into the Graseby detection and identification system for defense and civilian customers. The Biotrace-patented Adenosine triphosphate (ATP) luminescence system/Smiths Group Detection and Protection System is a novel two-step method for the detection of harmful biological agents.

The system is able to detect harmful biological agents in less than 2 min. The first step involves the 'nonspecific' detection of potentially harmful agents (Biotrace). This is followed by a 'trigger'-specific identifier system (Smiths Group) that pinpoints the particular agent. Graseby emphasizes that its technology is rugged and is able to identify harmful biological agents. Besides, its technology may be integrated into weapons of mass destruction (WMD) detection systems.

Another harmful biological agent detection system is the TIGER biosensor technology being developed by ISIS Pharmaceuticals. TIGER is an acronym for Triangulation Identification Genetic Evaluation of Risks. The company indicates that its Ibis Therapeutics has received \$65 million from different governmental agencies to develop this technology. Ibis Therapeutics further indicates that the development of the TIGER technology has been supported by the following governmental agencies for specific purposes:

Governmental agency	*Purpose*
Defense Advanced Research Projects Agency	Bioweapons defense
Centers for Disease Control and Prevention (CDC)	Epidemiological surveillance
Federal Bureau of Investigation (FBI)	Microbial agent database
National Institute of Allergy and Infectious Diseases (NIAID)	Biological products screening

Ibis emphasizes that the TIGER biosensor is capable of simultaneously identifying infectious agents. This includes previously unknown and newly emerging organisms. Furthermore, using its expertise Isis Pharmaceuticals is also developing novel biosensors to identify small molecule antibacterial and antiviral drugs that bind to RNA.

Other companies have also announced partnerships or collaborations to facilitate the detection of harmful biological agents. Innovative Biosensors Inc. (2004) located in

Gaithersburg, MD have indicated a licensing agreement with MIT (Cambridge, USA) to use the CANARY™ technology for use not only in detecting harmful biological agents but also in food testing and in human and clinical diagnostics. The above-mentioned company indicates that the CANARY biosensor researched initially at MIT permits the binding of the antigen to the engineered antibody on a cell surface. This produces light emission which is not only easily detectable in a quick fashion, but also permits high levels of sensitivity and specificity.

Similarly, QTL Biosystems, LLC (2004) located in Santa Fe, NM is collaborating with United First Responders, LLC of Fayette County, PA to manufacture a hand-held biosensor to detect harmful biological agents such as anthrax and ricin. The two companies indicate that about 6 months are required to manufacture the prototype and high-quality units. QTL's expertise lies in integrating chemistry, molecular and cell biology, and in instrumentation. United First Responders provides products and training services with regard to chemical and biological threat detection.

Cross and Freeman (2004) indicate the development of a biosensor based on feeding alternate orthogonal polarization states of light to a chip using a fast liquid crystallization switch. The authors indicate that their method overcomes the restrictions on using *one* polarization of light at a time for measurement. These authors also indicate that cheap (low cost) biosensor systems are required with a long service between scheduled maintenance. Their method provides extreme surface sensitivity and low chip cost. The authors further indicate that it took 6 years of R&D (research and development) investment to bring their product to the market. No numbers were provided as to the total expense required to bring the biosensor to the market. Their technique is called Dual Polarization Interferometry (DPI) and is capable of subatomic resolution measurement in 'real time.' Changes in thickness of molecular layers of interest may be detected by DPI. These authors emphasize that developments in software should provide upgrades for the biosensor as well as increase the user base. The authors emphasize that market penetration is significantly dependent on the ease of use. Finally, the authors indicate that based on the development of their biosensor they have launched Fairfield Photonics, Ltd. This would facilitate developing more products based on their technology.

NVE Corporation (2004) in Minnesota, USA is developing a biosensor to detect biological warfare agents, real-time DNA testers, and laboratory-on-a-chip diagnostic systems. It will use its working BioMagnetic Interfacing Concepts (BioMagnetICs) biosensors. These materials are a few atoms thick. The company estimates the commercial market for spintronics at about $100 billion per year. The company has been awarded a $1.2 million to develop this concept from the DARPA of the Defense Services office (DSO). The company indicates that the long-term goals are to provide a low-cost hand-held device that is as easy to use as a digital thermometer. Furthermore, accurate results will be provided in minutes rather than in hours.

An area of interest where biosensor may be used as an 'electronic nose' is in demining operations, in other words, to detect land mines. Berg (2003) indicates a 2-year project funded at 4,028,408 Euros (equal to $5,333,612, exchange rate 1 Euro = $1.324) by the European Community (EC). It is multi-organization project that includes the Swedish Rescue Service Agency (SRSA), Biosensor Applications, Sweden, and the Norwegian Peoples Aid (NPA). The goal of the project is to find the smallest quantity of explosives in mines and to reduce the area where mines are located. Furthermore, the project aims to build a prototype 'electronic nose' to verify that a certain area has been cleared of mines. The intention is to be able to detect TNT, RDX, and PETN. One of the goals of the project is to decrease the size of the sensing system so that it may go to areas inaccessible by vehicles. Finally, another goal of the project is to test the system in a simulated mine field, and in a real mine field.

O'Neil (2004) describes an effective biosensor using an air-sampling method to detect drugs and explosives at airports. This author indicates the development of a biosensor for amphetamine and for the detection of cocaine and cannabis. A biosensor for TNT has also been developed, and biosensors for the detection of RDX and PETN are under development. Their biosensor is capable of detecting the harmful analyte in under 3 min. The intention is to reduce the response time to under 30 s.

O'Neil (2004) emphasizes that present-day screening methods are costly, inefficient, and often unreliable. Furthermore, there is considerable inconvenience in (a) the long lines at airport check-in prior to entering the gate area for departure and (b) in the 'pat down' procedures (often sensed as overtly intrusive by passengers and the subsequent changes of procedures after considerable passenger complaints). Their company biosensor is an extremely sensitive piezo-electric quartz microbalance system. This author emphasizes that their biosensor is very accurate, is able to sample over a large volume (their system is able to concentrate the sample), and can detect analytes at low concentrations. Their receptor is an antibody, which is very specific for the analytes. Finally, O'Neil (2004) indicates that their biosensor fills the present-day critical need for making the airport screening process more reliable, efficient, and less intrusive. Their biosensor should prove to be cost effective as it may help speed up the flow of passengers, and also help minimize the costs of delay and lost business.

Another area of interest is biofabrication. DARPA, Defense Science Office (http://www.darpa.mil/dso/future/biofab.htm) has initiated a program to examine the possibility for utilizing biological processes to manufacture materials for defense. These processes may perhaps provide opto-electronic materials and photonic devices that may prove useful in biosensor development.

DuPont (2004) has introduced a line of screen-printable thick film conductive materials. These can be used for biosensors in medical monitoring, diagnostics, food and beverage testing, and environmental monitoring. Each material is specifically designed for each application. The company indicates that its materials may be used as both active and passive materials for biosensors. Furthermore, their printable materials may be used cost-effectively in high-volume manufacturing processes for biosensors. The company states that it has been able to make these advanced materials for biosensor applications due to its strengths in materials science, chemistry, and fine particle technology. For example, their platinized carbon may be used as working electrodes in amperometric sensors for analyte detection.

As far as materials are concerned van Hoof *et al.* (2004) indicate that due to its large processing power per unit area, silicon will presumably remain the material of choice, generally used for smart sensors. The large processing power per unit area permits a decrease in area and a lower cost for the sensor. Finally, these authors indicate that future cheap materials for sensors may probably be polymers, metals, textile materials, or even paper (Ender *et al.*, 2004).

Biowise (2000) too has also mentioned the disadvantages of the receptor molecule used in biosensors. It has classified receptors into: enzymes, antibodies, and microorganisms. Other types of receptors for biosensors are also available. However, at present, we only analyze enzymes, antibodies, and microorganisms. These three types of receptors have advantages also, but here we concentrate only on the bottlenecks/disadvantages. As far as enzymes are concerned these receptors have limited shelf life, they are susceptible to inhibition by other (than the analyte) substances present in the sample, and ambient conditions. The shelf life of these types of biosensors may be increased by improving the way by which the enzyme(s) (or other receptors) are deposited on the biosensor surface (Biowise, 2000). Adverse pH and temperature conditions may quickly inactivate the enzyme making it useless as a receptor. Though significant advances are being made in antibody technology, this is still classified as an emerging technology, and target analytes may thus be limited (Biowise, 2000). Furthermore, antibodies are very specific and cannot detect unknown substances. The speed of response also needs to be improved when using antibodies in an ELISA format. The response time may generally decrease from hours (ELISA format) to minutes when they are used in a biosensor format. When microorganisms are used as receptors in a biosensor format, the results are often variable (Biowise, 2000), and the response time is larger than when other types of receptors are used on biosensor surfaces. Biowise (2000) further indicates that the response time needs to be decreased to the order of 15–60 min when microorganisms are used as receptors on biosensor surfaces for them to become competitive.

Raghupathy (2002) briefly mentions two requirements or performance parameters for effective biosensor operations. These are sensitivity and a fast response. The detection of analyte concentrations at nanomolar or lower is considered as highly sensitive, and the fast response has a time frame of milliseconds.

TFS Sensor Technology (2004a,b) indicates five reasons that are hindering evanescent biosensors from entering the market. These include: nonspecific binding, biological fluorescence, inadequate sensitivity, reproducibility of measurement, and fabrication. This company has attempted to overcome or mitigate these limitations. They have a patented process that treats the optic fiber used for biosensors. The process minimizes nonspecific binding. The company indicates that previously proteins in biological fluids of interest such as blood, serum, and urine have prevented fiber optic biosensors from operating successfully. Furthermore, the natural fluorescence of biological substances present in serum, urine or in environmental samples has prevented fiber optic biosensors to be used in sensitive measurements. Three-fold sensors (TFS) has managed to use sensors wherein the fluorescence characteristics are clearly different from the biological fluorescence exhibited by the compounds that are not of interest. In other words, this type of interference is minimized, and more sensitive measurements may be performed.

TFS Sensor Technology (2004a) further indicates that the signal losses from optical fibers characteristically decrease the sensitivity of fiber optic biosensors. TFS' biosensor design minimizes these losses, besides improving the collection efficiency for the fluorescence from the biosensor surface. The company emphasizes that their biosensor design permits the detection of analytes at the subpicogram level. A reason that hinders reproducible results is the variations on the fiber optic surface. This points to differing degrees of heterogeneity (different fractal dimensions, D_f) (Sadana, 2003). The present author is not surprised with this statement, since in this whole chapter we have indicated that, in general, the binding and the dissociation coefficient(s) are very sensitive to the degree of heterogeneity that exists on the biosensor surface. TFS, however, uses a unique self-calibrating method that minimizes the ill effects of these variations on the fiber optic surface on the reproducibility. Finally, tapered optical fibers result in yielding a better sensitivity. However, previous fabrication techniques were not suitable to manufacture these types of fibers. The TFS design process does overcome this limitation to a large degree.

Ambri located in Chatswood, New South Wales, Australia (Ambri Annual Report, 2004) has developed the Ion Channel Switch (ICM™) Technology and its successor, SensiDx™ biosensor. This is aimed at POC assays that may be used not only in the Hospital Critical Care market but also in medical clinics, doctors' offices, retail pharmacies, and at home.

This company emphasizes that its main operational focus is to move both of the systems towards market readiness. It uses a novel, commercially viable technology (CVT) milestone to get its biosensor market ready. It uses the following key targets for CVT evaluation:

Tests. Commercial levels of sensitivity demonstrated
Analyzer. Connect Australian hospitals to match demand
Regulatory. Use good manufacturing principles (GMP) to develop manufacturing
Research. Use next generation membrane chemistry for sensitivity and stability
Validation. External evaluation of developments

For example, under the criteria for tests its biosensor system demonstrated the detection of hCG (for pregnancy) in blood of about 15 mIU/ml and below range with improved reproducible results. Their system is also able to detect troponin (a clinical marker for heart attacks). Under the analyzer criteria the company was able to connect their Sensidex™ system successfully with the Australian Hospital system. Other criteria were also similarly attained or satisfied.

This systematic (step-wise) procedure would facilitate the fast track development of their patented ICS™ Technology with a new silicon chip platform. Their miniature biosensor for POC testing combines the following attributes which include high volume fabrication at low cost and a tiny disposable biosensor to do several tests simultaneously, and it is a simple hand-held device.

Ambri emphasizes that due to efficient restructuring and financial discipline procedures, it has decreased its operating expenses by 42.1%, from \$9.5 million to \$5.5 million (Australian dollar) for the months July 2002–December 2002 and July 2003–December 2003, respectively. Note that one Australian dollar is equal to 0.7586 US dollar (exchange rate December 14, 2004). Their financial report indicates that they have \$16 million in reserve and expect to get their next-generation biosensor to the market in about a year's time. Furthermore, it has reallocated \$8 million to fast track their next-generation biosensor. This provides one with the perspective of time and financial resources required to get a competitive, cutting-edge biosensor to the market. Finally, this aggressive forward-looking company, which has done some successful fund-raising (\$22.5 million), has set-up a POC advisory group, is minimizing investment risk, and attempting to enhance licensing opportunities for its technology, besides attempting to improve its bottom line via different avenues.

DuBois (2003) in a recent article emphasizes the importance of accelerating speed to market a medical product. He indicates that, and we quote, "the ability to shorten the developmental cycle is critical to financial reward." He emphasizes the 'Thinking-out-of-the-box approach.' Though unconventional, these types of approaches have a high reward along with a high risk factor. This author further indicates that a fundamental principle is to set a time frame, and then get the resources together to complete the project. If the resources are not available internally, then one should outsource. This way the company can remain focused without the need for (a) building staff and minimizing (b) overhead and (c) capital expenses. The author emphasizes that the outsourcing industry is estimated to be around \$8 billion, and is growing at a rate of 25%. Furthermore, this author emphasizes that in outsourcing communication and trust are the key. Any individual who has done some form of consulting can attest to this.

In chronological order, DuBois (2003) indicates the following steps that are involved in taking the product concept to the market. Though this has been proposed for IVD (*in vitro* device) medical devices, the procedure is presumably generic, and with suitable modifications may be applied to other concept-to-market ready products. These steps are: concept, preliminary engineering, model creation (efficacy of concept), development engineering (specifications, materials requirements), prototype, final engineering, and pilot release.

DuBois (2003) emphasizes that communication is the key during all the steps. Flaws may thus be corrected, before they become major problems. The author suggests that a quality control group can assist in this during different stages of development.

DuBois (2003) indicates that the market for Diabetes Mellitus (DM) is estimated to be \$4.6 billion, with more than 17 million Americans being affected. Out of these the author estimates that 1 million have type 1 DM, and 11 million have type 2 DM. Six million Americans are still undiagnosed. Approximately, one in every four Americans is estimated to have DM.

Newman *et al.* (2004) indicate that according to the World Health Organization (WHO) the number of diabetics will double worldwide from 150 to 300 million by the year 2025. This represents doubling of the number of diabetics in about 20–22 years. Using the rule of 72, this represents approximately a 3.27–3.6% increase. For all practical purposes, this is a 3% increase, since the rule of 72 does not strictly apply here. Clearly, there is a critical need for a less painful and more efficient glucose testing biosensor.

The measurement of glucose, as indicated elsewhere in this chapter is painful and time-consuming (DuBois, 2003). An accurate monitoring device is required to prevent the onset of other 'cluster' diseases, such as heart attack, blindness, stroke, etc. The poor management of DM increases the probability of these 'cluster' diseases. The author emphasizes the need for (a) a less painful device, (b) testing speed, and (c) accuracy and precision. As a practical fast-track example, DuBois (2003) indicates that Nova Medical and Becton Dickinson (BD) collaborated on a glucose biosensor for Diabetes Mellitus, and in 18 months BD was able to leverage Nova Biomedicals diagnostic to take it from the 'concept' stage to the market. The market for DM is large and expanding. In these 18 months, the hand-held device was designed and developed, the equipment was custom produced, and the manufacturing facilities were set up. Once again, communication and trust were the keys.

Sontra Medical Corporation (2004) indicates the co-development of the Symphony Diabetes Management System with Bayer Diagnostics. The company emphasizes the unfilled need for a noninvasive product to measure glucose levels. This, according to the company, is a \$5 billion home product market. Their management system consists of a hand-held SonoPrep® device. This device is able to permeate the skin, and the biosensor transmits the glucose levels wirelessly to a glucose meter.

The company estimates the worldwide sales of the market at \$5 billion in the year 2004, and the market is estimated to increase to \$8 billion by the year 2007. A noninvasive biosensor would significantly improve the compliance rate of measuring glucose levels (which are often painful), and minimize the occurrence of coronary and vascular diseases such as kidney failure, adult blindness, nontraumatic amputation, and nerve damage (Sontra Medical Corporation, 2004). Finally, the company indicates that Phase 1 clinical trials on patients with diabetes was completed in 2003. Sontra Medical Corporation (2004) emphasizes that the patients did not complain of either pain or irritation. However, the product still requires regulatory clearance and is still not commercially available.

Another study by Synthetic Blood International (SBI) (2004) located in Kettering, Ohio, USA indicates that diabetes affects men and women equally, though it affects the elderly more frequently. This company estimates that the direct cost of diabetes is around \$50 billion in the US. This is estimated to be around 6% of the total direct personal healthcare expense in the US which is about \$830 billion. These authors emphasize that the National Institute of Diabetes and Digestive and Kidney Diseases (NIDDK) (one of the National Institutes of Health located in Bethesda, Maryland, USA) recommends a very tight control of glucose levels to prevent the onset of complications that arise with diabetes.

There is a critical need as indicated elsewhere in this chapter for a less painful method of measuring glucose levels in the blood to increase the compliance rate. Estimates of the people afflicted with diabetes differ. SBI (2004) estimates that 16 million individuals in the US suffer from diabetes, and about 600,000–700,000 new cases of diabetes are diagnosed every year.

SYPD has developed an implantable glucose biosensor that eliminates the use of finger sticks to monitor glucose levels in the blood. The company indicates that this implantable form of the biosensor is also more accurate than the current portable measuring devices. The biosensor is about the size of a cardiac pacemaker. The company further states that the implant life of the biosensor is expected to be over a year. Finally, the company indicates the worldwide market for their implantable biosensor to be over \$1 billion.

Madou (2002) indicates that the high cost of disposable biosensors hinders the introduction of disposable biosensors in the POC market. This author is examining nonsilicon materials (biomaterials) for use in the manufacture of chemical and biosensors. This author indicates that the use of a polyimide material for biosensors has the potential of decreasing the biosensor cost by an order of magnitude than planar electrochemical sensors. Madou (2002) emphasizes that their polyimide biosensor is estimated to cost \$0.3–1 (this cost is comparable to glucose-measuring strips) rather than the \$3–10 cost for small biosensor arrays. The author emphasizes the additional advantage due to miniaturization, in that the biosensor size will be reduced approximately by 50% when compared with the present-day biosensors due to improved architectural design facilitated by placing the modular structure of the biosensor one on top of the other, rather than side by side. See Figure 13.1.

May (2004) in a recent article, 'building a better biosensor', indicates that even though quite a few biosensors have been prototyped, they may not make it into the field. Biosensors are required that would detect agents at very low levels, in real time, and over long periods (preferably unattended). The trick is to take the laboratory know-how and effectively translate it for use in real-life applications. Portable, inexpensive, and rapidly deployable biosensors are required which can be used on-site. This author emphasizes that problems and challenges remain in translating biosensor detection technologies from the laboratory to the market. Ideally, one would require a universal biosensor that would be able to detect all harmful biological agents.

In a program solicitation for proposals, the NSF (National Science Foundation, NSF 03-512; Sensors and Sensor Networks, 2003) has attempted to address some of the perceived obstacles facing biosensor development. Some of these include:

(a) robustness under anticipated manufacturing schemes
(b) quantification of limits of detection, calibration, and interferences
(c) verification of accuracy
(d) miniaturization
(e) manufacture
(f) human-in-the-loop and closed loop adaptive feedback

The document emphasizes the need for inclusion in sensor development of anticipated progress in adjacent technologies such as wireless communications, nanofabrication, biosystems, and ubiquitous computing.

In a later program solicitation for proposals the NSF (National Science Foundation, NSF 04-522: Sensors and Sensor Networks (Sensors), 2004) indicates that emerging technologies are anticipated to decrease the size, weight, and increase the accuracy of biosensors. Emphasis is placed on integrating large number of sensors into systems that would increase not only the performance but also the lifetime. Furthermore, life cycle costs would also be decreased. The document emphasizes the need to develop functionalized receptors and materials. These would result in next-generation devices, for example, materials of varying porosities that would permit the detection of a single toxic compound in complex mixtures.

In a more recent program solicitation announcement for proposals the NSF (National Science Foundation, NSF 05–526: Sensors and Sensor Networks (Sensors), 2005) is looking for proposals that would 'seek to advance fundamental knowledge in the area of sensor development.' The proposal emphasizes that newer and emerging technologies have the potential to decrease the size, weight, and costs of sensors and sensor arrays. It is estimated that these changes would be of an order of magnitude. Furthermore, spatial and temporal resolution will increase along with an increase in accuracy. The NSF is seeking proposals that also lead to sensor integration into engineered systems.

An example of sensor integration is provided by Godso (2002). This author indicates that there is no industry standard for sensor integration. Their company, RPI, provides a sensor integration platform that permits rapid integration and interoperability of different sensors and sensor networks. The SensorView™ system has the capability of integrating and managing a wide range of disparate and distributed sensors. Furthermore, command, control, and monitoring of the distributed sensors by the SensorView™ platform have been demonstrated.

Other interesting concepts mentioned in the NSF document (NSF 03-512) include liquid surfaces with molecular recognition properties and the requirement of new methods for sensor fabrication, manufacture, and encapsulation. The above document further adds that though false alarms do not have a direct economic impact on the development of a biosensor, they do significantly impact economic and other aspects indirectly, such as time wasted, nuisance factor, etc. Thus, the need to detect and identify false alarms from the sensor data itself.

Santana (2002) indicates the development of the Automated Bioaerosol Collection and Detection (ABCD) system. This system continuously monitors biological agents in carrier facilities. The system automatically conducts tests at intervals of 5 min for near real-time bioaerosol detection.

Carriveau (2002) mentions the application of combining several sensor technologies to produce the next generation of chemical vapor detectors that reduces the false alarm rates during the detection of live chemical warfare agents and difficult interferents. These authors emphasize that their advanced system minimizes both false positive and false negative alarm rates without compromising on overall sensitivity.

Finally, intense competition may be expected from nonbiosensor devices wherein significant improvements and progress have been made to provide a quick and accurate result with a minimum of cost. Companies may perhaps develop a simple strategy where they have a strong base, for example, in glucose monitoring. The profit made in this area may then be successfully re-invested in areas such as autoimmune diseases, for example, rheumatoid arthritis as done perhaps by Abbott Laboratories. Other more complex strategies may be devised by companies in helping to diversify in biosensor development. This will eventually show up in a positive sense in the bottom line.

13.3 SUCCESSFUL AND MODEL COMPANIES FOR BIOSENSOR RESEARCH AND DEVELOPMENT

In this section we will present, as examples, three companies and try to give an idea of their growth. Their market niche will be emphasized which includes their in-house growth, along with growth due to the acquisition of other smaller companies which fit in with their vision of projected growth, and where the biosensor market is heading. At the outset, it needs to be emphasized that in this section the author's view is presented, which may or may not be coincident with the companies' vision. This author is trying to present the material as best as can be assimilated from the information that is available in the open literature and from the internet sources.

The three companies that we will present are Biacore, Abbott, and Biosensor Applications AB. Mead (2001) indicates that Biacore International was founded in 1984. It is the leading provider of the surface plasmon resonance (SPR) biosensor. This instrument is continuously being upgraded by the company, and is very frequently used to analyze biomolecular reactions occurring on surfaces and interfaces. The advantage of the SPR biosensor is that it provides high-quality real-time data of these biomolecular interactions. The software that comes along with it does provide values of the binding and dissociation rate coefficients as well as affinity values. The SPR is an expensive piece of equipment and it generally costs around $300,000–400,000 depending on the model that one purchases. Unfortunately, and this may be the sole opinion of this author, the software program that comes along with the SPR biosensor does not take into account either the diffusional aspects or the degree of heterogeneity that exists on the SPR chip surface whilst analyzing the kinetics of biomolecular reactions. Other biosensor users are also gradually resorting to other types of biosensors to help analyze the kinetics of these biomolecular interactions. However, this point has been discussed quite a bit in the previous chapters of this book and will not be further discussed here. Over here, our main intention is to present an idea of the growth of Biacore from its inception in 1984.

Mead (2001) indicates that Biacore has a niche which is to provide an analytical tool (albeit expensive) to help analyze biomolecular interactions in real-time. Pharmaceutical companies can use this SPR biosensor to help in drug discovery. This is a very popular biosensor and is used frequently in both the industrial and academic environments. This biosensor, as indicated earlier in this book, is finding increasing usage albeit its presumed 'shortcomings.'

Biacore's President Ulf Jonsson indicates that the SPR technology was used to start the company in 1984 with an initial investment of $50 million. It took the company 10 years to break even. According to Mead (2001), the company had a revenue of $40.9 million in 2000. Assuming a very reasonable 10% growth rate per year, the revenues for the years 2001–2005 may be initially estimated to be 44.99, 49.49, 54.44, 59.88, 65.87 in million dollars, respectively. Actual numbers are presented in the next paragraph, and the sales growth is more than 10%. The Internet revealed figures for the years after 1999. The company is listed on the Stockholm Exchange as well as on Nasdaq (BCOR).

The annual sales figures for Biacore International AB (Neuchatel, Switzerland) are available. They are presented in Table 13.3a. Numbers for the years 1999, 2000, 2001, and 2002 were available from Hoovers online (http://www.hoovers.com/biacore/). At the outset, please note that the numbers from the Hoover (internet source) and that given in Mead's (2001) article differ by $5.7 million. The Mead number for the year 2000 is $40.9 million, and the number in the Hoover Internet source for the same year is $46.6 million. This, once again, underscores the 'unreliability' in these economic numbers. Nevertheless, we will try to make the best of the situation with whatever is available in either the open literature or Internet sources. We have modeled the data available, and our equation that estimates the projected annual sales till the year 2008 is also presented. Please note that the equation used to provide these projections was based on the sales figures available for 4 years only. Needless to say, market competition and other factors, including geopolitical events may change these numbers for the estimated annual sales for future years. One may note the sharp change in the annual sales figures from the years 2001 and 2002. This represents a net positive change of 36.4%. This may or may not be just a one-time event. Numbers for the years 2003 and 2004 were unavailable at the present time in the open literature.

Table 13.3a

Annual sales figures for Biacore International AB

Year	Annual sales, \$ (in million; reported[a])	Annual sales, \$ (in million; estimated by eq. 13.1a)
1999	39.9	38.04
2000	46.6	49.23
2001	51.9	57.26
2002	70.8	63.72
2003		69.24
2004		74.10
2005		78.48
2006		82.47
2007		86.17
2008		89.62

[a]Biacore International AB (2004).

The data presented in columns 1 and 2 of Table 13.3a were modeled to yield the following equation which is shown in Figure 13.2(a):

$$\text{Annual sales \$ million} = (38.036 \pm 4.597)(\text{year})^{0.3722 \pm 0.1095} \tag{13.1a}$$

Only four data points were available. The fit is quite good. The availability of more data points would lead to a more reliable fit. Please note that in Figure 13.2(a) year 1 corresponds to 1999, year 2 to 2000, year 3 to 2001, etc. Projections are made from 10 years starting from 1999 to 2008. As indicated above, these are just projections of sales, and geopolitical events and other factors may and will influence the above-mentioned numbers. Nevertheless, the above equation should prove useful to biosensorists, and presumably also to Biacore who are in a much better position to refine the (projected) numbers, since they are the ones who have access to the present-day numbers.

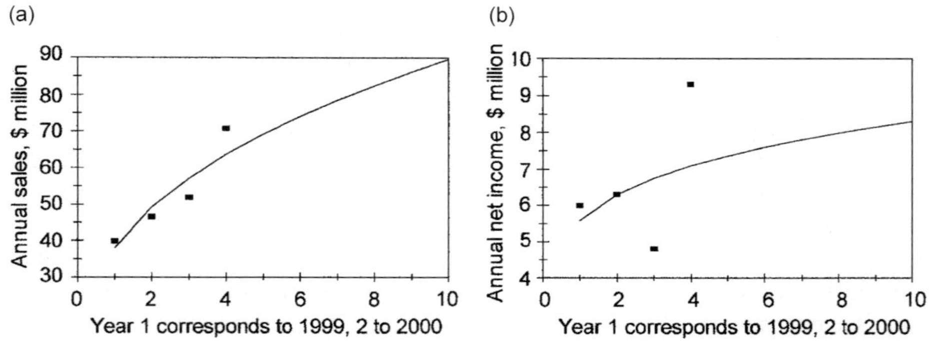

Figure 13.2 (a) Annual sales figures (in million dollars) for Biacore for the years 1999–2008 (Mead, 2001). Year 1 is 1999, and so on. (b) Annual net income (in million dollars) for Biacore for the years 1999–2008 (Mead, 2001). Year 1 is 1999 and so on.

The annual sales in million dollars is only mildly sensitive to the year (basis year 1999 equal to year 1), as noted by the less than one-half order (equal to 0.3722) of dependence on year. The company would, of course, like to increase the order of dependence on the year, which in other words means a sharper rise in annual sales. Any effort made by Biacore to improve its 'bottom line' would apparently lead to an increase in this order of dependence. Smart, appropriate, and timely acquisitions that significantly impact the bottom line (albeit in the future) could perhaps be one way of doing this. The share price would also be a good indicator of this.

Biacore is making attempts to team-up with key players. For example: *Development and commercialization of the Biacore SPR array technology*. A deal has been signed between Biacore and the Biological Information Research Center (BIRC) of the National (Japan) Institute of Advanced Industrial Science and Technology (AIST). This Japanese Institution specializes in proteonomics. This synergism is aimed at enhancing interactive proteonomics. Science Letter (November 11, 2004; 6:46 pm) (http://www.hoovers.com/free/co/news/detail.xhtml?COID = 52773&ArticleID = NR200411) indicates that this synergism would permit the faster generation of detailed proteonomic information by permitting the parallel analysis of biomolecules of interest against panels of proteins. Furthermore, the SPR technologies is ideally suited for drug discovery and high-throughput screening. For example, Mead indicates that the Biacore 3000 may be used upstream in the drug discovery process to help locate appropriate targets. The Biacore S51 may be used in the downstream drug discovery process to help better characterize potential pharmaceutical products.

The next table shows Biacore's annual income in million dollars. Projections of net annual income till the year 2008 are also provided as above for the annual sales figures.

The data presented in columns 1 and 2 of Table 13.3b were modeled to yield the following equation which is shown in Figure 13.2(b):

$$\text{Annual net income, \$ million} = (5.590 \pm 2.047)(\text{year})^{0.1720 \pm 0.2996} \tag{13.1b}$$

Table 13.3b

Annual net income for Biacore International AB

Year	Annual sales, $ (in million; reported[a])	Annual income, $ (in million; estimated by eq. 13.1b)
1999	6.0	5.59
2000	6.3	6.30
2001	4.8	6.753
2002	9.3	7.095
2003		7.373
2004		7.608
2005		7.812
2006		7.994
2007		8.158
2008		8.307

[a]Biacore International AB (2004).

Only four data points were available. There is quite a bit of scatter in the data which is reflected in the error of the coefficient that depicts the order (equal to 0.1720 ± 0.2996). There was a decrease in annual net income in the year 2001, and this is reflected in the error and in the figure. The availability of more data points would lead to a more reliable fit. Please note that in Figure 13.1(b) also year 1 corresponds to 1999, year 2 corresponds to 2000, year 3 corresponds to 2001, etc. Projections are made from 10 years starting from 1999 to 2008. As indicated above, these are just projections of annual net income sales, and geopolitical events and other factors may and will influence the above mentioned numbers. In this case, company decisions with regard to personnel and other items would also significantly impact net income figures. This author, as expected, is not privy to this type of confidential company information. Nevertheless, the above equation should prove useful to biosensorists, and presumably also to Biacore who are in a much better position to refine the (projected) numbers, since they are the ones, as mentioned previously, who have access to the present-day numbers.

Abbott Laboratories (Abbott Laboratories Fact Book, 2004) is a health care company that specialized in diagnostics and medical devices amongst other items. In the year 2003, it had a sales income of \$19.7 billion. Its two main divisions are the Pharmaceutical Products group with a sales of \$11.4 billion, and the Medical Products group with a sales of \$8.3 billion. The Medical Products group comprises of (a) US Hospital Products, (b) US Nutritionals, and (c) Worldwide Diagnostics. It is the growth of this Worldwide Diagnostics group that we are interested in here. The sales of the Worldwide Diagnostics group is estimated to be 14.5% of the total sales income of \$19.7 billion (Abbott Laboratories Fact Book, 2004; column 26 and 27) for the year 2003.

Table 13.4 shows the estimated Worldwide Diagnostics group from the years 1998 to 2003. In Table 13.4 both the total sales and the sales of the Worldwide Diagnostic group only are presented. The same 14.5% fraction is also used for the years 1998–2002. These numbers are very difficult to obtain, since the author is presumably not a shareholder in Abbott Laboratories. Thus, the same fraction is used, as an approximation, for the previous years.

Table 13.4

Abbott Laboratories Worldwide Diagnostic group sales

Year	Annual total sales, \$ (in million; reported[a])	Annual sales for Worldwide Diagnostic group, \$ (in million; estimated[a])	Annual sales for Worldwide Diagnostic group, \$ (in million; estimated from eq. 13.2
1998	12512.7	1814	1687.51
1999	13177.6	1910.7	2006.24
2000	13745.9	1993.15	2219.91
2001	16285.2	2361.35	2385.17
2002	17684.7	2564.28	2521.79
2003	19680.6	2853.69	2639.20

[a]Abbott Laboratories Fact Book (2004).

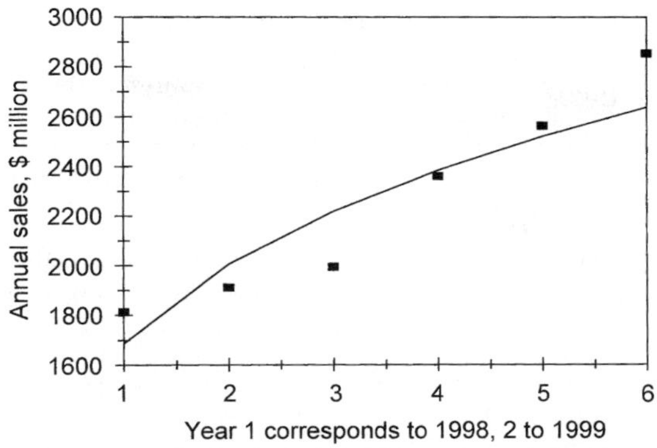

Figure 13.3 Annual sales (in million dollars) for Abbott Laboratories Diagnostic Worldwide group (for the years 1998–2003; Abbott Laboratories Fact Book, 2004). Year 1 is 1998 and so on.

Abbott Laboratories has a wide variety of diagnostic products (Abbott Laboratories Fact Book, 2004). Some of them are *Free Style* (blood glucose monitoring), *Abbott Prism* (screen donated blood for multiple viral assays), the *Architecht series* (t2000, i2000SR, etc.) for cancer, thyroid, fertility, Determine (HIV, hepatitis, and syphilis), *PathVision HER-2*, DNA probe Kit (HER-2/neu gene in breast cancer patients), and *Cell-Dyn 4000* (automated hematology analyzers).

Table 13.4 and Figure 13.3 show the increase in the Abbott Laboratories Worldwide Diagnostic group sales for the years 1998–2003.

The data presented in columns 1 and 3 of Table 13.4 were modeled to yield the following equation which is shown in Figure 13.3:

$$\text{Worldwide Diagnostic Group Sales, \$ million}$$
$$= (1687.51 \pm 140.83)(\text{year})^{0.2497 \pm 0.0541} \quad (13.2)$$

Only five data points are available. The fit is quite good considering the assumptions that have been made. Please note that in Figure 13.2, year 1 corresponds to 1999, year 2 to 2000, year 3 to 2001, and so on. These are just the estimates of the annual sales figures for the years 1998–2003 for the Abbott Laboratories Worldwide Diagnostic group. Once again, actual numbers are very difficult to get, and thus a 'real' comparison is not possible.

The above equation may be used to help provide initial projections of sales for the years 2004–2009. Needless to say, and as indicated above, market competition and other factors, including geopolitical events, may change these numbers for the estimated annual sales for the Worldwide Diagnostic group for future years.

Abbott is very aggressive, as it should be, in acquiring smaller companies and in collaborating with companies that fit its vision of growth in diagnostic products. Appropriate acquisitions would continually benefit its bottom line, and be reflected presumably in the order of dependence on year as shown in eq. 13.2. The present order of dependence is 0.2497. This is 45.17% higher than the order of dependence exhibited in eq. 13.1b for Biacore (equal to 0.1720) that manufactures only the SPR biosensor. Note that in the year 2000, the annual sales figures for Abbott Worldwide Diagnostics were about

Table 13.5

Influence of change in order on sales ratio for Abbott Laboratories Worldwide Diagnostic group

Year	Sales ratio
1 (2004)	1
2 (2005)	1.017
3	1.027
4	1.035
5	1.041
6	1.0456
7	1.0496

Abbott Laboratories Fact Book (2004).

300 times (316.4) more than that of Biacore. This factor (level of scale) may be responsible for the other higher order of dependence exhibited in Abbott's equation (13.2) when compared with that of Biacore's equation (13.1b). Surely, other factors are also responsible.

It is of interest to see how much the order of dependence on year affects the annual sales. For the sake of comparison let us compare the annual sales given by the following two expressions. The first one is eq. 13.2 with an order of dependence equal to 0.2497 (Abbott's figures), and the second one assumes a 10% increase in the order of dependence to 0.2746. The numbers are provided in Table 13.5. The first column is the year, with year 1 representing the year 2004, year 2 representing 2005, etc. Please remember these are projected annual sales for the coming years. The second column is the ratio of the numbers obtained using the orders 0.2746 and 0.2497, respectively. In both the cases, the number 1687.51 from eq. 13.2 was used.

This represents a 5% (4.96% actually) change in annual sales in 7 years with a 10% change in the order of dependence. The 5% change is not much, but the base on which it is based is very large (equal to \$19786.6 million). A 5% change represents \$984 million. This would be over and above the regular growth in sales estimated by eq. 13.2.

As indicated earlier, Abbott Laboratories is acquiring small, more-focused companies at an aggressive pace, that may help increase the previously mentioned influence of the order of dependence on its Worldwide Diagnostic sales by more than 10%. An Abbott Strategic Alliances and Acquisitions document (http://www.abbottdiagnostics.com/Abbott_Us/Alliances.cfm) indicates some of the following alliances and acquisitions. These acquisitions are presented in Table 13.6.

We now present the growth of a small company (less than 24 employees) which up until now has not turned a profit, but is expected to turn a profit in the next few years. Biosensor Applications AB in Sundbyberg, Sweden was sold to private investors by Bofors in 1998 (Annual Report, January 1 to December31, 2003; http://www.biosensor.se/eng-ekinfo-nyhetsbrev_q1-2003.asp). Up until the year 2003, it had accumulated a loss of 170,947,236 Swedish Kroner. One Swedish Kroner equals 0.150 US dollar (www.exchangerate.com), December 6, 2004; 11:00 am, Central Standard Time). This is equivalent to US \$25.6 million. However, the losses incurred by the company shows

Table 13.6

Abbott Laboratories Strategic Alliances and Acquisitions

Company	Diagnostic test
Artus GmbH	Detect a form of coronavirus suspected of causing severe acute respiratory syndrome (SARS)
Celera Diagnostics	*In vitro* molecular diagnostic products for disease detection
i-STAT Corporation	Point-of-care (POC) testing
Therasense Inc.	Glucose monitoring systems to reduce the pain for testing for glucose
Promega Corporation	Nucleic acid product testing to automate testing for infectious diseases

http://www.abbottdiagnostics.com/Abbott_Us/Alliances.cfm

a decreasing trend, and very shortly the company is expected to show a profit. The losses incurred in the years 2002 and 2003 were 34,245,000 and 27,618,00 Swedish Kroner, respectively (Annual Report, January 1 to December 31, 2003; http://www.biosensor.se/eng-ekinfo-nyhetsbrev_q1-2003.asp). This is equivalent to $5.136 and 4.14 million, respectively. Its losses are decreasing by about $1 million per year. The company is making smart alliances worldwide (North and South America, Europe, the Middle East, India, and Asia). It has also sold its first biosensor system to Japan and has carried out successful tests of its drug detection systems with the Canadian and American Customs authorities. In the year 2003, the company introduced its BIOSENS-D to worldwide markets. Once again, it may be noted that presumably around the year 2008 (10 years after it became independent from Bofors in 1998), Biosensor Applications AB, Sweden will start turning a profit. Though only one example of a small biosensor company is given, these numbers do appear to be typical with regard to investment and time of what it takes to make a biosensor company profitable.

Some examples of the investment and time involved in setting up a biosensor company were given above. Perhaps, it is instructive to provide some real life numbers if one wants to set up a stand-alone biosensor industry. The numbers presented here are different from the ones if one were to just expand or diversify into biosensors. In the second case, biosensors are not the only product that one manufactures and attempts to sell in the market. For much better, and a more detailed, analysis on the commercial applications of biosensors one may refer to *Biosensors: A Clearer View* by Newman *et al.* This is an expensive book that has recently come out in 2004, but is presumably well worth its price since it is written and compiled with the leading authority in biosensors.

Different types of biosensors may be made. Initially, one should leverage as much as one can based on one's expertise. For example, if one has experience with fiber-optic biosensors (for example, worked for a company before, or did a thesis, preferably doctoral), then one can use some, if not all of this knowledge that one already possesses. Money has to be raised to start and run the industry, even before one starts selling one's product. That too takes money, and needs to be considered as an 'expense.' One may offer an Initial Public Offering (IPO), for example. One may also make appropriate collaborations or strategic partnerships. These aspects are not considered in the numbers given below, since they are very variable, do contribute quite a large fraction of the total costs, and behoove one to explore different avenues to raise the capital required to run the industry, in this case biosensors.

The numbers presented below are for a small biosensor industry that employs 15–20 individuals. Assume that it takes 10–12 years to get the biosensor to the market. We are starting from scratch. This time may be minimized considerably if one were to acquire the technology from some other source, say for example, from university sources, or a 'specialized boutique' that has shown that the biosensor runs efficiently in the laboratory, and is looking for commercial partners to bring the biosensor to the market. There are quite a few people like these who are looking for investors. Also, it is not unusual to see investors or investment firms at scientific meetings. This author has attended biotechnological and biomedical meetings. These people were there at these meetings. They may also be present at other types of meetings.

Number of scientists, administrators, employees	15–20
Number of years required to develop the biosensor	10–12
Number of man years required	150–240
Cost of man-year (scientist, administrator, employee)	\$50,000–70,0000
Total personnel cost	\$7.5–16.8 million
Overhead (includes cost of financing project, equipment, supplies) at 200% of personnel cost	\$15–33.6 million
	\$22.5–50.4 million

This is an expensive project and ranges from approximately \$22 to around 50 million. The numbers are very general, and one can use better numbers that are more specific to one's project. For example, if one is able to use one's resources more efficiently, by selecting and employing more focused and experienced individuals on the project, then one may save both considerable time and money to successfully launch the biosensor in a '*highly competitive*' market. Needless to say, a 'niche market' would be excellent, but on the other hand its volume in all probability will be very small. Considerable time and money may also be saved by collaborating with like-minded companies, or mutually complementary industries, or build, as indicated earlier, strategic partnerships. However, one needs to carefully balance the risk-gain inherently associated with each choice. Continued improvements in the process, no matter how small, should be the *mantra* since large sums of money or investment are involved.

The above example indicates the general level of investment required to start-up a biosensor company. It is perhaps useful to do one's 'homework' or research the issue intensively, from different aspects, preferably by yourself, before one starts to invest time, money, and effort into a cost-intensive long-term project. Most people would consider a 'business model' prepared by consultants or experts in this field. Here we will attempt to provide just the basic overall perspectives and guidelines. The details may be filled out later, by interested individual(s), and depending on the type of biosensor project one is looking to establish. We will also attempt to provide a real-life example that perhaps fills the three criteria outlined below. This is just an example; more detailed criteria will surely be required:

Criteria A. Identify a market, preferably a large on, even if it is highly competitive. The glucose monitoring market for diabetes quickly comes to mind. Another market, though presumably small, but does exhibit the potential to grow is the market for the diagnosis and monitoring of autoimmune diseases, such as cancer, SLE, and arthritis. Serious consequences are involved if the ailment is not monitored and treated properly.

Criteria B. Identify a need in the competitive market that is not yet filled, or the need is filled, but there is considerable room for improvement, for example, the pain and inconvenience for the monitoring of glucose. Another example could be the inconvenience of monitoring different analytes such as auto-antibodies, creatinine, etc for SLE, to help predict a 'flare' for the diseases so that it may be better managed. Recognize that one generally resorts to drugs such as steroids to better manage diseases such as SLE. Steroids have very significant side effects.

Most of these drugs have serious side effects, as noted by the intense scrutiny on drugs such as Vioxx (removed from the market by Merck) and on Celebrex (still on the market: Pfizer; December 21, 2004). It may perhaps be useful to have a detection system that helps measure these side effects.

Criteria C. Attempt to fill the need with advanced technology and newer approaches. Also, possess plans for future expansion.

Pelikan Technologies in South San Francisco, California (2004) is attempting to introduce a hand-held biosensor that would specifically assist in the monitoring of glucose in the management of diabetes.

Criteria A. This is a highly competitive market. This company estimates that there are 18.3 million diabetics in the United States alone. Each year 1.3 million diabetics are added to this number. Most reports, along with the American Diabetic Association (ADA), indicate that this autoimmune disease is reaching epidemic proportions. There are quite a few big name companies that have diagnostic devices and POC monitoring devices for determining glucose levels in blood. If the disease is not treated properly or glucose levels are not monitored carefully, then heart disease, blindness, kidney damage, and other medical conditions may result.

Criteria B and C. Measuring of blood glucose levels is extremely painful and inconvenient. This often leads to negligence in glucose monitoring which is undesirable. Better, less invasive and painful measurement techniques would promote the compliance of measuring glucose levels. Pelikan's innovative and novel technology is aimed to address this issue of inconvenience and pain, and help provide an efficient blood glucose monitoring sensor. Pelikan technology is aimed at transforming the multi-step, painful, and 'difficult to use' present-day glucose monitoring systems to a one-step, simple to use and automated, and much less painful procedure which is also reliable. Furthermore, the company hopes to use their technology platform for the monitoring of analytes in blood for the detection of other ailments. This aspect cannot be over-emphasized, that is making the POC test painless. Even at a higher cost, it is safe to presume that if the test is painless, most individuals would purchase the POC test, and this would lead to a significantly higher compliance rate.

REFERENCES

Abbott Laboratories Fact Book, 2004, available online.

Agricultural Research Service, United States Department of Agriculture, Annual Report, Biosensor processes for detecting pathogenic bacteria in foods, 2002, http://www.ars.usda.gov/research/projects/projects.htm?

Ambri Financial Report, Increasing shareholder value by building businesses based on Ion Channel Switch™ Technology, 2004, http://www.ambri.com/Content/display.asp?name = Products%20and%Services

Applied Nanotech, Applied nanotech announces demonstration of improved biosensor using carbon nanotubes, 2003, http://www.tmcnet.com/usubunit/2003/Oct/1020765.htm

Berg, A, BIOSENS—Vapor detection—Area reduction in demining (EC Funded projects), 2003, http://www.eudem.vub.ac.be/projects/project.asp?project_id = 2

Berney, H, DNA biosensor for analytical investigation and labelling in food DNA-BAILIF, 2001, http://www.europa.eu.int/comm/research/quality-of-life/ka1/volume1/qlk1-2001-00876.htm

Biacore Drug Discovery and Development, 2004, http://www.biacore.com/drug/core.lasso

Biacore Food Analysis, 2004, http://biacore.com/food/core.lasso

Biacore International AB, Neuchatel, Switzerland, Hoover online, 2004, http://www.hoovers.com/biacore/

Biacore Online Catalog, 2004, http://biacore/catalog/core.lasso

Biosensor Applications AB, Sundbyberg, Sweden, Annual Report January 1 to December 31, 2003, http://www.biosensor.se/eng-ekinfo-nyhetsbrev_q1-2003.asp

Biotrace International, Plc, Collaboration with Smith Group's Graseby, May 1, 2002, http://www.biotrace.com/content.php?hID= 1&nhID = 21&nID= 36

Biowise, *Biosensors for Industrial Applications*, a review of biosensor technology, 2000.

Bohannon, B, On-site rapid detection of anthrax spores, *Detection Technologies—The Next Generation in Identification and Analysis*, December 5–6, Arlington, Virginia, USA, 2002.

Camilleri, D, Instant TB screening moves closer to reality, Rapid Biosensor Systems Ltd, 2004, http://www.rapidbiosensor.com/new_march_2004.htm

Carriveau, GW, The next generation in chemical vapor detectors: reducing the false alarm rate using complementary detection modes and data fusion, *Detection Technologies—The Next Generation in Identification and Analysis*, December 5–6, Arlington, Virginia, USA, 2002.

Channel CNBC News; USA, 9:30 am, December 28, 2004.

Chaplin, M, What are biosensors, London South Bank University, 2004, http://www.lsbu.ac.uk/biology/enztech/biosensors.html

Cross, GH and NJ Freeman, Commercialization of optical waveguide technology, Department of Trade and Industry, P.O. Box 83, Didicot, Oxfordshire OX 11 OBR, United Kingdom, 2004, http://www.fairfield-sensors.com

Crow, MK and J Wohlgemuth, Microarray analysis of gene expression in lupus. *Arthritis Research & Therapy*, 2003, **5**, 279–287.

Defense Advanced Research Projects Agency (DARPA) Defense Science Office—Future Areas of Interest—BioFabrication, http://www.darpa.mil/dso/future/biofab.htm

Drexel University, Applications of Biosensors, 2004, http://www.drexel.edu/academics/coe/ce/web_books/EngBio/Hidden/Sensr/ch1/1_3.htm

DuBois, J, *Medical Product Outsourcing*, Accelerating speed to market: from concept to finished device, development efforts can benefit from thinking-out-of-the box, October, 2003, http://www.mpo-mag.com?oct033.htm

DuBois, J, and RA Shaw, IR spectroscopy in clinical and diagnostic applications, *Analytical Chemistry*, October 1, 2004, pp. 361A–367A.

DuPont, Biosensor Materials, 2004, http://www.devicelink.com/company98/co/118/11807.html

Ender, F, et al. *Applied Physics*, 2004, **84**, 2673.

Evans, J, Shedding light on pathogens. *Chemistry World*, 2004, November, pp. 14–14.

Eysel, HH, M Jackson, A Nikulin, RL Somorjai, GTD Thomson, and HH Mantsch, A novel diagnostic test for arthritis: multivariate analysis of infrared spectra of synovial fluid, *Biospectroscopy*, 1997, **3**(2), 161–167.

Farrell, I, Scientific Security. *Chemistry World*, November, 2004, pp. 44–48.

Freedonia, Chemical Sensors to 2006: Liquid, Gas, and Biosensors, April 2002, http://freedonia.ecnext.com/coms2/summary_0001_000898_000000_000000_0001_

Freedonia, Chemical Sensors to 2008: Liquid, Gas, and Biosensors, March, 2004, http://freedonia.ecnext.com/coms2/summary_0001_001878_000005_000011_1

Fuji-Kerzai, Market Report: US and World-Wide Biosensor Market, R & D and commercial application, 207 pages, 2004, http://www.mindbranch.com/catalog/product.jsp?code=R98-105&prsc=i060704

GenomeCanada, Vitrek awarded $1.2 million to apply FONA to water quality testing, Investment from Ontario's Healthy Futures for Agriculture Program, 2004, http://www.genomecanada.ca/GCmedia/communiques Presse/indexDetails.asp?id=136&1=e

Georgia Tech, Innovative biosensor to undergo field trials at poultry processing plant, 2004, http://gatech.edu/pt11-3/11-3_pl.r

Godso, DW, SensorView™—A standard sensor interoperability and data fusion platform for command, control, monitoring, and management of heterogeneous distributed sensors, *Detection Technologies—The Next Generation in Identification and Analysis*, December 5–6, Arlington, Virginia, USA, 2002.

Hoovers online, Deal signed with Japanese organization for new SPR array technology, *Science Letters*, November 11, 6:46 pm, 2004, http://www.hoovers.com/free/co/news/detail?COID=52773&"ArticleD"=NR200411

Innovative Biosensors, Inc., and MIT (Massachusetts Institute of Technology), announce licensing agreement for biosensor technology, 2004, http://www.innovativebiosensors.com/Article5.html

ISIS Pharmaceuticals (ISIS) TIGER Program, 2004, http://www.biospace.com/news_stort.cfm?ID=17599320&full=1

Khandelwal, A, Innovating for a successful future. *Chemistry and Industry*, 2004, (16), pp. 13–14, 16 August.

Kissinger, PJ, Introduction to amperometric biosensor configurations, 2004, http://www.currentseparations.com/issues/16-3/cs16-3f.pdf

Knecht, B, A Strasser, R Dietrich, E Martlbauer, R Niessner, and MG Weller, Development of a biosensor for the rapid detection of antibiotics in milk, 2003, http://www.ch.tum.de/wasser/weller/DCHB2003P3.pdf

Kostel, K, Biodefense: challenges, priorities, threats, *The Scientist*, May 24, 2004.

Lab_Bell, inc., Diagnostic for toxic molecules, http://www.innovatechquebec.com/Client/detail.asp?id=122&type=2&language=2

Lin, C-T and S-M Wang, Biosensor commercialization strategy: a theoretical approach. *Frontiers in Bioscience*, 2005, **10**, 99–106, January 1.

Madou, M, Planar electrochemical sensors, 2002, http://mmadou.eng.nci.edu/ResearchDevelopment/planar_electrochemical_sensors.htm

Mareno, S, Detecting legionaire's disease sooner, 2004, http://www.rmit.edu.au/browse?SIMID=r8ewqx7mpnuj&Status=A

May, M, Building a better biosensor. *The Scientist*, 2004, **18**(10), 36–38.

Mead, N, Pursuing excellence in research, Swedish Biotech Update, Biacore, 2001, http://www.scandinavianow.com/biotech/biacore.html

Medical Technologies, Biosensors: a brief tutorial. No. 4: Biosensor Markets—Opportunities and Obstacles, 1994, http://www.fraserclan.com/biosens4.htm

MedMarket Diligence, Current medical technologies, applications & market opportunities, Nanotechnology and MEMS, 2003, http://www/mediligence.com/rpt-t601.htm

Miller, JD, US Government launches biolab building spree, May 24, 2004, 48–49.

Mukhopadhay, R, Portable FTIR spectrometers get moving, *Analytical Chemistry*, October 1, 2004, pp. 369A–372A.

National Institute of Allergy and Infectious Diseases, NIAID funds construction of biosafety facilities, press release, September 30, 2003, available online at www2.niaid.nih.gov/Newsroom/Releases/nblscorrect21.htm

National Science Foundation, NSF 03-512, Program Solicitation, Sensors and Sensor Networks, Proposals due March 6, Arlington, Virginia, USA, 2003.

National Science Foundation, NSF 04-522, Program Solicitation, Sensors and Sensor Networks (Sensors), Proposals due February 26, Arlington, Virginia, USA, 2004.

National Science Foundation, NSF 05-526, Program Solicitation, Sensors and Sensor Networks (Sensors), Proposals due March 03, Arlington, Virginia, USA, 2005.

Newman, JD, LJ Tigwell, APF Turner and PJ Warner, Biosensors—An Inside View, may be purchased from the Institute of BioScience and Technology, 2004, http://www.cramfield.ac.uk/ibst/commercial/biosensor_report2004.htm

Niedbala, RS, Rapid multiplex detection of biological warfare agents using Up-converting phosphor receptors, *Detection Technologies—The Next Generation in Identification and Analysis*, December 5–6, Arlington, Virginia, USA, 2002.

Normile, D, First US case of mad cow disease sharpens debate over testing, *Science*, January 9, 2004, pp. 156–157.

Obducat, Biosensors—opening up a new world, 2004, http://www.obducat.com/pdf/biopaper.pdf

O'Neil, KE, The Bio-sensor: An effective drugs and explosives detection system, *Advanced Aviation Technology*, Compton, Surrey, UK, 2004, http://www.aatl.net/publications/Bio-sensor.htm

Pelikan Technologies, 2004, http://www.pelikantechnologies.com/cgi/content.php4?rubrik+aboutus

Poultry Tech, Development of innovative biosensor makes progress, 2003, http://atrp.gatech.edu/pt15-3/15-3_p2.html

QTL Biosystems, LLC, QTL Biosystems, LLC and United First Responders to manufacture handheld biosensor in Pennsylvania, 2004, http://www.biospace.com/news_story.cfm?storyID=16290720&full=1

Raghupathy, R, Biosensor Technology, *Science Technology*, November, 2002, http://www.techno-preneur.net/timeis/technology/Sci TechNov02/DecBiosensor.html

Rajan, M, US market for biosensors and bioelectronics to cross $1.81 billion by 2006, Report RB-159, Business Communications Company, Inc., Norwalk, CT, 2002, http://www.bccresearch.com/editors/RB-159.html

Research and Markets, Biosensor market, R & D and Commercial Implication: biosensor market projected to grow to about $10.8 billion with a growth rate of 10.4%, 2004, http://www.tmcnet.com/usubmit/2004/apr/1036069.htm

Ruzgas, T, P Skaldal, D Pfeiffer, D Leech, E Dominguez, J Kulys, U Wollenberger, M Winther-Nielsen, L Norgaard, and T Welander, Intelligent signal processing of biosensor array using pattern recognition for characterization of wastewater: aiming towards alarm systems (INTELLISENS), adapted from Abstract, 2000, http://europa.eu.int/comm/research/quality-of-lif/cell-factory/volume/projects/qlk3-200

Sadana, A, *Biosensors: Kinetics of Binding and Dissociation using Fractals*, Elsevier, Amsterdam, 2003.

Santana, TA, An automated bioaerosol collection and detection system for the detection of biological agents in carrier facilities, *Detection Technologies—The Next Generation in Identification and Analysis*, December 5–6, Arlington, Virginia, USA, 2002.

Singapore's Institute of Bioengineering and Nanotechnology, Singapore's Institute of Bioengineering and Nanotechnology develops ultrasensitive biosensors for breast cancer and blood sugar diagnostics, 2003, http://www.biomed-singapore.com/bms/sg/en_uk/index/newsrrom/pressrelease/0/singapore

Sontra Medical Corporation, The new wave of transdermal science, 2004, http://sontra.com/productapplications/glucosemonitoring/

Strategic Alliances and Acquisitions, http://www.abbottdiagnostics.com/About_Us/alliances.cfmk.

Synthetic Blood International, Glucose biosensor, 2004, http://daibet.chat.ru/english/research/glucose_sensor_eng.html

Talukder, G, Biosensors & Bioelectronics, Business Communications Company, Inc., Norwalk, Connecticut, USA, Report B-159, September, 2002, http://www.buscom/biotech/B159.html

TFS Sensor Technology, 2004a http://mywebpages.comcast.net/tfs-jdownward/web_pages/TFS_Sensor_Technology.html

TFS Sensor Technology, 2004b http://mywebpages.comcast.net/tfs-jdownward/web_pages/TFS_Fabrication_Technology.html

Trimark Publications, Report: Point of Care Diagnostic Testing World markets, May 1, 689 pages, 2004.

University of Delaware, Graduate Class notes for Biosensors and BIOMems, Newark, Delaware, USA, 2004.

van Hoof, C, K Baert and A Witvrou, The best materials for tiny, clever, sensors. *Science*, 2004, **306**, 986–987, November 5.

Walsh, M, Nano and MEMS Technologies for Chemical Biosensors, *Advanced Technology Program, National Institute of Standards and Technology*, 2003, http://atp.nist.gov/atp/focus/98wp-nan.htm

Zarkoff, J, Promising growth in biosensor and antibody markets, 2002, http://www.devicelink.com/ivdt/archive/02/11/008.html (IVDT archive November 2002).

Index